Functional Nanoparticles for Environmental Contaminants Removal and Agricultural Application

Functional Nanoparticles for Environmental Contaminants Removal and Agricultural Application

Editor

Yu Shen

MDPI

Basel • Beijing • Wuhan • Barcelona • Belgrade • Novi Sad • Cluj • Manchester

Editor
Yu Shen
College of Biology and the Environment
Nanjing Forestry University
Nanjing
China

Editorial Office
MDPI
St. Alban-Anlage 66
4052 Basel, Switzerland

This is a reprint of articles from the Special Issue published online in the open access journal *Coatings* (ISSN 2079-6412) (available at: www.mdpi.com/journal/coatings/special_issues/func_nano_enviro_contam_remov_agri_app).

For citation purposes, cite each article independently as indicated on the article page online and as indicated below:

Lastname, A.A.; Lastname, B.B. Article Title. *Journal Name* **Year**, *Volume Number*, Page Range.

ISBN 978-3-0365-8975-6 (Hbk)
ISBN 978-3-0365-8974-9 (PDF)
doi.org/10.3390/books978-3-0365-8974-9

Contents

Preface

Sustainably providing a safe and secure food supply will be a significant challenge in the coming years, and the field of nanotechnology is well positioned to have a tremendous positive impact on these efforts. The potential benefits are numerous, including the enhancement of food quality and safety, the reduction of environmental waste and agricultural inputs, and the absorption of nanoscale nutrients from soil. Nanotechnology could also better enable plants to respond to a changing environment. It can also play an important role in crop productivity through controlling nutrient availability and utilization, as well as through mediating nutritional status in the presence of biotic and abiotic stress. Nanoscale platforms can also be used to monitor water quality parameters, including pesticide residue presence, in efforts to promote the sustainable development of agriculture. Nanomaterials have tremendously diverse functionality, making it difficult to perform a general assessment of crop health and environmental risks. For example, although dose is clearly one of the most important factors differentiating benefit from toxicity, factors such as environmental conditions and species sensitivity can dramatically shift that dividing line. In addition, the properties (other than size) of nanoparticles that influence toxicity include chemical composition, shape, surface structure, surface charge, and the extent of particle aggregation (clumping) or disaggregation. For this reason, even nanomaterials of the same chemical composition with different sizes or shapes can exhibit different toxicities or benefits.

There is also a general lack of understanding of how transformation processes in biotic and abiotic media influence overall nanomaterial activity. Regardless, nanotechnology applications in the agricultural sector have been a topic of intense interest, and many argue that it is a key factor for sustainable development. To explore this important topic, we have assembled a Special Issue in *Coatings* to encourage researchers and to provide them with a platform to publish their novel studies on the subject of "Functional Nanoparticles for Environmental Contaminants Removal and Agricultural Application".

Yu Shen
Editor

coatings

Editorial

Special Issue: Functional Nanoparticles for Environmental Contaminants Removal and Agricultural Application

Ran Tao [1,2], Mengmeng Kong [3], Yinlong Zhang [1] and Yu Shen [1,2,*]

1 Co-Innovation Center for Sustainable Forestry in Southern China, College of Biology and the Environment, Nanjing Forestry University, Nanjing 210037, China; nature19980626@gmail.com (R.T.); ecoenvylz@163.com (Y.Z.)
2 National Positioning Observation Station of Hung-tse Lake Wetland Ecosystem in Jiangsu Province, Nanjing 210037, China
3 School of Biological Science and Technology, University of Jinan, Jinan 250022, China; bio_kongmm@ujn.edu.cn
* Correspondence: yushen@njfu.edu.cn or sheyttmax@hotmail.com

Citation: Tao, R.; Kong, M.; Zhang, Y.; Shen, Y. Special Issue: Functional Nanoparticles for Environmental Contaminants Removal and Agricultural Application. *Coatings* **2023**, *13*, 1633. https://doi.org/10.3390/coatings13091633

Received: 5 September 2023
Accepted: 13 September 2023
Published: 18 September 2023

Currently, increasing demands are being placed on agricultural production, presented with the challenge of finding sustainable ways to meet the needs of the world's growing population [1]. Global warming and climate change have become a "wifi" contributing to agricultural and environmental issues, with rising global temperatures and extreme weather events leading to crop failures and impacting food security [2]. Furthermore, intensive agricultural practices and unsustainable land use are causing soil erosion, a loss of fertility, and reduced productivity [3]. Limited access to water, over-extraction, and pollution pose major challenges to agriculture, particularly in arid and semi-arid regions [4]. Additionally, the spread of pests and diseases has become increasingly challenging to control due to factors such as global trade and climate change [5].

Nanotechnology, and particularly the field of nanomaterials, holds the potential to revolutionize agriculture and the environment [6]. It can be utilized to enhance crop production efficiency [7], reduce the use of pesticides and fertilizers [8], and improve food quality [9]. Nanotechnology can also contribute to the development of novel materials that enhance irrigation efficiency, crop storage, and various other aspects of farming [10]. Moreover, nanotechnology has the potential to create sensors capable of detecting changes in soil quality, enabling farmers to make more informed decisions about land management [11]. Encouragingly, significant progress has been made in the field of nano-agriculture and environmental applications, as highlighted in our Special Issue "Functional Nanoparticles for Environmental Contaminants Removal and Agricultural Application" in *Coatings*.

Nanotechnology has played a positive role in agriculture. Guo et al. (2022) observed that silver nanoparticles outperformed other nanoparticles in terms of final crop seed germination percentages, and zinc nanoparticles were found to be the most effective in promoting root length growth during seed germination in a meta-analysis conducted between 1950 and 2021 [12]. Hexaconazole is a widely used, broad-spectrum, and highly efficient triazole fungicide, but its extensive use can lead to environmental disasters [13]. In response, Pan et al. (2021) applied azobenzene-modified bimodal mesoporous silica nanoparticles (BMMs-Azo) in conjunction with β-cyclodextrin to control the release of hexaconazole, reducing its environmental impact [14]. Biochar is known for its numerous surface functional groups and porous structure, which can reduce nutrient loss and enhance nutrient uptake by plants [15]. Combining Methylotrophic bacillus, colloidal biochar (containing dissolved nanoparticles), and organic fertilizer significantly increased the contents of lycopene, vitamin C, total sugar, and soluble sugar in tomato fruits by 58.40%, 46.53%, 29.45%, and 26.65%, respectively [16]. Additionally, Yang et al. (2022) demonstrated that biochar nanoparticles, acting as nanocarriers, substantially increased the fertilizer utilization of Chinese cabbage and promoted plant growth by over 50% [17]. Copper nanoparticles have widespread applications in various industries, but their effects in agriculture and the environment vary. Yang et al. (2022) observed that 5 ppm copper nanoparticles caused the slight swelling of epithelial cells in the epididymal duct of Chinese soft-shelled turtles, while higher copper ion concentrations severely damaged the epithelial structure of the epididymal tube [18]. Faraz et al. (2022) reported on the foliar application of 2–16 ppm copper oxide nanoparticles (CuO NPs) on Brassica juncea 25 days after sowing, whereby 8 ppm resulted in optimal chlorophyll content, net photosynthetic rate, leaf proline content, and antioxidant enzyme activity [19]. However, the effectiveness of this concentration may vary for other plants. Surface coating is another factor affecting the interaction of CuO NPs with plants. Deng et al. (2022) found that the foliar spraying of 300 ppm bare NPs yielded similar results to 75 ppm citric-acid-coated NPs, with Cu NPs increasing the yield by approximately 170% compared to the control [20]. Interestingly, soil amendment with both bare and coated NPs did not significantly impact the plant mass relative to untreated plants. The use of manufactured nano-objects (MNOs), including carbon nanotubes (CNTs), nanoparticles (NPs), and nanopesticides, raises concerns, as these substances can impact the life cycle and not only accumulate in soils but also in other environmental components, negatively affecting the soil biota and processes. MNOs can interfere with soil physicochemical properties and microbial metabolic activity in rhizospheric soils [21]. Notably, negatively charged CuO NPs were found to significantly reduce disease progression and increase biomass, whereas positively charged NPs and a CuSO4 salt control had little impact on plants [22]. Based on element characteristics, it has been reported that magnesium nanoparticles (Mg-NPs) are more effective in enhancing the development of 'Superior Seedless' berries during various growth stages (flowering, fruit set, version, and harvest) compared to sulfate magnesium ($MgSO_4 \cdot 7H_2O$) and magnesium disodium EDTA (Mg-EDTA) [23]. As we move forward in the realm of coated nanoparticles for environmental remediation and agricultural use, recent studies have shown promise. It has been reported that β-carotene-coated chitosan nanoparticles (CNPs) have the potential to block polycyclic aromatic hydrocarbons (PAHs) and protect crops in PAH-contaminated soil. Under specific conditions of 20 °C, pH 6, and 10 mg/mL TPP, spinach biomass significantly increased, and the transfer of PAHs from the soil to the roots was reduced [24]. In agricultural activities, insecticides and pesticides can have unintended effects, and the nanofeature of higher efficiency is utilized to reduce ecotoxicity. Alginate CNPs have been employed in combination with DMT to reduce insecticide toxicity in zebrafish larvae while reducing the required dose [25]. Furthermore, CNPs can serve as a core for novel SnS_2 quantum dots with Azadirachta indica leaf extract, with coated CNPs exhibiting an enhanced removal of crystal violet dye [26]. Overall, our Special Issue reflects the cutting-edge trends in nanotechnology applications in agriculture and the environment.

Environmental catalysis has garnered significant attention for its clean methods of producing useful chemicals and facilitating various chemical processes. This approach can decompose and eliminate organic pollutants from aqueous environments while enabling the production of valuable chemicals [27]. Liu et al. (2023) synthesized novel N-doped biochar nanoparticles through the one-step pyrolysis of algal sludge without external nitrogen sources, yielding a highly active and cost-effective carbon-based catalyst capable of activating new oxidants for contaminant removal [28]. Additionally, Liu et al. (2022) found that Sedum plumbizincicola nanoparticles can effectively remove Bisphenol A (BPA) from complex wastewater, demonstrating stable and efficient reactions [29]. Nanoscale schwertmannite (nano-SWT) was prepared using an indoor temperature synthesis method with the assistance of polyvinylpyrrolidone, and nano-SWT was found to be effective in reducing sulfamethoxazole in the presence of Fenton-like catalysts using hydrogen peroxide (H_2O_2) [30]. Currently, Fenton catalysts are widely employed in conjunction with nanoparticles for environmental contaminant removal. Manganese tetroxide (Mn3O4) nanoparticles have been reported to simultaneously degrade estriol and 17α-ethinylestradiol (E3/EE2) with removal efficiencies of 97.5% and 96.4% for E3 and EE2, respectively, using Fenton-like catalysts [31]. Copper–iron peroxide nanoparticles (CFp

NPs) have been designed for tumor-microenvironment-mediated synergistic therapy in a heterogeneous chemodynamic therapy system. CFp NPs generate oxygen during catalysis and exhibit a tumor-microenvironment-responsive T1 magnetic resonance imaging contrast enhancement, aiding in tumor oxygenation and in vivo tumor monitoring [32]. Furthermore, it has been reported that S-doped carbon and Fe7S8 nanoparticles interact effectively for the high-efficiency removal of antibiotics through a Fenton-like degradation process. Within 40 min under neutral pH conditions, amoxicillin, norfloxacin, and tetracycline hydrochloride were removed at rates of 98.9%, 97.8%, and 99.3%, respectively, with the catalyst demonstrating excellent cycle stability [33]. At the same time, with the oxygen atoms in the sulfonic acid group cooperating with Ag+ to form a synergistic complexation, a novel magnetic fluorescent nanoprobe (Fe_3O_4@ZnS@MPS(MFNPs)) was designed for Ag+ detection in aquatic media [34]. These findings underscore the significant progress made in leveraging nanomaterials and catalytic processes for cleaner and more efficient environmental solutions, offering promise for a sustainable future.

In conclusion, the advancements in nano-agriculture and environmental applications are evident from the research covered in the Special Issue "Functional Nanoparticles for Environmental Contaminants Removal and Agricultural Application" in *Coatings*. Noteworthy examples include the use of silver nanoparticles to enhance crop seed germination and the application of azobenzene-modified bimodal mesoporous silica nanoparticles to control the release of the fungicide hexaconazole, thereby reducing environmental impacts. While nanotechnology presents exciting prospects, it is crucial to consider its potential ecological effects and develop responsible applications. Overall, the progress in nanotechnology applications in agriculture and the environment underscores the field's potential to contribute to a more sustainable future.

Funding: This research was funded by the National Science Foundation, grant number 31872415.

Conflicts of Interest: The authors declare no conflict of interest.

References

1. Duarte, C.M.; Bruhn, A.; Krause-Jensen, D. A seaweed aquaculture imperative to meet global sustainability targets. *Nat. Sustain.* **2021**, *5*, 185–193. [CrossRef]
2. Kazemi Garajeh, M.; Salmani, B.; Zare Naghadehi, S.; Valipoori Goodarzi, H.; Khasraei, A. An integrated approach of remote sensing and geospatial analysis for modeling and predicting the impacts of climate change on food security. *Sci. Rep.* **2023**, *13*, 1057. [CrossRef] [PubMed]
3. Wang, X. Managing Land Carrying Capacity: Key to Achieving Sustainable Production Systems for Food Security. *Land* **2022**, *11*, 484. [CrossRef]
4. Mahlknecht, J.; Aguilar-Barajas, I.; Farias, P.; Knappett, P.S.K.; Torres-Martinez, J.A.; Hoogesteger, J.; Lara, R.H.; Ramirez-Mendoza, R.A.; Mora, A. Hydrochemical controls on arsenic contamination and its health risks in the Comarca Lagunera region (Mexico): Implications of the scientific evidence for public health policy. *Sci. Total Environ.* **2023**, *857*, 159347. [CrossRef]
5. Brennan, F.P.; Alsanius, B.W.; Allende, A.; Burgess, C.M.; Moreira, H.; Johannessen, G.S.; Castro, P.M.L.; Uyttendaele, M.; Truchado, P.; Holden, N.J. Harnessing agricultural microbiomes for human pathogen control. *ISME Commun.* **2022**, *2*, 44. [CrossRef]
6. Fu, X.; Zheng, Z.; Sha, Z.; Cao, H.; Yuan, Q.; Yu, H.; Li, Q. Biorefining waste into nanobiotechnologies can revolutionize sustainable agriculture. *Trends Biotechnol.* **2022**, *40*, 1503–1518. [CrossRef]
7. Guleria, G.; Thakur, S.; Shandilya, M.; Sharma, S.; Thakur, S.; Kalia, S. Nanotechnology for sustainable agro-food systems: The need and role of nanoparticles in protecting plants and improving crop productivity. *Plant Physiol. Biochem.* **2023**, *194*, 533–549. [CrossRef]
8. Beig, B.; Niazi, M.B.K.; Sher, F.; Jahan, Z.; Malik, U.S.; Khan, M.D.; Américo-Pinheiro, J.H.P.; Vo, D.-V.N. Nanotechnology-based controlled release of sustainable fertilizers. A review. *Environ. Chem. Lett.* **2022**, *20*, 2709–2726. [CrossRef]

9. Sompura, Y.; Sangwan, D.; Barupal, T. Impact of nanotechnology on environment and their applica-tion in agronomy and food stuffs production: An Overview. *Biomater. J.* **2022**, *1*, 1–4.
10. Oladosu, Y.; Rafii, M.Y.; Arolu, F.; Chukwu, S.C.; Salisu, M.A.; Fagbohun, I.K.; Muftaudeen, T.K.; Swaray, S.; Haliru, B.S. Superabsorbent Polymer Hydrogels for Sustainable Agriculture: A Review. *Horticulturae* **2022**, *8*, 605. [CrossRef]
11. Silva, L.; Rodriguez-Sedano, F.; Baptista, P.; Coelho, J.P. The Digital Twin Paradigm Applied to Soil Quality Assessment: A Systematic Literature Review. *Sensors* **2023**, *23*, 1007. [CrossRef] [PubMed]
12. Guo, H.; Liu, Y.; Chen, J.; Zhu, Y.; Zhang, Z. The Effects of Several Metal Nanoparticles on Seed Germination and Seedling Growth: A Meta-Analysis. *Coatings* **2022**, *12*, 183. [CrossRef]
13. Syrgabek, Y.; Alimzhanova, M. Modern Analytical Methods for the Analysis of Pesticides in Grapes: A Review. *Foods* **2022**, *11*, 1623. [CrossRef] [PubMed]
14. Pan, H.; Li, W.; Wu, L.; Huang, W.; Zhang, F. β-Cyclodextrin-Modified Mesoporous Silica Nanoparticles with Photo-Responsive Gatekeepers for Controlled Release of Hexaconazole. *Coatings* **2021**, *11*, 1489. [CrossRef]
15. Rombel, A.; Krasucka, P.; Oleszczuk, P. Sustainable biochar-based soil fertilizers and amendments as a new trend in biochar research. *Sci. Total Environ.* **2022**, *816*, 151588. [CrossRef]
16. Gu, S.; Lian, F.; Yang, H.; Han, Y.; Zhang, W.; Yang, F.; Gao, J. Synergic Effect of Microorganism and Colloidal Biochar-Based Organic Fertilizer on the Growth and Fruit Quality of Tomato. *Coatings* **2021**, *11*, 1453. [CrossRef]
17. Yang, R.; Shen, J.; Zhang, Y.; Jiang, L.; Sun, X.; Wang, Z.; Tang, B.; Shen, Y. The Role of Biochar Nanoparticles Performing as Nanocarriers for Fertilizers on the Growth Promotion of Chinese Cabbage (*Brassica rapa* (Pekinensis Group)). *Coatings* **2022**, *12*, 1984. [CrossRef]
18. Yang, L.; Wei, Y.; Gao, S.; Wang, Q.; Chen, J.; Tang, B.; Bian, X. Effect of Copper Nanoparticles and Ions on Epididymis and Spermatozoa Viability of Chinese Soft-Shelled Turtles *Pelodiscus sinensis*. *Coatings* **2022**, *12*, 110. [CrossRef]
19. Faraz, A.; Faizan, M.; Hayat, S.; Alam, P.; Shahzadi, I. Foliar Application of Copper Oxide Nanoparticles Increases the Photosynthetic Efficiency and Antioxidant Activity in *Brassica juncea*. *J. Food Qual.* **2022**, *2022*, 5535100. [CrossRef]
20. Deng, C.; Wang, Y.; Cantu, J.M.; Valdes, C.; Navarro, G.; Cota-Ruiz, K.; Hernandez-Viezcas, J.A.; Li, C.; Elmer, W.H.; Dimkpa, C.O.; et al. Soil and foliar exposure of soybean (*Glycine max*) to Cu: Nanoparticle coating-dependent plant responses. *NanoImpact* **2022**, *26*, 100406. [CrossRef]
21. Mubeen, B.; Hasnain, A.; Wang, J.; Zheng, H.; Naqvi, S.A.H.; Prasad, R.; Rehman, A.U.; Sohail, M.A.; Hassan, M.Z.; Farhan, M.; et al. Current Progress and Open Challenges for Combined Toxic Effects of Manufactured Nano-Sized Objects (MNO's) on Soil Biota and Microbial Community. *Coatings* **2023**, *13*, 212. [CrossRef]
22. Borgatta, J.; Shen, Y.; Tamez, C.; Green, C.; Hedlund Orbeck, J.K.; Cahill, M.S.; Protter, C.; Deng, C.; Wang, Y.; Elmer, W.; et al. Influence of CuO Nanoparticle Aspect Ratio and Surface Charge on Disease Suppression in Tomato (*Solanum lycopersicum*). *J. Agric. Food Chem.* **2023**, *71*, 9644–9655. [CrossRef] [PubMed]
23. El-Ezz, S.F.A.; Lo'ay, A.; Al-Harbi, N.A.; Al-Qahtani, S.M.; Allam, H.M.; Abdein, M.A.; Abdelgawad, Z.A. A Comparison of the Effects of Several Foliar Forms of Magnesium Fertilization on 'Superior Seedless' (*Vitis vinifera* L.) in Saline Soils. *Coatings* **2022**, *12*, 201. [CrossRef]
24. Zhang, J.; Cui, M.; Tao, R.; Yao, Y.; Han, J.; Shen, Y. The Application of Carotenoid-Coated Chitosan Nanoparticles to Reduce the PAHs Stress on Spinach Growth. *Coatings* **2023**, *13*, 1404. [CrossRef]
25. Vallim, J.H.; Clemente, Z.; Castanha, R.F.; Pereira, A.d.E.S.; Campos, E.V.R.; Assalin, M.R.; Maurer-Morelli, C.V.; Fraceto, L.F.; Castro, V.L.S.S.d. Chitosan nanoparticles containing the insecticide dimethoate: A new approach in the reduction of harmful ecotoxicological effects. *NanoImpact* **2022**, *27*, 100408. [CrossRef] [PubMed]
26. Gadore, V.; Mishra, S.R.; Ahmaruzzaman, M. Green and environmentally sustainable fabrication of SnS_2 quantum dots/chitosan nanocomposite for enhanced photocatalytic performance: Effect of process variables, and water matrices. *J. Hazard. Mater.* **2023**, *444*, 130301. [CrossRef]
27. Yang, Y.; Wu, M.; Zhu, X.; Xu, H.; Ma, S.; Zhi, Y.; Xia, H.; Liu, X.; Pan, J.; Tang, J.-Y.; et al. 2020 Roadmap on two-dimensional nanomaterials for environmental catalysis. *Chin. Chem. Lett.* **2019**, *30*, 2065–2088. [CrossRef]
28. Liu, C.; Chen, Z.; Kang, R.; Wang, J.; Lu, Q.; Wang, T.; Tian, D.; Xu, Y.; Wang, Z.; Ding, H. N-Rich Algal Sludge Biochar for Peroxymonosulfate Activation toward Sulfadiazine Removal. *Coatings* **2023**, *13*, 431. [CrossRef]
29. Liu, C.; Chen, Z.; Kang, R.; Niu, Y.; Su, W.; Wang, X.; Tian, D.; Xu, Y. Sedum Plumbizincicola Derived Functional Carbon for Activation of Peroxymonosulfate to Eliminate Bisphenol A: Performance and Reaction Mechanisms. *Coatings* **2022**, *12*, 1892. [CrossRef]
30. Meng, X.; Wang, L.; Yang, Y.; Song, Y.; Yuan, C. Heterogeneous Fenton-like Catalyzation of Nanoscale Schwertmannite for Sulfamethoxazole Degradation. *Coatings* **2023**, *13*, 1097. [CrossRef]
31. Liu, M.; Lin, M.; Owens, G.; Chen, Z. Fenton-like oxidation mechanism for simultaneous removal of estriol and ethinyl estradiol by green synthesized Mn_3O_4 NPs. *Sep. Purif. Technol.* **2022**, *301*, 121978. [CrossRef]
32. Koo, S.; Park, O.K.; Kim, J.; Han, S.I.; Yoo, T.Y.; Lee, N.; Kim, Y.G.; Kim, H.; Lim, C.; Bae, J.-S.; et al. Enhanced Chemodynamic Therapy by Cu–Fe Peroxide Nanoparticles: Tumor Microenvironment-Mediated Synergistic Fenton Reaction. *ACS Nano* **2022**, *16*, 2535–2545. [CrossRef] [PubMed]

33. Zhang, X.; Wang, J.; Wang, Y.; Yao, Z.; Guo, W.; Xu, H.; Jiang, Z. Boosting electron transport process over multiple channels induced by S-doped carbon and Fe_7S_8 NPs interface toward high-efficiency antibiotics removal. *J. Hazard. Mater.* **2023**, *442*, 130115. [CrossRef] [PubMed]
34. Gao, Y.; Chen, X.; Xu, P.; Chen, J.; Yu, S.; Liu, Z.; Zeng, X. A Novel Magnetic Fluorescent Fe_3O_4@ZnS@MPS Nanosensor for Highly Sensitive Determination and Removal of Ag^+. *Coatings* **2023**, *13*, 1557. [CrossRef]

coatings

Article

The Effects of Several Metal Nanoparticles on Seed Germination and Seedling Growth: A Meta-Analysis

Huanhuan Guo [1], Yong Liu [1,*], Jidai Chen [2], Yan Zhu [3,*] and Zihan Zhang [3]

1 Key Laboratory for Silviculture and Conservation, Ministry of Education, Beijing Forestry University, Beijing 100083, China; ghh@bjfu.edu.cn
2 Key Laboratory of Digital Earth Science, Aerospace Information Research Institute, Chinese Academy of Sciences, Beijing 100094, China; chenjidai@aircas.ac.cn
3 Research Institute of Forestry, Chinese Academy of Forestry, State Key Laboratory of Tree Genetics and Breeding, Beijing 100091, China; bobzhang_2007@hotmail.com
* Correspondence: lyong@bjfu.edu.cn (Y.L.); zhuyanzi522@163.com (Y.Z.)

Abstract: Using the proper means to improve seed germination is of great significance in agriculture and forestry. Here, a meta-analysis was used to examine whether metal nanoparticle treatments have a specific effect on the seed germination and seedling growth of agricultural species. Using the Web of Science (1950–2021), PubMed (1950–2021), and Scopus (1950–2021) databases, a paper search was conducted using the following items ("nanoparticles" and "seed germination", "nanomaterials" and "seed germination") to filter the references in the title, abstract, and keywords of the published articles. The results indicated that nanoparticle (NP) treatments had a significantly positive effect on the final germination percentage (FGP), with a mean difference (MD) (that is, the overall effect) of 1.97 (0.96, 2.98) for the silver (Ag)-NP subgroup, 1.21 (0.34, 2.09) for the other-NP subgroup, 1.40 (0.88, 1.92) for the total based on the NP types, 1.47 (0.85, 2.09) for the "Concentrations: <50 mg/L" subgroup, and 1.40 (0.88, 1.92) for the total based on the NP concentrations. Similarly, root length (RL) was positively and significantly affected by NP treatment, with an MD (95% CI) of 0.92 (0.76, 1.09) for the zinc (Zn)-NP subgroup, 0.79 (0.65, 0.92) for the other-NP subgroup, 0.82 (0.72, 0.93) for the total based on the NP types, 0.90 (0.77, 1.04) for the "Concentrations: ≤50 mg/L" subgroup, 0.80 (0.60, 0.99) for the "Concentrations: >50 mg/L" subgroup, and 0.82 (0.72, 0.93) for the total based on the NP concentrations. However, there was no statistical correlation between the nanoparticle concentrations and shoot length (SL), due to the inclusion of zero in the 95% CI of the overall effect. Therefore, Ag-NPs could increase the FGP more than other-NPs, while Zn-NPs enhanced RL more. Moreover, NPs at lower concentrations could improve the FGP and RL of crop species to a larger extent than NPs at higher concentrations. This meta-analysis can provide a reference for the nanoparticle treatment technology utilization in agricultural and forest seeds.

Keywords: metal nanoparticles; final germination percentage; root length; shoot length; meta analysis

Citation: Guo, H.; Liu, Y.; Chen, J.; Zhu, Y.; Zhang, Z. The Effects of Several Metal Nanoparticles on Seed Germination and Seedling Growth: A Meta-Analysis. *Coatings* **2022**, *12*, 183. https://doi.org/10.3390/coatings12020183

Academic Editor: Aivaras Kareiva

Received: 4 January 2022
Accepted: 30 January 2022
Published: 31 January 2022

1. Introduction

Currently, advancements in manufacturing have led to the fabrication of nanoparticles with various sizes and shapes in large quantities; as a result, scientific studies have been conducted that investigate the environmental risks and toxic effects of nanoparticles [1,2]. Nanoparticles (NPs), with a particle size less than 100 nm in at least one dimension, can modify the physicochemical properties of a material compared with the corresponding bulk material, such as the reduction ability and conductivity, allowing them to efficiently enhance catalysis and to adsorb and deliver substances of interest [3–5]. Due to the continuously increasing production and use of nanoparticles in a variety of instruments and goods, plants are more prone to exposure to nanoparticles, which may accumulate in the living organs and cells of plants through direct exposure, contaminated soil, or air pollution [6]. In plant science, the most commonly used NPs are metal-based NPs, such as silver, titanium, zinc, and gold NPs, which were selected by certain researchers to better understand their effects on plants; therefore, mainly metal-based NPs (hereafter "metal NPs") are discussed in this study. Moreover, it is well-known that the effects induced by these materials are determined by the NP-type, the plant species, and the growth media, which are inconsistent among the various studies [7]. The effects of metal NPs on plants remain unresolved.

The phytotoxicity of metal nanoparticles has been studied for the plant species in seed germination and root elongation tests with the goal of promoting their use for agricultural applications in recent years, and the germination percentage (GP) and seedling vigor index (SVI) calculated from the root length and the shoot length as indicators are commonly used in seed germination and root elongation studies, as they provide a good estimate of potential field performance [1,2,4,8]. Data from limited studies have reported both positive and negative effects of metal NPs on higher plants (mostly agricultural species), which are therefore mainly discussed in this study [4,9]. López-Moreno et al. (2010) showed that the germination of corn, tomato, and cucumber seeds was decreased significantly (approximately 30%, 30%, and 20%, respectively) by nanoceria at 2000 mg/L [10]. For root growth, cucumber and corn root elongations were improved by nanoceria, while alfalfa and tomato root elongations were inhibited. Moreover, nanoceria improved the shoot length in the four plant species at approximately all concentrations. Feizi et al. (2013) found that nanosized TiO_2 with low and intermediate concentrations increased germination indicators, such as the germination value, vigor index, and mean daily germination, of fennel (*Foeniculum vulgare* Mill) seeds [11]. Moreover, the results found by Kumar et al. (2013) indicated that GNP (gold nanoparticle) exposure at both 10 and 80 μg/mL concentrations has significantly enhanced the seed germination rate, vegetative growth, and free radical scavenging activity of *Arabidopsis thaliana* [12]. However, seed germination, emergence, and the lengths of plumules and the principal and seminal roots of maize (*Zea mays*), were significantly inhibited by ZnO-NPs and CuO-NPs [13]. In addition, seed priming utilizing different types of metal NPs has been proven to enhance the seed germination and seedling vigor of agricultural species, which may be because NP treatment stimulates several metabolic mechanisms associated with seed germination, including the upregulation of aquaporin genes, α-amylase activity, reactive oxygen species (ROS) production, and antioxidant systems [8,14].

Thus, understanding the phytotoxic behavior of the metal NPs is necessary before using them under field conditions [2]. There were some reviews and systematic assessments concerning the metal NP effects on crop growth, while few meta-analyses could be found. Meta-analysis, mainly used in medical science, is a useful approach to explore whether metal NPs can promote seedling growth or not [15,16]. Therefore, the aim of this study was to explore the effects of metal NPs on the final germination percentage (FGP), root length (RL), and shoot length (SL) of the agricultural species through a meta-analysis to provide a meaningful reference for the wider utilization of the metal NPs, such as with forest species.

2. Materials and Methods

2.1. Data Collection

The data search was carried out in the Web of Science (https://www.webofscience.com/wos/alldb/basic-search), PubMed (https://pubmed.ncbi.nlm.nih.gov), and Scopus (https://www.scopus.com/search/form.uri?display=basic#basic) databases and was concluded on 3 September 2021. First, the keywords "nanoparticles and seed germination" were used to search for the articles via the three databases, and data visualization using social network analysis was conducted by means of the information (all records and cited references) exported from the databases (Figure 1). Studies reflecting the relationship between nanoparticles and growth, germination, oxidative stress, and antioxidant enzymes are abundant, indicating that nanoparticles may affect seed germination and seedling growth. Then, based on the data visualization, the search keywords ("nanoparticles" and "seed germination", "nanomaterials" and "seed germination") were utilized to screen the references according to the title, abstract, and keywords of the articles published without language restriction and without excluding any botanical family [17]. A selection process was then carried out to eliminate duplicate articles and those with topics not related to the objective of the meta-analysis. The papers that passed the first filter were submitted to a full text review by establishing selection criteria that each one had to meet to be accepted into the meta-analysis. The inclusion criteria of studies were as follows: (1) the seeds other than the seedlings were treated with nanoparticles; (2) both the control group (seeds treated with water) and the experimental group (seeds treated with metal nanoparticles) with three or more replicates were included in each study; (3) final germination percentage, root length, and shoot length were used as germination response variables in each study; (4) the mean and standard deviation (SD) for each variable and number of seeds used in the treatments were provided; and (5) only metal nanoparticle-involved studies were retained.

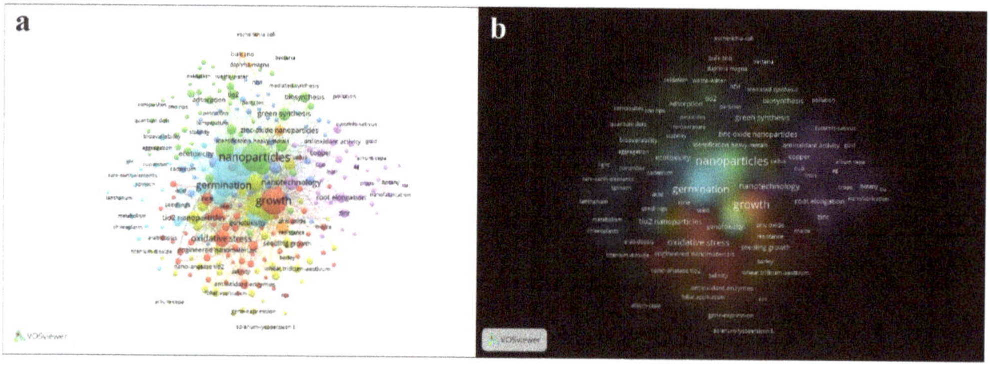

Figure 1. (**a**) Keywords of "nanoparticles/nanomaterials and seed germination" co-occurrence network visualization; (**b**) keywords of "nanoparticles/nanomaterials and seed germination" density visualization. The "curves" are the representatives of co-occurrence between the connected "nodes", and each node denotes a "keyword". Different keywords were segregated into different clusters, shown in different colors.

2.2. *Risk of Bias and Publication Bias Assessment of the Included Studies*

The qualitative aspects of the methodological process to evaluate the risk of bias for the included studies were as follows: sterilizing the seeds used in all treatments; seed germination and seedling growth under controlled conditions; seed exposure to a direct nanoparticle suspension; uniformity of the seeds; and cultivation of the seeds on the filter papers inside Petri dishes. The results were illustrated graphically in the forest plots of three indices, in which it was specified whether or not the inclusion criterion was met (a green color represents the study meeting the criterion with a low risk of bias, the yellow represents the study not meeting the criterion with an unclear risk of bias, and red represents the study not meeting the criterion with a high risk of bias). Publication bias was assessed through a funnel plot.

2.3. *Germination and Seedling Growth Response Variables*

We used the final germination percentage (FGP) (Equation (1)), root length (RL), and shoot length (SL) at optimal nanoparticle treatment concentrations as the main response variables for the germination treatments:

$$\mathrm{FGP} = S_g / S_s \times 100 \tag{1}$$

where FGP is the final germination percentage, S_g is the number of seeds germinated, and S_s is the number of seeds sown. The three outcome indicators were selected because they can reflect the effects of nanoparticle treatment on seed germination and seedling growth [8]. The outcome index values at the best concentration were selected, which can reflect the potential of nanoparticles to promote seed germination and seedling growth with little random error. Why not choose the value under the same concentration treatment or that under the same treatment time? The reason is that the variation in outcome indicators with nanoparticle concentrations does not show the same trend in each study. In addition, although the seeds used in each study belong to the crop species, there are differences in seed germination ability and germination time of different species or different varieties of the same species. Selecting these values would disturb the results of this meta-analysis and decrease its reliability.

2.4. Meta-Analysis and Statistical Analyses

The meta-analysis, an analytical method integrating the different results of various studies into a single common result, was developed through the international standard called Preferred Reporting Items for Systematic Meta-analyses (PRISMA) in this study [17]. In our study, Review Manager (Version 5.4.1) software was used to analyze the data for performing the meta-analysis to explore the effect of metal NP treatments on seed germination and seedling growth by calculating the mean difference (MD):

$$MD_{FGP} = T_{FGP} - C_{FGP} \tag{2}$$

$$MD_{RL} = T_{RL} - C_{RL} \tag{3}$$

$$MD_{SL} = T_{SL} - C_{SL} \tag{4}$$

where MD_{FGP}, T_{FGP}, and C_{FGP} represent the mean differences in FGP, FGP of the treatment group, and FGP of the control group, respectively. MD_{RL}, T_{RL}, and C_{RL} represent the mean differences of RL, RL of the treatment group, and RL of the control group, respectively, and MD_{SL}, T_{SL}, and C_{SL} represent the mean differences of SL, SL of the treatment group, and SL of the control group, respectively. Therefore, MD more than zero favors the treatment group. The chi-square (Q) test was selected to evaluate the study, subgroup, and total heterogeneity [18]. The inconsistency index (I^2) could be computed from the Q value and degree of freedom (df) according to Higgins et al. (2003) as follows [19]:

$$I^2\ (\%) = 100 \times [(Q - \mathrm{df})/Q] \tag{5}$$

A larger I^2 value indicated a high extent of heterogeneity or inconsistency among the data of the selected studies, and whenever the I^2 value was negative, it was set to zero, indicating no observed heterogeneity at all. In this meta-analysis, I^2 values lower than 25%, 25–50%, and above 50% represent low, moderate, and high heterogeneity, respectively [17]. All figures exported from Review Manager (Version 5.4.1) software in our study were processed using Adobe Photoshop software to obtain clear figures with proper sizes.

3. Results

3.1. Description of the Dataset

According to the social network analysis, the topics such as germination, root elongation, and growth included in this study were present in a large proportion. Then, by searching the databases, a total of 1860 articles were found, of which 884 were from the Web of Science, 271 were from the PubMed, and 705 were from the Scopus database. After removing the duplicates, screening for the title and the abstract, and selecting full-text articles with reliability and eligibility, and which met the inclusion criteria, 29 original articles were found to be eligible for the meta-analysis (Figure 2).

All of the included studies evaluated the effect of metal nanoparticle treatment on seed germination or seedling growth. Based on Table 1, all the species used by these studies belong to agricultural seeds, such as maize (*Zea mays*), radish (*Raphanus sativus*), barley (*Hordeum vulgare* L., cv. Annabell), rice (*Oryza sativa* L., cv. Swarna), and lettuce (*Lactuca sativa*), among others. There were various nanoparticle types, including Ag-NPs, Zn-NPs, Al-NPs, Ce-NPs, Fe-NPs, Cu-NPs, and Ti-NPs, with the exposure mode of suspension or suspension mixture with soil. The best concentrations of these metal NPs varied among different metal NP types, different indicators, and different species. In terms of the growth media, the seeds were mostly cultivated on filter papers in Petri dishes, followed by pots with soil, semisolid agar medium, and MS medium. Moreover, FGP, RL, and SL were evaluated as the outcome indices (Table 1).

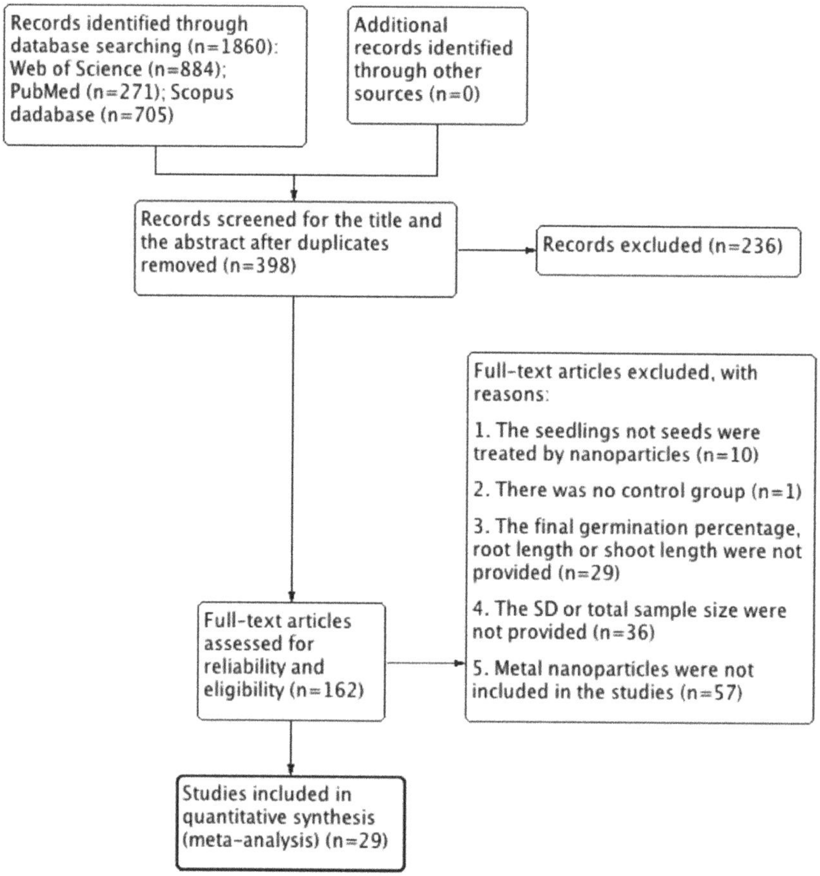

Figure 2. Flow chart of the study selection in the meta-analysis.

3.2. Risk of Bias and Publication Bias Assessment of the Included Studies

The funnel plot can be used to intuitively show the publication bias. The studies for the final germination percentage and root length analysis presented a roughly uniform distribution based on the overall effect lines (Figure 3a,b), while the studies for the shoot length analysis did not (Figure 3c). Among the three outcome indicators, studies with large sample sizes were in the majority, distributed at the top of the diagram, whereas those with small sample sizes were in the minority, accounting for fewer weight values in the heterogeneity analysis, which were distributed in the lower part. Overall, the studies for the shoot length analysis might exhibit a certain publication bias.

Table 1. Literature review of the studies included in the meta-analysis.

No.	Study	Seed Species	Metal Nanoparticle Treatment			Cultivation Media	Sample Size	Indicators
			Type	Exposure Mode	Best Concentration			
1	Acharya et al., 2020 [5]	Watermelons (*Citrullus lanatus*)	nano-Ag	Suspension	31.3 ppm	Filter papers in Petri dishes	400	FGP
2	Acharya et al., 2020a [20]	Onion (*Allium cepa*)	nano-Ag	Suspension	31.3 ppm	Filter papers in Petri dishes	300	FGP
3	Ahmed et al., 2021 [13]	Maize (*Zea mays*)	nano-Zn	Suspension	0.05 mg/mL	Semisolid agar in Petri dishes	90	SL
4	Belhamel et al., 2020 [21]	*P. vulgaris* var. "Piattelli"	nano-Al	Suspension	1.6 mg/mL	Filter papers in Petri dishes	50	RL
5	Corral-Diaz et al., 2014 [22]	Radish (*Raphanus sativus*)	nano-Ce	Mixture with soil	62.5 mg/kg [b], 125 mg/kg [c]	Pots with loamy sand soil	32	RL, SL
6	Duran et al., 2018 [23]	*Phaseolus vulgaris*	nano-Fe	Suspension	10 mg/L	Filter papers in Petri dishes	100	RL
7	El-Temsah and Joner, 2012 [9]	Barley (*Hordeum vulgare* L., cv. Annabell)	nano-Ag	Suspension	10 mg/L	Filter papers in Petri dishes	30	SL
8	Gupta et al., 2018 [24]	Rice (*Oryza sativa* L., cv. Swarna)	nano-Ag	Suspension	20 ppm	0.8% agar medium	30	RL
9	Kasote et al., 2019 [25]	Watermelon (*Citrullus lanatus*)	nano-Fe	Suspension	160 mg/L	Filter papers in Petri dishes	60	SL
10	Li et al., 2016 [26]	Corn (*Zea mays*)	nano-Fe	Suspension	50 mg/L	Filter papers in Petri dishes	30	FGP
11	Li et al., 2021 [27]	Fragrant rice varieties, Xiangyaxiangzhan and Yuxiangyouzhan (*Oryza sativa*)	nano-Zn	Suspension	50 mg/L [a], 25 mg/L [b]	Filter papers in Petri dishes	100 [a], 40 [b]	FGP, RL
12	Lin and Xing, 2007 [4]	Lettuce (*Lactuca sativa*)	nano-Al	Suspension	2000 mg/L	Filter papers in Petri dishes	30	FGP
13	Liu et al., 2018 [28]	Rice (*Oryza sativa japonica*)	nano-Cu	Mixture with soil	100 mg/L	Growth containers with soil	200	FGP
14	López-Moreno et al., 2017 [29]	Corn (*Zea mays*)	nano-Zn	Suspension	800 ppm	Filter papers in Petri dishes	30	RL
15	Mahakham et al., 2017 [14]	Jasmine rice (*Oryza sativa* L. cv. KDML105)	nano-Ag	Suspension	20 mg/L	Filter papers in Petri dishes	30	FGP

Table 1. *Cont.*

No.	Study	Seed Species	Metal Nanoparticle Treatment			Cultivation Media	Sample Size	Indicators
			Type	Exposure Mode	Best Concentration			
16	Nguyen et al., 2021 [30]	Green and red beans	nano-Zn	Suspension	10 mg/L ^{a,b,c}	Filter papers in Petri dishes	60	FGP, RL, SL
17	Saquib et al., 2016 [31]	Radish (*Raphanus sativus*)	nano-Fe	Suspension	0.25 mg/L	Filter papers in Petri dishes	60	RL
18	Segura et al., 2020 [32]	Radish (*Raphanus sativus*)	nano-Ag	Suspension	500 μg/mL	Filter papers in Petri dishes	30	RL
19	Singh et al., 2016 [33]	Tomato (*Solanum lycopersicum*)	nano-Zn	Suspension	1.2 mM ^{b,c}	Filter papers in Petri dishes	30	RL, SL
20	Singh et al., 2019 [34]	Wheat (*Triticum aestivum*)	nano-Zn	Suspension	250 mg/L ^{b}, 15 mg/L ^{c}	Filter papers in Petri dishes	12	RL, SL
21	Singh et al., 2020 [35]	Wheat (*Triticum aestivum*)	nano-Ag	Suspension	25 mg/L	Filter papers in Petri dishes	75	FGP, RL, SL
22	Song et al., 2013 [36]	Tomato (*Lycopersicon esculentum*)	nano-Ag	Suspension	100 mg/kg	Filter papers in Petri dishes	50	FGP
23	Subpiramaniyam et al., 2021 [37]	Mung bean (*Vigna radiata*)	nano-Cu	Mixture with soil	1 mg/kg ^{b,c}	Glass beakers with soil	30	RL, SL
24	Sun et al., 2019 [38]	Mung bean (*Vigna radiata*)	nano-Fe	Suspension	450 mg/L ^{a}, 150 mg/L ^{b}	Filter papers in Petri dishes	30	FGP, RL
25	Tan et al., 2017 [39]	Basil (*Ocimum basilicum*)	nano-Ti	Mixture with soil	750 mg/kg	Pots with topsoil	16	RL
26	Trujillo-Reyes et al., 2013 [40]	Radish (*Raphanus sativus*)	nano-Ce	Suspension	50 ppm ^{b,c}	Filter papers in Petri dishes	30	RL, SL
27	Wang et al., 2016 [41]	*Arabidopsis thaliana*	nano-Cu	Suspension	20 mg/L	MS media	60	FGP
28	Yadu et al., 2018 [42]	*Cajanus cajan*	nano-Ag	Suspension	1.2 nM	Filter papers in Petri dishes	350	FGP
29	Zuverza-Mena et al., 2016 [43]	Radish (*Raphanus sativus*)	nano-Ag	Suspension	125 mg/L	Filter papers in Petri dishes	120	FGP

Note: FGP—final germination percentage; RL—root length; SL—shoot length. ^{a} value for FGP; ^{b} value for root length; ^{c} value for shoot length.

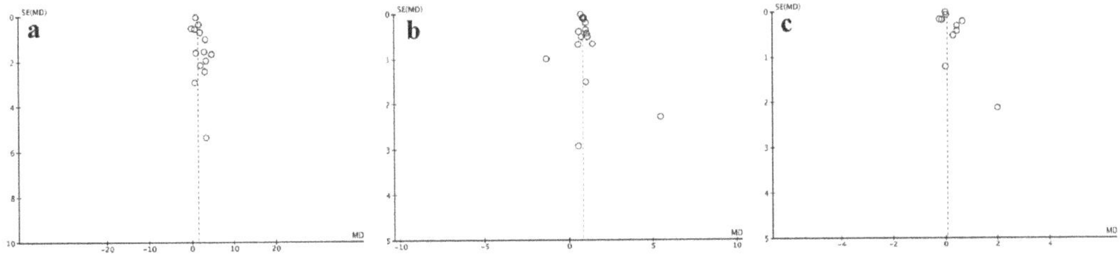

Figure 3. (**a**) Funnel plot for FGP; (**b**) RL; and (**c**) SL. The dotted lines represent the overall effects; hollow circles represent the studies.

The risk of bias assessment (Figure 4c, Figure 5c, Figure 6c) indicated that 64.29% of the studies for the final germination percentage analysis, 75% for the root length analysis, and 60% for the shoot length analysis performed the sterilization of the seeds used in all treatments, representing a low risk, whereas 35.71%, 6.25%, and 10%, respectively, did not illustrate whether the seeds were sterilized or not, representing an unclear risk; moreover, 18.75% and 30% for the latter two did not sterilize the seeds, thereby indicating a high risk. For the conditions utilized for seed germination and seedling growth, 71.43% of the studies for the final germination percentage, 62.5% for the root length, and 50% for the shoot length showed the detailed conditions, such as light intensity, photoperiod, temperature, etc., with a low risk, while 28.57%, 37.5%, and 50%, respectively, did not indicate the details, with an unclear risk. In 92.86% of the studies for the final germination percentage, 87.5% for the root length and 90% for the shoot length, the seeds were exposed to a direct nanoparticle suspension, showing a low risk, while in the remaining 7.14%, 12.5%, and 10% studies, respectively, the seeds were not, showing a high risk. All of the included studies for the three indicators reported the uniformity of the seeds with a low risk. The seed germination trial was carried out on filter papers inside Petri dishes in 85.71% of the studies for the final germination percentage, 75% for the root length, and 70% for the shoot length with a low risk, whereas 14.29%, 25%, and 30% of the studies had a high risk, respectively. Overall, 82.86% of the studies for the final germination percentage analysis, 80% for the root length analysis, and 74% for the shoot length analysis corresponded with all the evaluation criteria; however, 17.14%, 20%, and 26%, respectively, showed no accord with one or more criteria.

3.3. *Effect of Metal Nanoparticles on the Final Germination Percentage*

Based on the forest plot, a two-type subgroup analysis was carried out on the basis of nanoparticle types or nanoparticle concentrations to explore the effect of the two factors on the final germination percentage of seeds (Figure 4a,b). First, in the nanoparticle type-based subgroup analysis, two subgroups of Ag nanoparticles (Ag-NPs) and other-NPs were included, because other-NPs, such as Fe-NPs, Zn-NPs, Cu-NPs, and Al-NPs, involved studies in a few quantities, which would result in a decrease in meta-analysis reliability. There was no heterogeneity between the two subgroups (p = 0.27, I^2 = 19%), while moderate heterogeneity existed in each subgroup (I^2 = 36% for the Ag-NP subgroup; I^2 = 48% for the other-NP subgroup). The overall effects of the two subgroups were significant (p < 0.001), with a mean difference (MD) of 1.97 for the Ag-NP subgroup, 1.21 for the other-NP subgroup, and 1.40 for the total NP-type subgroups; the 95% CI was 0.96 and 2.98 for the Ag-NP subgroup, 0.34 and 2.09 for the other-NP subgroup, and 0.88 and 1.92 for the total NP-type subgroups (Figure 4a), showing a significant effect of nanoparticle type on the final germination percentage. After dividing the subgroups, the heterogeneity of each subgroup did not show significant variation compared with the total, indicating that the nanoparticle type might not have been the major source of heterogeneity in this research.

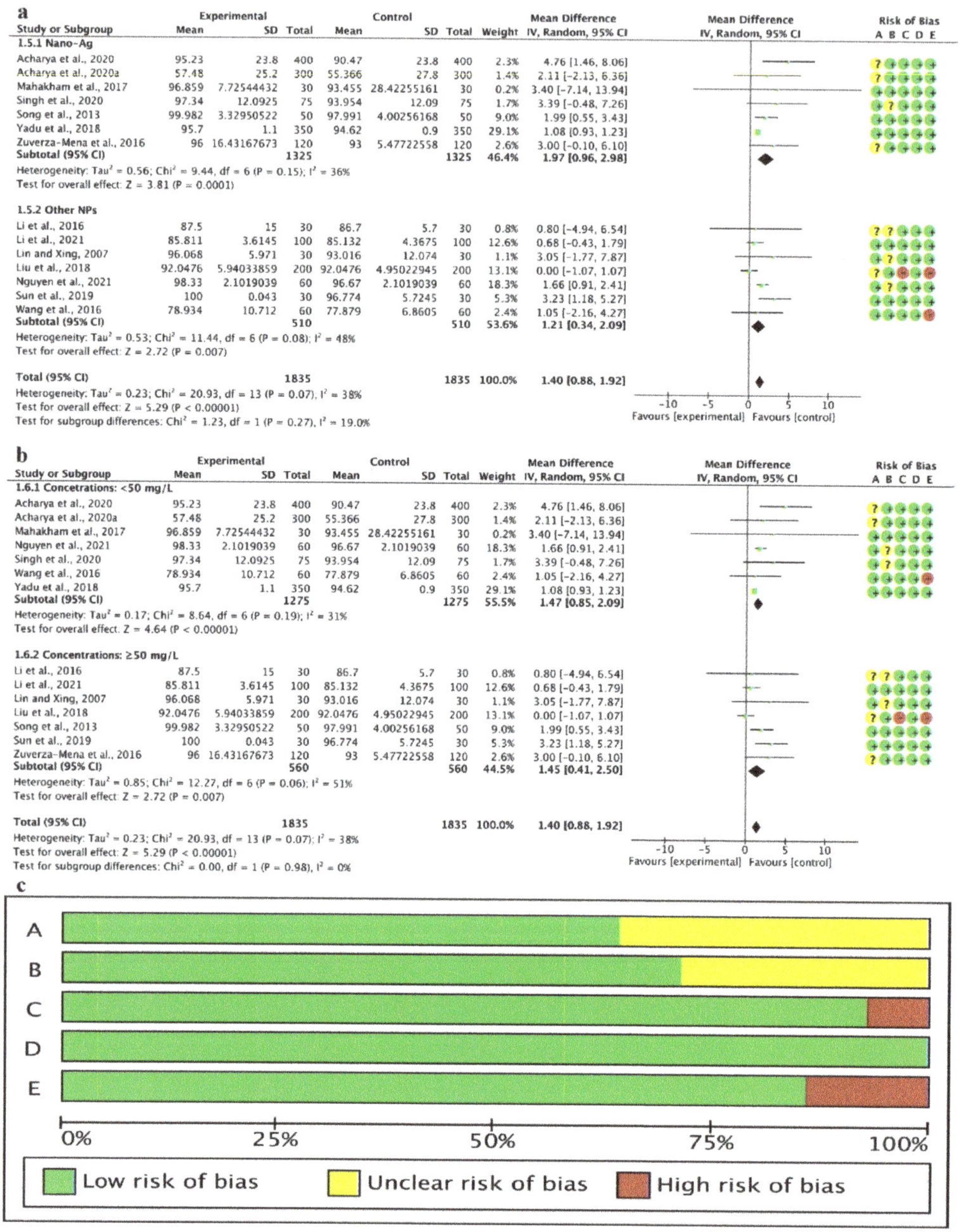

Figure 4. Heterogeneity evaluation, subgroup analysis, and risk of bias assessment for FGP. (**a**) Subgroup analysis for NP types; (**b**) subgroup analysis for NP concentrations; (**c**) risk of bias assessment, A—sterilizing the seeds used in all treatments, B—seed germination and seedling growth under controlled conditions, C—seed exposure to a direct nanoparticle suspension, D—uniformity of the seeds, E—cultivation of the seeds on the filter papers inside Petri dishes (the same below).

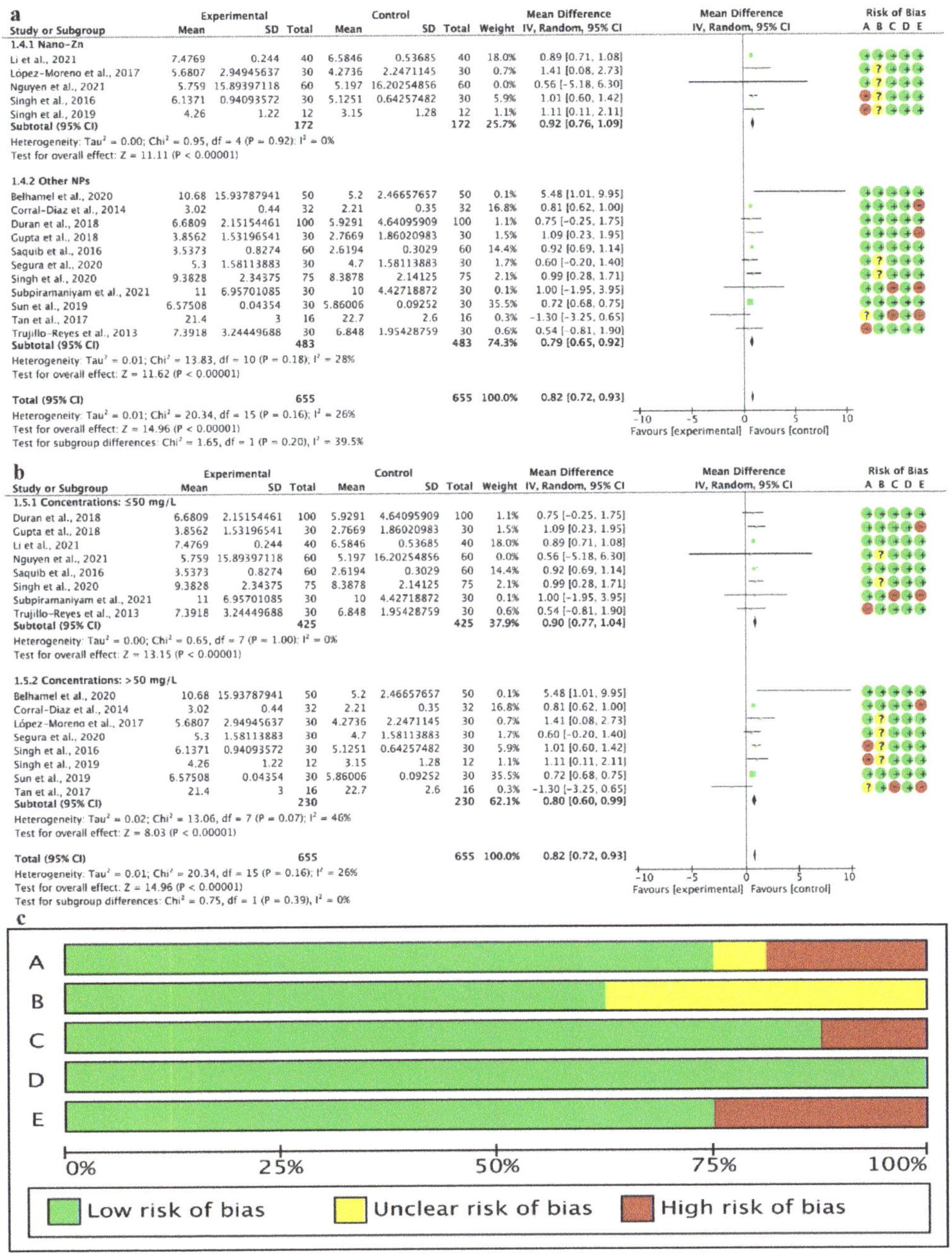

Figure 5. Heterogeneity evaluation, subgroup analysis, and risk of bias assessment for RL. (**a**) Subgroup analysis for NP types; (**b**) subgroup analysis for NP concentrations; (**c**) risk of bias assessment.

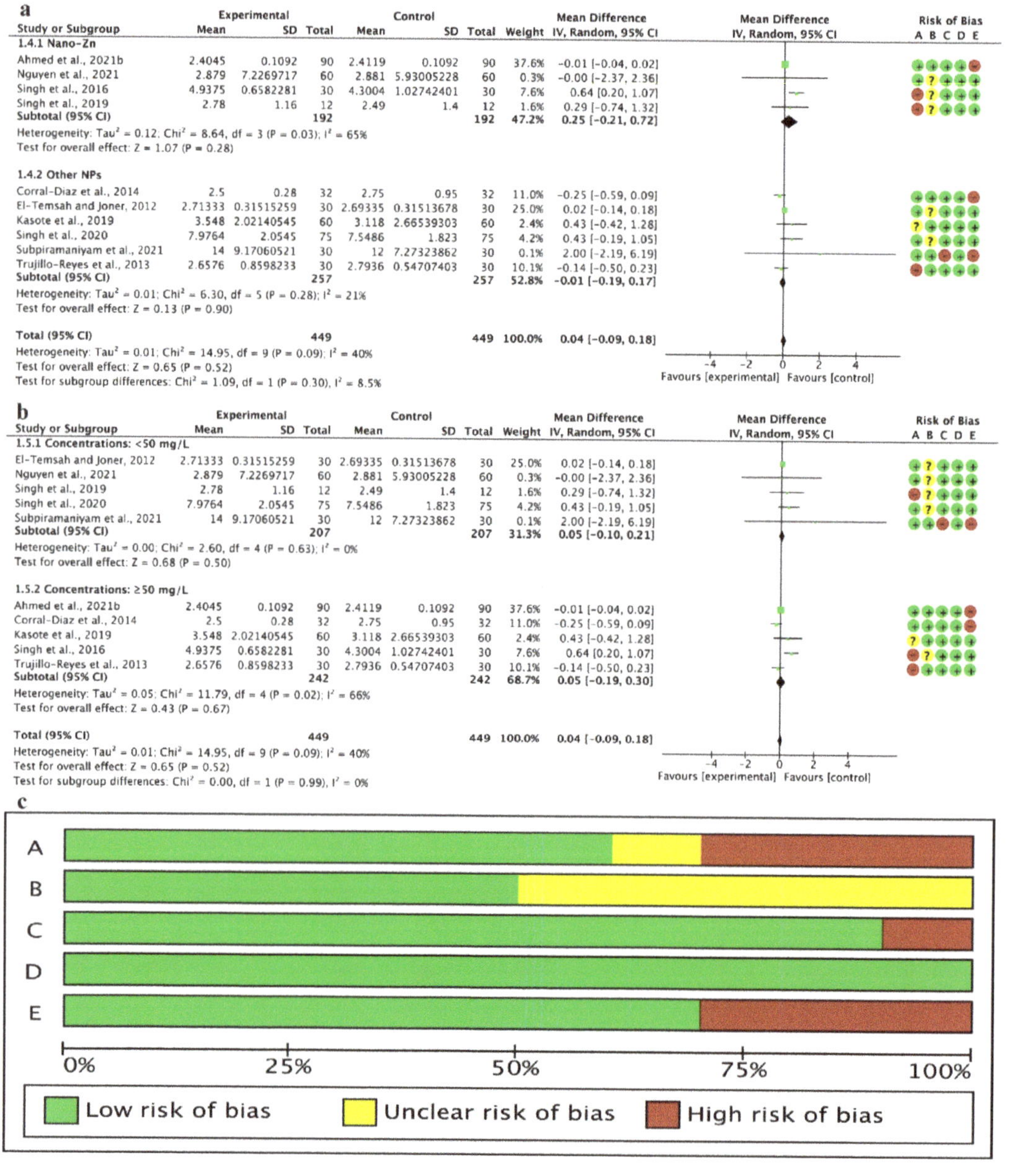

Figure 6. Heterogeneity evaluation, subgroup analysis, and risk of bias assessment for SL. (**a**) Subgroup analysis for NP types; (**b**) subgroup analysis for NP concentrations; (**c**) risk of bias assessment.

In the nanoparticle concentration-based subgroup analysis, there were two subgroups consisting of "Concentrations: <50 mg/L" and "Concentrations: ≥50 mg/L", for which approximately similar numbers of studies were included in each subgroup in this way. There was no heterogeneity for the two subgroups (p = 0.98, I^2 = 0%), whereas moderate heterogeneity existed in the "Concentrations: <50 mg/L" subgroup (I^2 = 31%) and the total (I^2 = 38%), and greater heterogeneity existed in the "Concentrations: ≥50 mg/L" subgroup (I^2 = 51%) with an unreliable overall effect. Therefore, the overall effects of the

"Concentrations: <50 mg/L" subgroup and the total were effective and reliable, with MD (95% CI) values of 1.47 (0.85, 2.09) and 1.40 (0.88, 1.92), respectively (Figure 4b), showing the significant effect of nanoparticle concentrations on the final germination percentage. Dividing the subgroups in this way did not markedly affect the heterogeneity in each subgroup compared with the total; as a result, the nanoparticle concentration may be not the major heterogeneity source herein.

3.4. Effect of Metal Nanoparticles on Root Length

Two-type subgroup analysis was also carried out based on the nanoparticle types (Figure 5a) or nanoparticle concentrations to explore the effect of the two factors on the root length of the seedlings (Figure 5b). In the nanoparticle type-based subgroup analysis, two subgroups of Zn-NPs and other-NPs were included, for which the reason was that other-NPs, such as Fe-NPs, Cu-NPs, Ag-NPs, Ce-NPs, etc., involved fewer than three studies, causing an unreliability of this meta-analysis. There was moderate heterogeneity between the two subgroups ($p = 0.20$, $I^2 = 39.5\%$), and for the other-NP subgroup ($p = 0.18$, $I^2 = 28\%$) and the total ($p = 0.16$, $I^2 = 26\%$), whereas no heterogeneity existed in the Zn-NP subgroup ($p = 0.92$, $I^2 = 0\%$), indicating that the nanoparticle types might be a major source of the heterogeneity. The overall effects of the two subgroups were significant ($p < 0.001$), with an MD of 0.92 for the Zn-NP subgroup, 0.79 for the other-NP subgroup, 0.82 for the total, and the 95% CIs of (0.76, 1.09), (0.65, 0.92), and (0.72, 0.93), respectively (Figure 5a), showing a significantly positive effect of nanoparticle types on the root length.

Additionally, in the nanoparticle concentration-based subgroup analysis, two subgroups consisting of "Concentrations: ≤50 mg/L" and "Concentrations: >50 mg/L", with approximately similar numbers of studies each, were included in this study. There was no heterogeneity between the two subgroups ($p = 0.39$, $I^2 = 0\%$) and for the "Concentrations: ≤50 mg/L" subgroup ($p = 1.00$, $I^2 = 0\%$), indicating the possibility of the concentration as the major heterogeneity source, whereas moderate heterogeneity existed in the "Concentrations: >50 mg/L" subgroup ($I^2 = 46\%$) and the total ($I^2 = 26\%$), showing that the concentration gradients of more than 50 mg/L may present different effects on root length, which requires more detailed division. Moreover, the overall effects were 0.90 (0.77, 1.04) for the "Concentrations: ≤50 mg/L" subgroup, 0.80 (0.60, 0.99) for the "Concentrations: >50 mg/L" subgroup, and 0.82 (0.72, 0.93) for the total, respectively, of the MD (95% CI) (Figure 5b), showing the significantly positive effect of nanoparticle concentrations on root length.

3.5. Effect of Metal Nanoparticles on Shoot Length

Similarly, two-type subgroup analysis was performed based on the nanoparticle types (Figure 6a) or nanoparticle concentrations to explore the effect of the two factors on the shoot length of the seedlings (Figure 6b). In terms of the nanoparticle type-based subgroup analysis, two subgroups of Zn-NPs and other-NPs were included, with a similar reason as the former two. Moderate heterogeneity existed in both subgroups ($p = 0.03$, $I^2 = 65\%$; $p = 0.28$, $I^2 = 21\%$) and the total ($p = 0.09$, $I^2 = 40\%$). The overall effects of the two subgroups and the total were not significant ($p > 0.05$), with an MD (95% CI) of 0.25 (−0.21, 0.72) for the Zn-NP subgroup, −0.01 (−0.19, 0.17) for the other-NP subgroup, and 0.04 (−0.09, 0.18) for the total (Figure 6a), which showed that there was no statistical correlation between the nanoparticle types and the shoot length due to the cross between the diamond (that is, the overall effect) and the invalid line.

In the nanoparticle concentration-based subgroup analysis, two subgroups consisting of "Concentrations: <50 mg/L" and "Concentrations: ≥50 mg/L", with approximately similar numbers of studies each, were included in this study. There was no heterogeneity for the "Concentrations: <50 mg/L" subgroup ($p = 0.63$, $I^2 = 0\%$), indicating the possibility of the concentration as the major heterogeneity source, while there was moderate heterogeneity for the "Concentrations: ≥50 mg/L" subgroup ($p = 0.02$, $I^2 = 66\%$) and the total ($p = 0.09$, $I^2 = 40\%$). The overall effects of subgroups and the total were ineffective because zero was included in the 95% CI, with an MD (95% CI) of 0.05 (−0.10, 0.21) for the "Concentrations: <50 mg/L" subgroup, 0.05 (−0.19, 0.30) for the "Concentrations: ≥50 mg/L" subgroup, and 0.04 (−0.09, 0.18) for the total (Figure 6b). Similarly, there was no statistical correlation between the nanoparticle concentrations and the shoot length due to the inclusion of zero in the 95% CI.

4. Discussion

4.1. FGP Increased More under Ag-NPs Treatment

NP types have a positively significant effect on FGP, with overall effect values (1.97 and 1.21 in each subgroup) and their confidence intervals greater than zero. It can be seen that Ag-NPs can increase the FGP more than other-NPs. The seed germination process culminates in the rupture of the seed coat and the emergence of the radicle, which enables direct contact with the NPs in the soil media and potentially impacts the development of the seed [37]. Ag-NP is a growing hot topic for researchers owing to its imperative physiochemical properties [42]. Mahakham et al. (2017) indicated that Ag-NPs can penetrate the seed coat and accelerate water uptake to promote seed germination and starch metabolism in rice; additionally, Ag-NP priming can upregulate the expression of aquaporin genes, thus facilitating water and H_2O_2 diffusion, and increased H_2O_2 may act as a signaling molecule for stimulating the germination process [14]. Acharya et al. (2020) found that the seeds treated with Ag-NPs can improve the seed germination of watermelons through an eco-friendly and sustainable nanotechnological approach because NPs have the advantage of being able to trigger certain metabolic processes (e.g., enhancing the levels of glucose and fructose, thus promoting glycometabolism) that are normally activated during the early phase of germination [5]. Moreover, Acharya et al. (2020a) found that NP treatments selectively modulated ZA and GABA levels in onion seed germination compared with the control, and then significantly affected germination inhibitors in onion seeds along with these germination stimulators [20].

Exogenous application of Ag-NPs can also alleviate the damage of adverse conditions to seed germination and enhance the stress resistance of plants. Yadu et al. (2018) showed that Ag-NP treatment promoted the germination percentage of *Cajanus cajan* under fluoride stress, and decreased reactive oxygen species (ROS) levels by suppressing the expression of the NADPH oxidase (NOX) gene and lipoxygenase (LOX) activity [42]. Due to their involvement in regulating ROS generation and its scavenging, Ag-NPs can affect antioxidant enzyme activity levels and gene expression patterns [24]. However, the chemical composition, particle size, shape, synthesis methods, and surface coating of Ag-NPs and their exposure form, and the plant species used, should be taken into consideration, due to their various effects on plants [34,44].

4.2. RL Enhanced More under Zn-NPs Treatment

Nutrient uptake improvement is an objective in crop breeding. The root system plays an essential role in nutrient and water acquisition, and its architecture is the spatial arrangement of roots (such as root length) that affects the capacity of plants to access nutrients [45,46]. Guo et al. (2017) found that root proliferation in response to external nitrate is a behavior which integrates local N availability and the systemic N status of the plant [46]. In this study, Zn-NPs could enhance the root length more than other-NPs. As an important transitional metal, zinc is the only metal present in all six classes of enzymes that acts as a cofactor for many essential enzymes of plants at below threshold levels, and a functional component for several transcription factors [13,47]. In eukaryotes, zinc is mainly dominated by 10% of zinc-binding proteins and 36% of zinc proteins, which are involved in gene expression. Thus, zinc should be used to regulate stress-related gene expression, especially in harsh environments [48]. Moreover, Zn is also an essential micronutrient that plays a vital role in the growth and yield of plants by maintaining cell membrane integrity and cell elongation, protein synthesis, and stress tolerance in plants [34].

Zn has potential to increase the biosynthesis of chlorophyll and carotenoids; enhance the contents of pigment, protein, and sugar; and thereby improve the photosynthetic capability of the species [33,49]. Photosynthesis, a fundamental process, plays a significant role in the growth, dry matter production, and yield of species [50]. Wu et al. (2020) also revealed that Zn is regarded as an important factor in plant photosynthesis; the chlorophyll concentration increased significantly with the application of ZnO NPs (10–50 mg/L) [51]. It has been reported that Zn nanofertilizers can improve the fruit yield and quality of pomegranate (*Punica granatum* cv. Ardestani) without affecting its physical characteristics, and ZnO-NPs could have been used as a fungicide in agriculture; thus, seed germination can be improved by treating the seeds with ZnO-NPs [29,30]. ZnO belongs to the class of metal oxides, which is characterized by photocatalytic and photo-oxidizing capacities against chemical and biological species [33]. Indeed, ZnO-NPs play a principal role in physiological and anatomical responses, as well as in hormone metabolism [30]. The root and shoot length of fragrant rice were substantially enhanced with ZnO-NP treatment, which induced modulations in physiological and biochemical attributes, e.g., the superoxide dismutase (SOD) activity, peroxidase (POD) activity, and metallothionein contents in roots, which were increased under low levels of ZnO-NPs [27]. In addition, ZnO-NPs can improve plant growth and induce resistance responses against *Sclerospora graminicola* in pearl millet by activating defense signaling pathways [48]. Notably, green ZnO-NPs could be used as a better material for agricultural products, such as nanofertilizers or nanopesticides, relative to their chemically synthesized counterparts [34].

4.3. FGP and RL Improved More in Lower Levels of NPs

In this study, NP concentrations also significantly affected FGP and RL, and the results indicated that NPs at lower concentrations could improve the FGP and RL of agricultural species to a larger extent. Several studies have obtained similar results; for example, the germination index of corn seeds under treatments with 20 and 50 mg/L γ-Fe_2O_3 NPs was 27.2% and 18.9% higher than that of the control, respectively [26]. Zuverza-Mena et al. (2016) showed that Ag-NPs increased the germination of radish seeds by 3% at the concentration of 125 mg/L, while they reduced the germination by 3% and 6% under 250 and 500 mg/L, respectively [43]. Therefore, the dosage represents a decisive factor influencing the ecological effects of NPs [38]. Higher Ag accumulation would change the structure of amino acids, nitrogenous bases, and nucleotides by forming complexes with them, interfering with the respiratory enzymes, and inducing oxidative stress in seeds, leading to a decrease in seed germination and seedling growth; in contrast, a low concentration of Ag-NPs with a short exposure time could not only reduce the toxicity of Ag-NPs but also enhance the germination and starch metabolism of aged rice seeds [14]. It is worth noting that the toxicity of NPs sometimes presents a nonlinear correlation with their dose, due to their aggregation with increasing density [38].

Similarly, the observation that NPs at lower concentrations can improve the RL of agricultural species to a larger extent has been evaluated in a variety of studies on crop species. Nanoparticle concentrations may affect the agronomic effectiveness of ZnO-NPs [33]. Under green ZnO-NPs at a moderate concentration (62 mg/L), the wheat seed samples presented the most significant enhancement ($p < 0.005$) in root length relative to other concentration levels by 50% [34]. In addition, ZnO-NP treatment at 50 ppm has been demonstrated to increase the seedling growth and reduce the excessive generation of ROS, while adverse effects on rice seedling growth have been observed at concentrations of 500 and 1000 ppm [27]. García-López et al. (2018) also found that ZnO-NP suspensions at 100, 200, and 500 ppm inhibited the seedling growth of *Capsicum annuum* and promoted the accumulation of phenolic compounds with phytotoxic effects [52]. Subpiramaniyam et al. (2021) found that the root length was slightly higher than that of the control (not significantly) under the lowest CuO-NP treatment (1 mg/kg) [37]. Moreover, γ-Fe_2O_3 NPs at a concentration of 20 mg/L significantly promoted root elongation by 11.5%, whereas those at 50 and 100 mg/L remarkably decreased root length by 13.5% and 12.5%, respectively [26]. Excess nZVI (zero-valent iron nanoparticles) accumulation in roots could not only block the uptake and transport of water but also disturb plant nutrient uptake and balance owing to the adherence of nZVI to the root surfaces and penetration into the root tissues, leading to reduced water flow and limited root hydraulic conductivity, thereby inhibiting the root elongation of the species [13,38].

However, there was no statistical correlation between the nanoparticle concentrations and the shoot length due to the cross between the diamond (that is, the overall effect) and the invalid line. Therefore, the variation trend of SL was more complex than that of FGP and RL in our study, possibly because in contrast to roots, which are likely to be most affected by NPs as the first organ to encounter soil-borne contaminants, shoots did not have direct contact with NPs, and the toxicity of NPs on SL depends not only on NP properties and environmental conditions but also on the test organisms and transportation capacity of the crop species [13,38,53]. Sun et al. (2019) illustrated that Fe-NPs accumulated in the roots are the nontransferable form with low mobility; thus, Fe-NP concentrations in shoots were not significantly affected [38]. However, there are similarities with FGP and RL, e.g., lower NP concentrations would promote shoot elongation more relative to the higher levels. The results of Li et al. (2021) indicated that the exogenous application of ZnO-NPs at a suitable concentration would be able to promote the growth of rice, whereas high levels could have inhibitory effects [27].

5. Conclusions

This meta-analysis presented the significant effects of nanoparticles on the final germination percentage and root length of crop species, based on the types and concentrations. Silver nanoparticles increased the final germination percentage more than other nanoparticles, while zinc nanoparticles enhanced the root length more. Moreover, nanoparticles at lower concentrations could improve the final germination percentage and root length of crop species to a larger extent than those at higher concentrations. The variation trend of the shoot length was more complex compared with the other two in our study, because the toxicity of nanoparticles on shoot length not only depended on nanoparticle properties and environmental conditions but also on the test organisms and transportation capacity of the crop species, and therefore, the heterogeneity, risk bias, and publication bias should be taken into consideration. In this global meta-analysis, nanoparticle effects on seed germination and seedling growth have been assessed mostly in agricultural species, whereas few studies have focused on tree seeds, which may show the more various dormancy types. One of the reasons why nanoparticles can promote seed germination and root growth is that they increase water permeability; moreover, the seed dormancy of some tree species is caused by difficulty in absorbing water. Therefore, nanoparticles may play a role in breaking this dormancy characteristic of certain tree seeds, indicating the application prospect of nanoparticle treatment technology in tree seeds in the future.

Author Contributions: Conceptualization, H.G.; formal analysis, J.C.; methodology, H.G.; supervision, Y.L. and Y.Z.; writing—original draft, H.G.; writing—review and editing, Y.L., Y.Z., and Z.Z. All authors have read and agreed to the published version of the manuscript.

Funding: This study was supported by "The Fundamental Research Funds for the Central Universities" (BFUKF202111/).

Institutional Review Board Statement: Not applicable.

Informed Consent Statement: Not applicable.

Data Availability Statement: Data sharing is not applicable to this article.

Conflicts of Interest: The authors declare no conflict of interest.

References

1. Seeger, E.M.; Baun, A.; Kästner, M.; Trapp, S. Insignificant acute toxicity of TiO_2 nanoparticles to willow trees. *J. Soil Sediment* **2008**, *9*, 46–53. [CrossRef]
2. Khot, L.R.; Sankaran, S.; Maja, J.M.; Ehsani, R.; Schuster, E.W. Applications of nanomaterials in agricultural production and crop protection: A review. *Crop Prot.* **2012**, *35*, 64–70. [CrossRef]
3. González-Melendi, P.; Fernández-Pacheco, R.; Coronado, M.J.; Corredor, E.; Testillano, P.S.; Risueño, M.C.; Marquina, C.; Ibarra, M.R.; Rubiales, D.; Pérez-De-Luque, A. Nanoparticles as smart treatment-delivery systems in plants: Assessment of different techniques of microscopy for their visualization in plant tissues. *Ann. Bot.* **2008**, *101*, 187–195. [CrossRef] [PubMed]
4. Lin, D.; Xing, B. Phytotoxicity of nanoparticles: Inhibition of seed germination and root growth. *Environ. Pollut.* **2007**, *150*, 243–250. [CrossRef] [PubMed]
5. Acharya, P.; Jayaprakasha, G.K.; Crosby, K.M.; Jifon, J.L.; Patil, B.S. Nanoparticle-mediated seed priming improves germination, growth, yield, and quality of watermelons (*Citrullus lanatus*) at multi-locations in Texas. *Sci. Rep.* **2020**, *10*, 5037. [CrossRef]
6. Rico, C.M.; Majumdar, S.; Duarte-Gardea, M.; Peralta-Videa, J.R.; Gardea-Torresdey, J.L. Interaction of nanoparticles with edible plants and their possible implications in the food chain. *J. Agric. Food Chem.* **2011**, *59*, 3485–3498. [CrossRef]
7. Arruda, S.C.; Silva, A.L.D.; Galazzi, R.M.; Azevedo, R.A.; Arruda, M.A.Z. Nanoparticles applied to plant science: A review. *Talanta* **2015**, *131*, 693–705. [CrossRef]
8. Arnott, A.; Galagedara, L.; Thomas, R.; Cheema, M.; Sobze, J.M. The potential of rock dust nanoparticles to improve seed germination and seedling vigor of native species: A review. *Sci. Total Environ.* **2021**, *775*, 145139. [CrossRef]
9. El-Temsah, Y.S.; Joner, E.J. Impact of Fe and Ag nanoparticles on seed germination and differences in bioavailability during exposure in aqueous suspension and soil. *Environ. Toxicol.* **2012**, *27*, 42–49. [CrossRef]
10. Lopez-Moreno, M.L.; Rosa, G.D.L.; Hernández-Viezcas, J.A.; Peralta-Videa, J.R.; Gardea-Torresdey, J.L. X-ray absorption spectroscopy (XAS) corroboration of the uptake and storage of CeO_2 nanoparticles and assessment of their differential toxicity in four edible plant species. *J. Agric. Food Chem.* **2010**, *58*, 3689–3693. [CrossRef]
11. Feizi, H.; Kamali, M.; Jafari, L.; Moghaddam, P.R. Phytotoxicity and stimulatory impacts of nanosized and bulk titanium dioxide on fennel (*Foeniculum vulgare* Mill). *Chemosphere* **2013**, *91*, 506–511. [CrossRef] [PubMed]
12. Kumar, V.; Guleria, P.; Kumar, V.; Yadav, S.K. Gold nanoparticle exposure induces growth and yield enhancement in *Arabidopsis thaliana*. *Sci. Total Environ.* **2013**, *461–462*, 462–468. [CrossRef] [PubMed]
13. Ahmed, B.; Rizvi, A.; Syed, A.; Elgorban, A.M.; Khan, M.S.; AL-Shwaiman, H.A.; Musarrat, J.; Lee, J. Differential responses of maize (*Zea mays*) at the physiological, biomolecular, and nutrient levels when cultivated in the presence of nano or bulk ZnO or CuO or Zn^{2+} or Cu^{2+} ions. *J. Hazard. Mater.* **2021**, *419*, 126493. [CrossRef] [PubMed]
14. Mahakham, W.; Sarmah, A.K.; Maensiri, S.; Theerakulpisut, P. Nanopriming technology for enhancing germination and starch metabolism of aged rice seeds using phytosynthesized silver nanoparticles. *Sci. Rep.* **2017**, *7*, 8263. [CrossRef] [PubMed]
15. Jamari, J.; Ammarullah, M.I.; Saad, A.P.M.; Syahrom, A.; Uddin, M.; van der Heide, E.; Basri, H. The Effect of Bottom Profile Dimples on the Femoral Head on Wear in Metal-on-Metal Total Hip Arthroplasty. *J. Funct. Biomater.* **2021**, *12*, 38. [CrossRef]
16. Ammarullah, M.I.; Afif, I.Y.; Maula, M.I.; Winarni, T.I.; Tauviqirrahman, M.; Akbar, I.; Basri, H.; van der Heide, E.; Jamari, J. Tresca Stress Simulation of Metal-on-Metal Total Hip Arthroplasty during Normal Walking Activity. *Materials* **2021**, *14*, 7554. [CrossRef]
17. Ureta-Leones, D.; García-Quintana, Y.; Vega-Rosete, S.; Pérez-Morell, L.; Bravo-Medina, C.A.; Arteaga-Crespo, Y. Effect of pre-germination treatment with direct magnetic field exposure: A systematic review and meta-analysis. *Eur. J. Forest Res.* **2021**, *140*, 1029–1038. [CrossRef]
18. Soltani, E.; Baskin, J.M.; Baskin, C.C.; Benakashani, F. A meta-analysis of the effects of treatments used to break dormancy in seeds of the megagenus *Astragalus* (Fabaceae). *Seed Sci. Res.* **2020**, *30*, 224–233. [CrossRef]
19. Higgins, J.P.T.; Thompson, S.G.; Deeks, J.J.; Altman, D.G. Measuring inconsistency in meta-analyses. *Br. Med. J.* **2003**, *327*, 557–560. [CrossRef]
20. Acharya, P.; Jayaprakasha, G.K.; Semper, J.; Patil, B.S. 1H nuclear magnetic resonance and liquid chromatography coupled with mass spectrometry-based metabolomics reveal enhancement of growth-promoting metabolites in onion seedlings treated with green-synthesized nanomaterials. *J. Agric. Food Chem.* **2020**, *68*, 13206–13220. [CrossRef]
21. Belhamel, C.; Boulekbache-Makhlouf, L.; Bedini, S.; Tani, C.; Lombardi, T.; Giannotti, P.; Madani, K.; Belhamel, K.; Conti, B. Nanostructured alumina as seed protectant against three stored-product insect pests. *J. Stored Prod. Res.* **2020**, *87*, 101607. [CrossRef]
22. Corral-Diaz, B.; Peralta-Videa, J.R.; Alvarez-Parrilla, E.; Rodrigo-García, J.; Maria Isabel Morales, J.; Osuna-Avila, P.; Niu, G.; Hernandez-Viezcas, J.A.; Gardea-Torresdey, J.L. Cerium oxide nanoparticles alter the antioxidant capacity but do not impact tuber ionome in *Raphanus sativus* (L.). *Plant Physiol. Biochem.* **2014**, *84*, 277–285. [CrossRef] [PubMed]
23. Duran, N.M.; Medina-Llamas, M.; Cassanji, J.G.B.; de Lima, R.G.; de Almeida, E.; Macedo, W.R.; Mattia, D.; de Carvalho, H.W.P. Bean seedling growth enhancement using magnetite nanoparticles. *J. Agric. Food Chem.* **2018**, *66*, 5746–5755. [CrossRef] [PubMed]
24. Gupta, S.D.; Agarwala, A.; Pradhan, S. Phytostimulatory effect of silver nanoparticles (AgNPs) on rice seedling growth: An insight from antioxidative enzyme activities and gene expression patterns. *Ecotoxicol. Environ. Saf.* **2018**, *161*, 624–633. [CrossRef] [PubMed]
25. Kasote, D.M.; Lee, J.H.J.; Jayaprakasha, G.K.; Patil, B.S. Seed priming with iron oxide nanoparticles modulate antioxidant potential and defense-linked hormones in watermelon seedlings. *ACS Sustain. Chem. Eng.* **2019**, *7*, 5142–5151. [CrossRef]

26. Li, J.; Hu, J.; Ma, C.; Wang, Y.; Wu, C.; Huang, J.; Xing, B. Uptake, translocation and physiological effects of magnetic iron oxide (γ-Fe_2O_3) nanoparticles in corn (*Zea mays* L.). *Chemosphere* **2016**, *159*, 326–334. [CrossRef]
27. Li, Y.; Liang, L.; Li, W.; Ashraf, U.; Ma, L.; Tang, X.; Pan, S.; Tian, H.; Mo, Z. ZnO nanoparticle-based seed priming modulates early growth and enhances physio-biochemical and metabolic profiles of fragrant rice against cadmium toxicity. *J. Nanobiotechnol.* **2021**, *19*, 75. [CrossRef]
28. Liu, J.; Simms, M.; Song, S.; King, R.S.; Cobb, G.P. Physiological effects of copper oxide nanoparticles and arsenic on the growth and life cycle of rice (*Oryza sativa japonica* 'Koshihikari'). *Environ. Sci. Technol.* **2018**, *52*, 13728–13737. [CrossRef]
29. López-Moreno, M.L.; Rosa, G.D.L.; Cruz-Jiménez, G.; Castellano, L.; Peralta-Videa, J.R.; Gardea-Torresdey, J.L. Effect of ZnO nanoparticles on corn seedlings at different temperatures; X-ray absorption spectroscopy and ICP/OES studies. *Microchem. J.* **2017**, *134*, 54–61. [CrossRef]
30. Nguyen, D.T.C.; Le, H.T.N.; Nguyen, T.T.; Nguyen, T.T.T.; Bach, L.G.; Nguyen, T.D.; Tran, T.V. Multifunctional ZnO nanoparticles bio-fabricated from *Canna indica* L. flowers for seed germination, adsorption, and photocatalytic degradation of organic dyes. *J. Hazard. Mater.* **2021**, *420*, 126586. [CrossRef]
31. Saquib, Q.; Faisal, M.; Alatar, A.A.; Al-Khedhairy, A.A.; Ahmed, M.; Ansari, S.M.; Alwathnani, H.A.; Okla, M.K.; Dwivedi, S.; Musarrat, J.; et al. Genotoxicity of ferric oxide nanoparticles in *Raphanus sativus*: Deciphering the role of signaling factors, oxidative stress and cell death. *J. Environ. Sci.* **2016**, *47*, 49–62. [CrossRef]
32. Segura, R.; Vásquez, G.; Colson, E.; Gerbaux, P.; Frischmon, C.; Nesic, A.; García, D.E.; Cabrera-Barjas, G. Phytostimulant properties of highly stable silver nanoparticles obtained with saponin extract from *Chenopodium quinoa*. *J. Sci. Food Agric.* **2020**, *100*, 4987–4994. [CrossRef] [PubMed]
33. Singh, A.; Singh, N.B.; Hussaina, I.; Singh, H.; Yadava, V.; Singh, S.C. Green synthesis of nano zinc oxide and evaluation of its impact on germination and metabolic activity of *Solanum lycopersicum*. *J. Biotechnol.* **2016**, *233*, 84–94. [CrossRef] [PubMed]
34. Singh, J.; Kumar, S.; Alok, A.; Upadhyay, S.K.; Rawat, M.; Tsang, D.C.W.; Bolan, N.; Kim, K.H. The potential of green synthesized zinc oxide nanoparticles as nutrient source for plant growth. *J. Clean. Prod.* **2019**, *214*, 1061–1070. [CrossRef]
35. Singh, Y.; Kaushal, S.; Sodhi, R.S. Biogenic synthesis of silver nanoparticles using cyanobacterium *Leptolyngbya* sp. WUC 59 cell-free extract and their effects on bacterial growth and seed germination. *Nanoscale Adv.* **2020**, *2*, 3972. [CrossRef]
36. Song, U.; Jun, H.; Waldman, B.; Roh, J.; Kim, Y.; Yi, J.; Lee, E.J. Functional analyses of nanoparticle toxicity: A comparative study of the effects of TiO_2 and Ag on tomatoes (*Lycopersicon esculentum*). *Ecotoxicol. Environ. Saf.* **2013**, *93*, 60–67. [CrossRef]
37. Subpiramaniyam, S.; Hong, S.C.; Yi, P.I.; Jang, S.H.; Suh, J.M.; Jung, E.S.; Park, J.S.; Cho, L.H. Influence of sawdust addition on the toxic effects of cadmium and copper oxide nanoparticles on *Vigna radiata* seeds. *Environ. Pollut.* **2021**, *289*, 117311. [CrossRef]
38. Sun, Y.; Jing, R.; Zheng, F.; Zhang, S.; Jiao, W.; Wang, F. Evaluating phytotoxicity of bare and starch-stabilized zero-valent iron nanoparticles in mung bean. *Chemosphere* **2019**, *236*, 124336. [CrossRef]
39. Tan, W.; Du, W.; Barrios, A.C.; Armendariz, R., Jr.; Zuverza-Mena, N.; Ji, Z.; Chang, C.H.; Zink, J.I.; Hernandez-Viezcas, J.A.; Peralta-Videa, J.R.; et al. Surface coating changes the physiological and biochemical impacts of nano-TiO_2 in basil (*Ocimum basilicum*) plants. *Environ. Pollut.* **2017**, *222*, 64–72. [CrossRef]
40. Trujillo-Reyes, J.; Vilchis-Nestor, A.R.; Majumdar, S.; Peralta-Videa, J.R.; Gardea-Torresdey, J.L. Citric acid modifies surface properties of commercial CeO_2 nanoparticles reducing their toxicity and cerium uptake in radish (*Raphanus sativus*) seedlings. *J. Hazard. Mater.* **2013**, *263*, 677–684. [CrossRef]
41. Wang, Z.; Xu, L.; Zhao, J.; Wang, X.; White, J.C.; Xing, B. CuO nanoparticle interaction with *Arabidopsis thaliana*: Toxicity, parent-progeny transfer, and gene expression. *Environ. Sci. Technol.* **2016**, *50*, 6008–6016. [CrossRef] [PubMed]
42. Yadu, B.; Chandrakara, V.; Korramb, J.; Satnami, M.L.; Kumar, M.; Keshavkant, S. Silver nanoparticle modulates gene expressions, glyoxalase system and oxidative stress markers in fluoride stressed *Cajanus cajan* L. *J. Hazard. Mater.* **2018**, *353*, 44–52. [CrossRef] [PubMed]
43. Zuverza-Mena, N.; Armendariz, R.; Peralta-Videa, J.R.; Gardea-Torresdey, J.L. Effects of silver nanoparticles on radish sprouts: Root growth reduction and modifications in the nutritional value. *Front. Plant Sci.* **2016**, *7*, 90. [CrossRef]
44. Homaee, M.B.; Ehsanpour, A.A. Silver nanoparticles and silver ions: Oxidative stress responses and toxicity in potato (*Solanum tuberosum* L.) grown in vitro. *Hortic. Environ. Biotechnol.* **2016**, *57*, 544–553. [CrossRef]
45. Matthias, W.; Tobias, K.; Rose, T.J. From promise to application: Root traits for enhanced nutrient capture in rice breeding. *J. Exp. Bot.* **2016**, *12*, 3605.
46. Guo, Q.; Jonathan, L.; Song, J.; Jessica, R.; Turnbull, M.H.; Jameson, P.E. Insights into the functional relationship between cytokinin-induced root system phenotypes and nitrate uptake in *Brassica napus*. *Funct. Plant Biol.* **2017**, *44*, 832–844. [CrossRef]
47. Subbaiah, L.V.; Prasad, T.N.V.K.V.; Krishna, T.G.; Sudhakar, P.; Reddy, B.R.; Pradeep, T. Novel effects of nanoparticulate delivery of zinc on growth, productivity, and zinc biofortification in maize (*Zea mays* L.). *J. Agric. Food Chem.* **2016**, *64*, 3778–3788. [CrossRef] [PubMed]
48. Nandhini, M.; Rajini, S.B.; Udayashankar, A.C.; Niranjana, S.R.; Lund, O.S.; Shetty, H.S.; Prakash, H.S. Biofabricated zinc oxide nanoparticles as an eco-friendly alternative for growth promotion and management of downy mildew of pearl millet. *Crop Prot.* **2019**, *121*, 103–112. [CrossRef]
49. Kasivelu, G.; Selvaraj, T.; Malaichamy, K.; Kathickeyan, D.; Shkolnik, D.; Chaturvedi, S. Nano-micronutrients [γ-Fe_2O_3 (iron) and ZnO (zinc)]: Green preparation, characterization, agro-morphological characteristics and crop productivity studies in two crops (rice and maize). *New J. Chem.* **2020**, *44*, 11373. [CrossRef]

50. Sun, D.; Hussain, H.I.; Yi, Z.; Rookes, J.E.; Kong, L.; Cahill, D.M. Mesoporous silica nanoparticles enhance seedling growth and photosynthesis in wheat and lupin. *Chemosphere* **2016**, *152*, 81–91. [CrossRef]
51. Wu, F.; Fang, Q.; Yan, S.; Pan, L.; Ye, W. Effects of zinc oxide nanoparticles on arsenic stress in rice (*Oryza sativa* L.): Germination, early growth, and arsenic uptake. *Environ. Sci. Pollut. Res.* **2020**, *27*, 26974–26981. [CrossRef] [PubMed]
52. García-López, J.; Zavala-García, F.; Olivares-Sáenz, E.; Lira-Saldívar, R.; Barriga-Castro, E.D.; Ruiz-Torres, N. Zinc oxide nanoparticles boosts phenolic compounds and antioxidant activity of *Capsicum annuum* L. during germination. *Agronomy* **2018**, *8*, 215. [CrossRef]
53. Lei, C.; Sun, Y.Q.; Tsang, D.C.W.; Lin, D.H. Environmental transformations and ecological effects of iron-based nanoparticles. *Environ. Pollut.* **2018**, *232*, 10–30. [CrossRef] [PubMed]

coatings

MDPI

Review

Current Progress and Open Challenges for Combined Toxic Effects of Manufactured Nano-Sized Objects (MNO's) on Soil Biota and Microbial Community

Bismillah Mubeen [1,2,†], Ammarah Hasnain [1,2], Jie Wang [1,*,†], Hanxian Zheng [1], Syed Atif Hasan Naqvi [3,*], Ram Prasad [4], Ateeq ur Rehman [3], Muhammad Amir Sohail [5], Muhammad Zeeshan Hassan [3], Muhammad Farhan [3], Muhammad Altaf Khan [6] and Mahmoud Moustafa [7,8]

1 Tobacco Research Institute, Chinese Academy of Agricultural Sciences, Qingdao 266101, China
2 Institute of Molecular Biology and Biotechnology, The University of Lahore, Lahore 54000, Pakistan
3 Department of Plant Pathology, Faculty of Agricultural Sciences and Technology, Bahauddin Zakariya University, Multan 60800, Pakistan
4 Department of Botany, Mahatma Gandhi Central University, Motihari 845401, India
5 Hubei Key Laboratory of Plant Pathology, Huazhong Agricultural University, Wuhan 430070, China
6 Legume Research Group, Plant Production Department, College of Food and Agricultural Sciences, King Saud University, Riyadh 11451, Saudi Arabia
7 Department of Biology, Faculty of Science, King Khalid University, Abha 62529, Saudi Arabia
8 Department of Botany and Microbiology, Faculty of Science, South Valley University, Qena 83523, Egypt
* Correspondence: wangjie@caas.cn (J.W.); atifnaqvi@bzu.edu.pk (S.A.H.N.)
† These authors contributed equally to this work.

Abstract: Soil is a porous matrix containing organic matter and minerals as well as living organisms that vary physically, geographically, and temporally. Plants choose a particular microbiome from a pool of soil microorganisms which helps them grow and stay healthy. Many ecosystem functions in agrosystems are provided by soil microbes just like the ecosystem of soil, the completion of cyclic activity of vital nutrients like C, N, S, and P is carried out by soil microorganisms. Soil microorganisms affect carbon nanotubes (CNTs), nanoparticles (NPs), and a nanopesticide; these are called manufactured nano-objects (MNOs), that are added to the environment intentionally or reach the soil in the form of contaminants of nanomaterials. It is critical to assess the influence of MNOs on important plant-microbe symbiosis including mycorrhiza, which are critical for the health, function, and sustainability of both natural and agricultural ecosystems. Toxic compounds are released into rural and urban ecosystems as a result of anthropogenic contamination from industrial processes, agricultural practices, and consumer products. Once discharged, these pollutants travel through the atmosphere and water, settling in matrices like sediments and groundwater, potentially rendering broad areas uninhabitable. With the rapid growth of nanotechnology, the application of manufactured nano-objects in the form of nano-agrochemicals has expanded for their greater potential or their appearance in products of users, raising worries about possible eco-toxicological impacts. MNOs are added throughout the life cycle and are accumulated not only in the soils but also in other components of the environment causing mostly negative impacts on soil biota and processes. MNOs interfere with soil physicochemical qualities as well as microbial metabolic activity in rhizospheric soils. This review examines the harmful effect of MNOs on soil, as well as the pathways used by microbes to deal with MNOs and the fate and behavior of NPs inside the soils.

Keywords: manufactured nano-objects (MNOs); nano-agrochemicals (NAgs); nanoplastics (NPs); nano-scale materials (NSMs)

Citation: Mubeen, B.; Hasnain, A.; Wang, J.; Zheng, H.; Naqvi, S.A.H.; Prasad, R.; Rehman, A.u.; Sohail, M.A.; Hassan, M.Z.; Farhan, M.; et al. Current Progress and Open Challenges for Combined Toxic Effects of Manufactured Nano-Sized Objects (MNO's) on Soil Biota and Microbial Community. *Coatings* **2023**, *13*, 212. https://doi.org/10.3390/coatings13010212

Academic Editors: Yu Shen and Manuel Miguel Jordan-Vidal

Received: 5 December 2022
Revised: 3 January 2023
Accepted: 6 January 2023
Published: 16 January 2023

1. Introduction

According to some scientists, nanoparticles and nanostructured materials may have been formed during the Big Bang and brought to Earth via meteorites. The term "nanotechnology" gained widespread attention in the 1990s due to advances in imaging technologies that enabled practical applications in various industries. Seashells, skeletons, and other Nanostructures developed later in nature. Early people used fire to create nano-scaled smoke particles. The scientific story of nanomaterials, on the other hand, started considerably later. The colloidal gold particles created by Michael Faraday in 1857 are one of the first scientific reports. Nanostructured catalysts have also been researched for almost 70 years. By the early 1940s, precipitated and fumed silica nanoparticles were produced and supplied in Germany and the United States as alternatives to ultrafine carbon black for rubber reinforcements [1].

Nano sized objects have piqued people's curiosity for more than 30 years. These items are recognized in environmental sciences as having a major role in the biogeochemical cycles of chemical elements and organisms in their colloidal condition [2]. This is a result of their small size, commonplaceness, enormous specific surface area, and capacity to cling to the surface of aqueous solutions for extended periods of time. Analytical methods for characterizing naturally occurring colloidal objects have been developed as a result of the need to comprehend the behavior and fate of trace components [3]. The ability to synthesize nano-scale items has vastly increased since the early 2000s. The development of nanotechnologies has been aided by this increased knowledge. The qualities of the end goods generated are intimately tied to the properties of their nano-components. As a result, approaches for nano-scale analytical control has to be developed as well [4].

Nanotechnology advancements have driven the production of MNOs with fascinating and valuable material features, like nanoparticles (NPs) and nano-fibers (NFs) are defined as particles with diameters ranging from 1–100nm and many industrial products contain harmful substances like heavy metals. MNOs are widely used in medicine [5] and have a number of well-known benefits, like improving H_2O management [6], preservation of food and enhancement in agricultural production [7], preserving storage of energy [8], and a variety of other applications in working for the environment and improving it [9]. Unintentional releases of MNOs into the environment grow as production quantities increase, and this can happen throughout the life of goods containing manufactured nano-objects [10,11]. Furthermore, MNOs employed in agriculture as pesticides or fertilizers (nano-agrochemicals) are viewed as a viable approach for ensuring future food supply, although they may penetrate the environment in enormous numbers [12]. Several research initiatives have evaluated the environmental health and human safety consequences during the previous two decades [13]. Due to their tiny size, needle-like morphologies reduce the stability and enhance the area of surface-to-volume ratio, MNOs have nanoscale-specific detrimental effects when non-NSMs come in comparison with them [14]. MNOs, on the other hand, offer beneficial nano-specific physiochemical features that might be advantageous in applications like agriculture. Nano-agrochemicals, which include nanopesticides and nanofertilizers, are new nanoformulations that mix various surfactants, polymers, and inorganic NPs to enhance the dissolving power of substances that cannot dissolve in water to provide a managed and slower delivery [15]. Among nano-bio confluence, manufactured nano-objects are very helpful for plant defense because they can be employed to segregate phyto viruses, serve as a release system of organic nutrition, or enhance the role of enzymes as antioxidants [16]. If the application of manufactured nano-objects is excessive and straight in agriculture, current research focuses on enhancing MNO efficacy as well as risk evaluation. Along with physical and chemical changes, biological transformation occurs when MNOs and nano-agrochemicals are added to the soil. The pH, pore water, electrolyte, and organic matter in the soil all have an impact on these processes. The characteristics of the soil, such as pH, the make-up of the pore water and electrolyte, the quantity of natural organic matter (NOM), and other elements have a significant impact on these activities. During nano-bio interaction heteroaggregation, dissolution and oxidation-reduction takes place. For instance, the adsorption of protein NOM onto the surface of MNO might result in the creation of a corona, which increases particle mobility in the environment, as opposed to the adsorption of electrolytes (such as Ca^{2+}), which decreases particle mobility [17].

It is important to investigate the transformation mechanisms of MNOs, including through bioassays, in order to understand exposure routes and potential absorption by biota in risk assessments. Invertebrates, which play a crucial role in maintaining and improving soil fertility and structure through their participation in various biochemical and biological processes in agri-soils, are particularly important to consider. Invertebrates in the soil are significant indicators of soil characteristics and are crucial in assessing the risk of potentially contaminating substances. According to literature, among the negative impacts of manufactured nano-objects that are composed of basically carbon and metal, include effects on the community and morphology of soil, manufactured nano-objects completely change both of these and can also have negative effects on the environment and living things present there [18,19]. Scientists and industry are becoming increasingly aware of nanomaterial impacts while also realizing the enormous potential of nanomaterials. As a result, they're attempting to strike a balance between the two, such as using the "safe by design method." The physico-chemical properties of nanoparticles are investigated in this method. Then, by experimenting with various physicochemical factors, they strive to discover a strategy to minimize nanomaterial toxicity to the lowest possible level [20]. The objective of this review is to give a succinct overview of the existing knowledge on MNO behavior and its impact on soil biota (Figure 1).

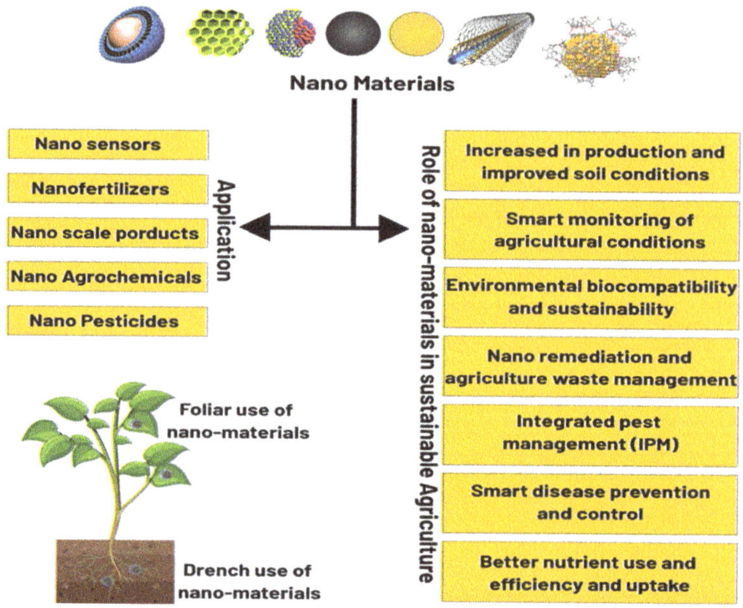

Figure 1. Possible role of nanomaterials in sustainable agriculture.

2. Occurrence of Manufactured Nano-Objects in Soil

Soil is a porous matrix containing organic and mineral substances as well as living creatures that are well organized and dynamic in all aspects, physically, geographically, and temporally. Soil microbiota, for example, serve as a reservoir in which plants choose a certain microbiome, which helps them develop and stay healthy. Microorganisms in the soil also take part in several ecosystemic functions in agrosystems, such as nutrient recycling in the soil ecosystem. Nano-agrochemicals are active ingredients developed using nanotechnologies and nanoformulations to enhance the features and qualities of active molecules used in pesticides in agriculture, such as biocides, herbicides, and nutrients. The promise of using nanotechnologies to improve pesticide, nutrient, and delivery efficiency has caused an explosion in the field of agronomy and will likely result in a reduction in the amount of input used in farming. However, the effect of these nanopesticides as a non-target organism on the soil microbiota has been underappreciated until recently [21,22].

3. Existence of MNOs in Environment

Nanotechnology enables the manipulation of matter, atoms, and molecules to change the properties of materials. This vitally important emerging technology has made it possible to create nano crystalline semiconductors, nano-agrochemicals and nanotherapeutics for cancer cures and a number of other purposes. Between 267,000 and 318,300 tonnes of MNOs, including Ag, Al_2O_3, CNTs, CeO_2, Cu, Fe, SiO_2, ZnO, TiO_2, and nanoclays are produced globally [22,23]. According to one market study, the United States generated 50% of MNOs, with the EU producing 19%, China producing 12%, Korea 6%, Japan 4%, Canada 3%, Taiwan 2%, and other nations producing 4%. According to a recent market assessment, worldwide MNO production volume in 2020 will range between 400,000 and 3,150,000 tonnes for nano-SiO_2 and 2 to 4 tonnes for nano-Ags [24].

MNO production volumes, according to the authors of the reference [24], represent a small portion of the mining industry's total ore production, such as 1% of the entire production of Silver or 0.000002% of the entire production of Fe, and as a result, has a negligible impact on the whole life of harmful elements in the extracting and synthesizing or production areas [24]. More research is needed to evaluate the mass concentrations of manufactured nano-objects in the universal cycle of handled or unprocessed materials because the quantities that have already been declared vary widely and occasionally are not accurate [18]. Using data and dynamic material flow models, it has been possible to predict the mass fluxes related to releases throughout the whole life in H_2O, soil, or the atmosphere [25,26]. According to the authors of the reference [25], by 2020 the use of nano-enabled products (such as building materials, packaging, medical products, etc.) was responsible for about 51% i.e., 12,200 tones of the worldwide release of Fe_2O_3, SiO_2, and TiO_2 while end-of-life releases accounted for 43% (9890 tones). The remaining leaks occurred during manufacture [25].

Release models were established by the authors of the reference [27,28], to predict MNO concentrations in urban, natural, and sludge-treated soils in Europe. Results show that most MNOs are converted or kept in operation through processes like wastewater treatment or rubbish incineration. Released nano-Ags, for instance, bind to biosolids which dissolve slightly in wastewater of acidic nature or convert into soluble Ag_2S or silver chloride types which flows with water or adheres to the wastewater and waste materials of sewage that is heated to remove contaminations and is applied to the agricultural lands to enhance the fertility of soil [27,28]. Al_2O_3, Fe_2O_3, SiO_2, or TiO_2, which are chemically more stable MNOs, bind to biosolids in the treatment time of wastewater and can build up in measureable concentrations in the soil of sludge treatment [27,28]. Literature estimates that the lowest MNO concentrations are of quantum dots. MNOs are approximately 8.4109 $mg{\cdot}kg^{-1}$ in the urban and natural soils, while the highest concentrations of nano-SiO_2 is anticipated to be around 4.9102 $mg{\cdot}kg^{-1}$ in landfills are predicted to be around that same level. The largest concentrations of MNOs are anticipated for the agricultural soils which are treated with sewage water. According to the authors of the reference [29,30], simulation of ten years of release of CeO_2 NPs, CuO NPs, SiO_2 NPs, and ZnO NPs in the San Francisco Bay region of California (The United States).

MNO release routes are influenced by the type of application, the product's lifespan of use, and the product's treatment for disposal. Soils treated with sludge in rural or urban areas are the most important MNOs sinks. Regarding risk evaluation and projected increase in MNO manufacturing volumes, adverse effects onland species cannot be ruled out anytime soon. Release models should be regularly updated with pertinent data because they serve as crucial guidelines for testing toxicity under realistic conditions and concentration ranges. Incorporating different MNO transformation processes into the environment is also essential for quantitative risk assessment [18,30]. Methods of expected toxicity tests or analyses of the relationship of structural activity suggest that dissolution of the particle is not the main cause of nano-toxicity. Other toxicological factors include membrane lysis, formation of ROS, redox activity, cationic stress, oxidative stress, and interference in embryo hatching and photoactivation [31]. To identify commonalities in toxicological effects among various MNO types, grouping approaches for regulatory testing have been developed in the last ten years. The connection of physiological and chemical characteristics with toxic effects is also being further investigated [32–34].

MNOs can be discharged accidentally into the soil or used purposely as biocides in agricultural applications, such as the usage of nanopesticides or nanofertilizers [35]. As a result of particle ingestion and transmission via trophic levels from bioaccumulation or biomagnifications, extensive ecosystem exposure may occur [36]. As a result of particle ingestion and transmission via trophic levels from bioaccumulation or biomagnification, extensive ecosystem exposure may occur. For instance, laboratory studies showed that the soil fungus *Penicillium solium* attracted the amino acid conjugated quantum dots; however, no MNO absorption was observed in the absence of this coating. According to the authors of the reference [37], invertebrates like *Eisenia fetida* (Earthworm) could be used to quantify the accumulation of MNOs by determining the ratios of MNOs present in tissues to the amounts present in water and in bio-contaminants, in which the ratio of MNOs is determined in prey and predator (Figure 2).

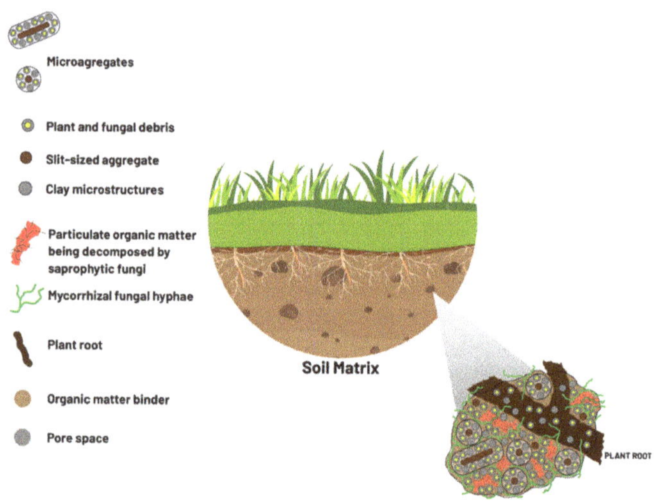

Figure 2. Soil matrix, aggregation and organic allocation induced by plant and microbial activity in the rhizosphere.

4. Nanoparticles and Mycorrhizas/Rhizobia Interactions

Certain nanoparticles like AgNPs, ZnO NPs, and TiO_2 have long been known to have antimicrobial effects against bacteria and fungi. Fungal hyphae and bacterial cells can be damaged by nanoparticles [38–40]. Limited or positive impacts of NPs on microbial populations of soil and function studies imply that NPs and microbe interactions depend on the environment [41–45]. Root and microbe symbioses occur in the rhizosphere, where factors like interactions between the soil biota, the complexity and availability of resources, and biophysical heterogeneities may affect how much NPs impact both free-living and root colonization microorganisms [46,47]. Overall, it appears that NP effects on the emergence of mycorrhizal (Arbuscular mycorrhiza (AM) is a common type of symbiotic relationship between plants and microbes. These fungi are found in many natural habitats and are known to offer a variety of ecological benefits, including improved plant nutrition and stress resistance, better soil structure and fertility, and increased tolerance to environmental challenges) and rhizobial symbioses are extremely context-specific. NPs have been shown in numerous studies to negatively affect mycorrhizal and rhizobial interactions [48–51]. Symbioses of mycorrhiza and rhizobia are the most significant relationship on the earth and have importance in the earth's ecosystem. It plays role in nutrient cycling in the soil, mineralization of the organic matter, modeling of microbial communities and plants, and eventually in the functioning of an ecosystem [52–54]. The characteristics of structure and plant fungus species involved have led to the reports of various advantageous root-fungal symbioses so far [55]. Although few of the studies have partly addressed other associations, the majority of research in the field of nanotechnology has focused on the common mycorrhiza, notably arbuscularmycorrhiza. The gymnosperms, angiosperms, bryophytes, and pteridophytes [56], associate with fungus from the Mucoromycota subphylum Glomeromycotina to establish arbuscularmycorrhizal (AM) symbioses. Rotating legume and non-legume crops, rhizobial symbioses, and resilience to environmental stresses like acidity, heavy metals, and organic pollutants are all essential aspects of the Fabaceae, the third-largest plant family [57]. Roots produce nodules to make room for N_2-fixing rhizobia at the symbiotic contact [58]. The overall impact of NPs on mycorrhizal colonization of nodule growth is influenced by the NP characteristics, concentration of fungal or bacterial species, and characteristics of the interaction matrix, where mycorrhizas and rhizobia reside and interact with plant roots. There is little study on how NPs affect these and how these symbiotic interactions function. However, available evidence indicates that studies on the toxicity of NPs against these advantageous root and microbe symbioses should concentrate on both structural and functional aspects [36].

Physical characteristics of NPs have a significant influence on the colonization or nodulation of root mycorrhizal fungi (e.g., kind, speciation, and size). AgNPs appear to be more toxic to mycorrhizas than ZnO NPs because of their negative effects on root colonization at approximately 5600 times lower soil concentrations [48,50,51,59,60]. In a pea of rhizobium leguminosarum symbiosis, exposure to Fe_2O_3 nanoparticles at 6 $g{\cdot}L^{-1}$ had no effect on nodulation. ZnO NPs and TiO_2 NPs both had unfavorable impacts on nodule growth 35 days after treatment, despite equal concentrations and exposure times [61]. In a soybean-*Bradyrhizobium japonicum* symbiosis, CeO NPs at 50 $g{\cdot}kg^{-1}$ had no effect on nodulation, but ZnO NPs at the same concentration increased nodulation [62]. Depending on the NP types and coating, the bioavailability and effects of NPs on mycorrhizas and rhizobia may differ. For instance, functionalized silver nanoparticles (PVP-AgNPs) were sown to reduce arbuscular mycorrhiza (AM) growth in tomato roots when applied at the same treatment rate of 100 $mg{\cdot}kg^{-1}$ in tomato roots when applied at the same treatment rate of 100 $mg{\cdot}kg^{-1}$ at the same treatment rate of 100 $mg{\cdot}kg^{-1}$ soil, however, silver sulphide NPs had no discernible effect [63]. The authors of the reference [64] used functionalized Fe_3O_4 NPs with positive and negative surface charges (carrying an amine and a carboxylic acid, respectively) to study NPs and rhizobia interactions and discovered that positively charged Fe_2O_4 NPs improved nodulation in soybean more than negatively charged Fe_2O_4 NPs. This demonstrates that NP surface coating and modification may affect their toxicity towards mycorrhizas and rhizobia before exposure, and as a result, NP physicochemical characteristics can be changed to achieve acceptable results or avoid unwanted results in NP interactions with rhizobia and mycorrhiza. It was also found that NP size affects interactions between NP and mycorrhiza. Tomato root colonization was reduced when exposed to 2 nm-AgNPs at the same dosage of 12 $mg{\cdot}kg^{-1}$ soil but was unaffected when exposed to larger AgNPs of 15 nm [64]. Moreover, soil spiked with TiO_2 NPs exhibited considerably greater Ti concentrations in the microcosm leachates than soil spiked with P25-TiO_2 NPs [65], indicating that NP shape and size may have an impact on the bioavailability. More research is necessary because our knowledge of how NP size influences NP–rhizobia interactions are currently relatively limited. Because partially or completely converted NPs may have a different toxicity potential than their pure counterparts, chemical alteration of NPs may have an effect on rhizobia [66,67]. The amount of NPs in the soil influences the interactions between mycorrhizas and rhizobia. No negative effects on Arbuscular mycorrhiza (AM) colonization in tomatoes were seen at low ZnO NP concentrations i.e., 25 and 400 mg·kg^{-1} soil, respectively [68] and maize plants [69], whereas, greater amounts (500 to 3200 mg·kg^{-1} soil) prevented colonization in maize [70]. In contrast to a low concentration of AgNPs higher concentrations i.e., 0.1 and 1 mg·kg^{-1} soil significantly enhanced root colonization in white clover (*Trifolium repens*) [53].

White clover grew AM fungi when exposed to a low concentration of FeO NPs i.e., 0.032 mg·kg^{-1} soil to a low concentration of FeO NPs (0.032 mg·kg^{-1} soil), but not when exposed to a concentration 100 times higher (3.2 mg·kg^{-1} soil) [62]. The concentration of NPs appears to play a role in its interaction with rhizobia, comparable to the responses observed for mycorrhizal colonization, despite the paucity of experimental evidence. The effects of ZnO NPs on nodulation in *Bradyrhizobium japonicum*-infected soybean plants changed from neutral to positive when the concentration was increased from 5 to 50 g·kg^{-1}. When exposed to Kocide, a fungicide made of copper that contains a substantial quantity of copper NPs [71], nodulation and N_2 fixation were unaffected at the authorized rate (1.7 mg·kg^{-1}), whereas nodulation was prevented at higher doses of 3.4 and 6.8 mg·kg^{-1} of treatment [72]. Negative NP and rhizobia interaction do not show a negative association between the concentration of NO and nodule growth [61,73]. The used concentration range may have exceeded the NPs' toxicity threshold in the relevant experimental circumstances (soil-less media), causing all concentrations to have a detrimental effect on the results [36]. According to the study, several mycorrhizal fungal species may respond to NPs differently. *Glomus caledonium*, an AM fungus, was shown to be more resistant to the toxicity of ZnO NPs than *G. versiforme* based on the extent of the unfavorable impact of ZnO NPs on root colonization [74]. This was attributed to the stronger resistance of *G. caledonium* to heavy metals like copper, zinc, cadmium, etc. [69] (Figure 3).

Figure 3. Toxic effects of manufactured nano-objects (MNOs) on earthworms.

5. NPs and Soil Biota: Mechanisms of Action

The knowledge gained about the toxicity mechanism of NPs will aid in the anticipated remodeling of nanoparticles to reduce their environmental impact. Non toxicity's main mechanisms include directly attaching to the external membrane surface of cells, dissolution of hazardous ions, and stimulation of oxidative stress [75]. Researchers clarified the influence of NPs on nanoparticles association with microorganisms and mitigation techniques for NP toxicity in the environment after elaborating on the nano-toxicity mechanism, providing an understanding of lowering toxicity and promoting long-term use of NPs [76].

Direct NP surface binding is a primary mechanism of inducing toxicity among several others and surprisingly, when NPs are found in near proximity to cells or animals, their interface is governed by electrostatic attraction. A bacterial cell's surface is usually found to have a negative charge [77]. According to the aforementioned finding, positively charged NPs are more closely linked to bacteria than negatively charged ones. A subset of cells in multicellular model organisms may take up NP as a result of such interactions [78]. When applied to bacterial culture, NPs persist on the surface of the cell, thus destruct membrane lipid, causing membrane loss or disruption [79]. The stimulation of an internal signaling cascade as a result of such a charge in membrane physical properties disturbs cells [80]. NPs then get dissolved in and release harmful ions on the surface of the cell which are cell permeable [81]. The authors of the reference [82] asserted that the relationship between NPs and gram-positive bacteria was likely caused by the attachment of negatively charged entities to lipopolysaccharides on the cell surface, which led to cell death. In another study, the presence of colloidal semiconductor nanocrystals made of poly diallydimethylammonium chloride covered with cadmium selenide (CdSe) quantum dots embedded in bilayer of lipid was demonstrated using tools like quartz crystal microbalance and atomic force microscopy [83]. When NPs bind to the cells; hydrophilic regions of the lipid bilayer collapse, disrupting cell membranes. Both eukaryotic and prokaryotic cells require these liquid domains for signaling and membrane transport [84–86].

The breakdown of harmful components from NPs, which causes an oxidative burst in affected organisms, is the most important mechanism of toxicity induced by NPs. Toxic ions can be released from NPs in a variety of ways, all of which are dependent on the identity of the released ions. A small number of ions attach to important proteins and enzymes, thus changing their metabolism and eventually causing major biological activities to be suppressed [87]. The hazardous ions are slowly released from metal oxides and get absorbed by membranes resulting in direct interactions with amino, mercapto, and carboxyl groups in nucleic acids and proteins. These interactions have a significant effect on the structure of cells and enzymatic activity, ultimately disrupting the exposed organisms overall physiology. Another method is to directly associate harmful ions with the damaged organism's phospholipid bilayer or even its genetic material [88]. As a result, metal ions cause an oxidative burst in organisms by increasing reactive oxygen species (ROS) levels [89]. The major form of hazardous ion dissolution in AgNPs is one of the primary sources of toxicity in organisms [90]. Due to the release of Ag and Pd ions into the solution, Pd nanolayers and nanowires with diameters ranging from 0.4 to 22.4 nm medium of polyethylene naphthalate were discovered to show antibacterial activity [91]. The authors of the reference [92] found that metaloxides had only minimal antibacterial activity when added to the culture, indicating that metal ions dissolution may not be the NPs antibacterial action. The toxicity of a few complex oxides including lithium (Li), nickel (Ni), manganese (Mn), and cobalt oxides to *Shewanella oneidensis* MR-1 bacterial cells was confirmed to be largely caused by the dissolution of ions [93]. To assess bacterial respiration, the researchers used a respirometer and optical density analysis to evaluate bacterial multiplication. According to this study, the complex oxides of Li, Ni, Mn, and Co were broken down to produce the ionic Li, Ni, Mn, and Co respectively. The continual dose and release of these nanomaterials, however, was straight linked to toxicity that was observed. An equi-stoichiometric NMC with different morphologies and bacterial toxicity was produced in order to highlight the importance of dissolution to Ni, Mn and Co oxide toxicity [94]. These were chosen because they show different crystal faces, demonstrating how variations in crystal faces affect the transition of metal coordination and how dissolution depends on them. Dissolution is connected to a material's exposed surface area, as was previously stated. Additionally, the surface area of the NMC was used to determine its toxicity rather than its mass. When using the surface area as the basis of dose, morphologies with different crystal faces demonstrated equal toxicity to the bacterial strain. As a result of the results made above, it was determined that neither crystal faces nor surface area significantly affects NMC toxicity to *Shewanella oneidensis* [76].

There are numerous studies that have been published recently that support the importance of ROS production and oxidative burst in relation to NP toxicity [95,96]. The four main kinds of ROS are singlet oxygen, hydrogen peroxide, hydroxyl ions, and superoxide anion, which are produced by short-term stress induced processes [97]. It has been determined that physiological harm is caused by singlet oxygen. Under typical environmental cues, exposed organisms maintain a balance between ROS generation and scavenging. On the other hand, when too much ROS is produced, the intercellular redox equilibrium is upset, which sets the stage for oxidation [98]. The main effects of ROS include lipid peroxidation and disruption of important enzymes such as mononuclear iron proteins [99]. Additionally, ROS production leads to the oxidation of DNA bases and deoxyribose, which causes mutations and DNA damage [95]. It is being determined whether NPs' electrical structures, ROS, and toxicity are correlated. Similar behaviors were also demonstrated in *E. coli* by the authors of the reference [100] using 7-metal oxide NPs with band edges close to the redox potential of reactive redox couples. It was discovered that the level of abiotic ROs produced in response to NPs exposure was directly correlated with the organism's level of toxicity. Another experimental study also showed the toxicity of 24 distinct metal oxide nanoparticles in *E. coli*. Only seven of the 24 metal oxides found in NPs were hazardous and increased the levels of intercellular ROS. It was established through the use of nanostructure activity association analysis that NPs' toxicity was directly related to (1) the conduction band energy of the nanomaterials and (2) the hydration enthalpy dictating their capacity to dissolve. Additionally, it was hypothesized that nanomaterials and biomolecules conduction bands might overlap if the materials were hazardous and disintegrated quickly. The NPs have also been shown to activate a chain of biological signals that results in an oxidative burst [66]. Another finding obtained by the authors of the reference [101] in the guts of *Daphia magna* was that when the guts were exposed to positively or negativity charged nano diamond particles of 5 nm to 15 nm size, the larger particle was observed to stimulate ROS production in a dose-dependent way in comparison to lesser sized NPs. The expression of genes related to oxidative burst was also confirmed to be decreased; indicating that cells were preventing ROS clarification as the basic mechanism of NPs-toxicity [101].

According to several pieces of research, NPs antifungal properties, which include the release of metal ions, have a detrimental effect on mycorrhizal colonization like Zn^{+2} and Ag [102]. The detrimental effects of NPs on mycorrhizal colonization may be directly related to their antifungal properties, which include adhering to cell surfaces and physically harming cell walls and membranes, increasing membrane permeability, obstructing water channels, and killing cells through NP penetration and deposition. Considering that NPs have sharp edges, they might also be able to cut through fungal structures and cell walls [101,103]. Inhibition of germination of spore due to formation of aggregations of NPs by Van der Waals forces [101] buildup of ROS by disruption of ROS scavenging defence mechanisms such as glutathione cycle and regulation [104]; ion discharge from metal-based NPs [105]; and capability of certain NPs, like TiO_2, of being photocatalytic are also related to negative effects of NPs on the colonization of mycorrhiza [106].

The mechanisms behind the beneficial NP-mycorrhiza interactions need to be studied more thoroughly. AgNPs have been discovered to promote AM fungal colonization [53]. Additionally, plants under heavy metal stress showed an increase in AM fungus colonization [107]. Plants exposed to NPs may see a decline in the growth of rhizobial nodules as a direct result of the NPs' antibacterial properties. The antibacterial effects of NPs may generally result from simultaneous oxidative stress induction and metal ion release processes such as cell membrane disruption [108–110]. Additionally, it has been verified that NP concentrations that are environmentally relevant have a significant impact on greenhouse gas emissions, crucial ecosystem services like nitrogen cycling, and soil microbial populations [111]. Applying any material to soil that is both persistent and immobile should generally be done with utmost caution. Due to their elemental nature, nanoparticles do not degrade in the environment. For instance, hazardous stages may be achieved in soil after ten years of continuous applications. Several NPs seem to be determined and mostly fixed in the soil, depending on the NPs and the soil properties. They also incorporate into plant tissues and the soil biota [111,112]. Therefore, it is necessary to undertake spatial and temporal trials for nano-agrochemicals and NP-containing amendments (such as AgNPs provided through sludge application) to assess the impact of repeated NP treatments on these significant root microbial symbioses. Cu concentrations in soil are up to an order of magnitude higher than in natural soils as a result of decades of historical use of Cu-based fungicides, which can have detrimental environmental consequences on soil fertility, water resources, and species [113] (Figure 4).

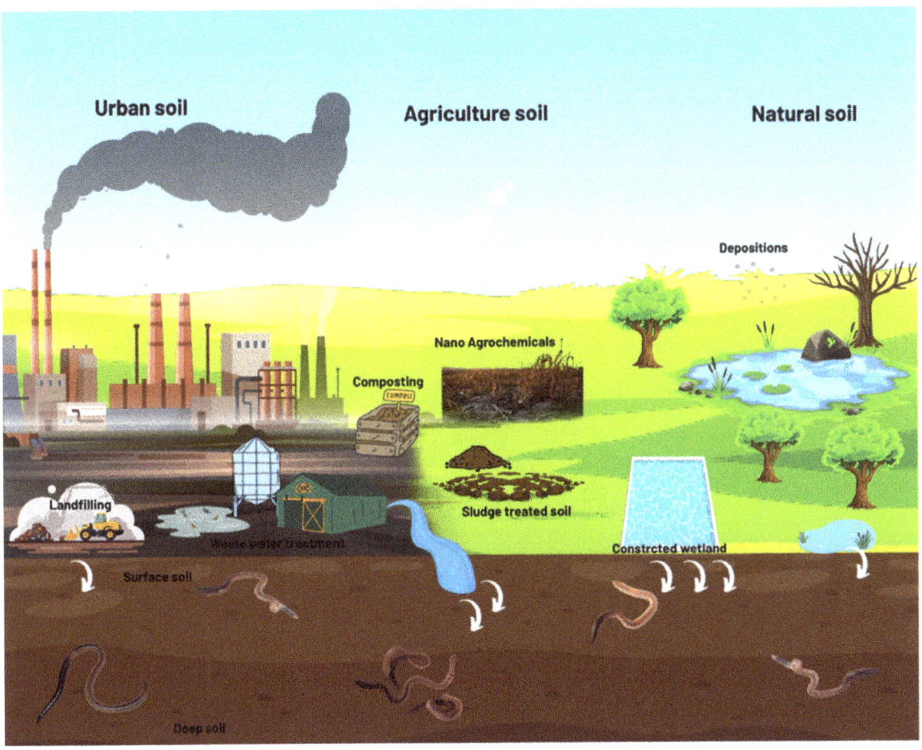

Figure 4. An outlook of MNOs possible pathways for earthworms.

5.1. Silver

Silver has long been used as a biocide. The antimicrobial/biocidal effects of silver-based nanopesticides are demonstrated against a variety of kinds of microorganisms, including bacteria, fungi, and viruses [114]. Ag-based nanopesticides, which have been commercialized and patented for use in plant protection, seed processing, and plant development improvement, have already been introduced in the plant protection measures. Even though these nanopesticides are advertised as having the ability to efficiently prevent phytopathogen illnesses in a variety of plants to boost the plant immune system and to lessen stress [115,116].

Due to their widespread use in commercial and industrial products, AgNPs may have accumulated in the soil accidently or on purpose, for example as fungicides and nano-agrochemicals. Predicted environmental concentration (PEC) values for Ag-NPs in soil vary by location around the world; in Denmark, native soil has a PEC value of 13–61 mg·kg^{-1}, whereas agricultural soil has a PEC value of 6–12 mg·kg^{-1} [117,118], while the expected values of Ag-NPs in American soils range from 6.6 to 29.8 ng·kg^{-1}, they are predicted to be between 17.4 and 58.7 ng·kg^{-1} in European soils [117].

The majority of the information that has been published in this field comes from studies about the effects of nanopesticides that are formed of silver on bacteria and microorganisms that are not intended targets. The authors of the reference [119] used biosolids and AgNPs to modify a clayey SiO_2 with low pH (SiO_2 73, Silt 22 and Clay 5%, pH 5.6, and small amounts of organic components) to a target concentration of 0.19 to 15 mg·kg^{-1} soil (particle size 15 nm). This mixture was kept at 22 °C in the dark for 30 days. Exoenzyme activity, soil respiration and potential ammonium oxidation (PAO) along with patterns of coming generations of bacteria were then carried out to assess the influence of AgNPs and examine bacterial diversity. Similar AgNP sensitivity was shown in these assays, with effects starting at levels of at least 1.67 mg·kg^{-1}. With an enrichment of proteobacteria, cytophagales, and spirobacteria, next-generation sequencing demonstrated a shift in the microbial population and variable sensitivity of bacterial groups. Some nitrifiers (nitrosomonadales) were adversely impacted, and this correlated with the reduction of PAO activity [119].

The authors of the reference [120] incubated AgNPs and added $AgNO_3$ to SiO_2 loamy soil with a hydrogen ion concentration of 5.61 and 0.93 percent organic content in order to study the effects of Silver nanoparticles on NH^{4+} oxidizing bacteria (AOB) for 140 days. Amounts of 0.56, 1.67, and 5 milligrams per killograms of silver nanoparticles kg^{-1} of $AgNO_3$ were applied. They showed relative hindrance of AOB with 1.67 and 5 milligrams per Kg's of silver nanoparticles starts at day fourteen and enhanced to 140 days, in contrast to silver ionic form, where inhibition begins on the first day and shows growth even during deficiency [121,122]. AgNPs impacts on a black poplar tree's associated phyllosphere and rhizosphere bacteria were examined by the authors of the reference [123]. Three years old poplar trees were chronically supplied with nano powder, amorphous carbon coated AgNPs of 25 nm size, and the dose applied was 1 mg·L^{-1} for a 10-week period, single supply was applied for 4 weeks and double supply was applied for 6 weeks. AgNPs foliar exposure was done with the soil covered, and no fertilizer was used during the experiment.

Using next-generation sequencing, the ITS 1 region and the V3-V4 section of the 16S rRNA, respectively, were used to study the bacterial and fungal microbiome. Root AgNPs treatment reduced the biodiversity of microorganisms like bacteria and fungi; application of AgNPs to leaves enhances the evenness of bacteria and fungi and discovered a significant decrease in both microbial groups. According to a bioinformatics functional analysis, the use of AgNPs enhanced the bacteria that do not require too much oxygen and can bear oxygen stress while lowering aerobic bacteria. However, the AgNPs treatments in this study mimicked the application of Ag nanopesticides on agricultural soil rather than a contaminated environment. For example, Zerebra®Agro, a commercial nanopesticides based on silver, has an Ag content of 0.5 g·L^{-1}. Additionally, 0.1 L·t^{-1} is the suggested amount for plant therapy, 0.1 L^{-1} and seed hectare-1 for use on cultivated crops from 1–3 times during the vegetative period as opposed to twenty grams per hectare [122].

The authors of the reference [38] investigated how citrate-coated AgNPs of 50 nm and their ions disturb the functioning of enzymes present in soil and the make-up of the microbial community in a bio-solid amended agricultural soil. During the treatment of wastewater solid part that was organic in nature was added to surface soil at the Macdonald campus of McGill University (soil/biosolid weight ratio: 50/1; depth: 35 cm; pH: 6.7). The amount of total AgNPs supplied to the soil was 1, 10, and 100 mg. The material disintegrated within the first two hours and maintained its stability for up to

30 days. AgNPs at one and ten mg per kg of enzymes that are present outside involved in phosphorous, carbon, and nitrogen cycling did not have any effect at short-term (2 h) concentrations. Because only 37% of the AgNPs were dissolved at 2 h, AgNPs had a milder effect on these enzymatic activities at 100 mg·kg^{-1} compared to Ag^{+}. After two hours and thirty days of exposure, the microbial population of the soil was tested using 16S rRNA gene amplicon sequencing. In comparison to all other treatments, the relative abundance of the Gamma proteobacteria category was remarkably greater for Ag^{+} ions and AgNPs at 100 mg·kg^{-1} of soil response to dissolve Ag and AgNPs [38]. According to the authors of the reference [123], decline in the abundance of AOB has been reported due to the application of 0.01 mg nano particles of silver per kilogram per year that showed bad effects on nitrogen fixation and soil microorganisms' environment, leucine amino peptidase activity and nitrogen-fixing microorganisms.

The authors of the reference [63] focused on distinct Ag speciation and NPs coating while examining the effects of AgNPs. They introduced 1, 10, or 100 mg of Ag_2S NPs, AgNPs coated in polyvinylpyrrolidone (PVP), and Ag^{+} to a sandy loam soil that had been modified with biosolids (pH 6.8). Before planting tomato seeds, the soil mixture was infected with an arbuscular mycorrhizal fungus (AMF) or a commercial inoculum (*Solanum lycopersicum*). The authors measured neutral lipid fatty acid (and phospholipids fatty acid) analyses, ammonium nitrate extractable Ag concentrations, and the overall microbial community structure in soil modified with biosolids. All silver applications at the rate of 1 mg·kg^{-1} and 10 mg·kg^{-1} did not substantially differ from the control, with the exception of three treatments i.e., 100 mg·kg^{-1} for Silver-PVP and Ag+ and 10 mg·kg^{-1} for Ag_2S NPs [63]. Ag-PVP, Ag+, and Ag_2S NPs had an effect on fungus and bacteria, including Actinomycetes and microbial community even at concentrations of 1 mg·kg^{-1} [124,125].

5.2. Copper

There has been a rise in the release of copper oxide NPs (CuO NPs) into terrestrial and aquatic environments as a result of their use in different sectors like goods against microorganisms and nano-agrochemicals [126]. About 79,000 tons of CuO-NPs are consumed annually in North America, which contributes about fifty percent of the world marketplace [127]. There has been extensive research on the effects of metal and metal oxide NPs on the soil and rhizosphere microbiome, mostly with an eye toward the effects of environmental pollution [128,129]. NPs of Silver, Zinc oxide, copper, and iron are the most extensively studied in toxicity studies. There are now two types of nanomaterials that have produced commercial agrochemicals that are nano-enabled and on the market: colloidal silver and copper nanoparticles for treating fungal infections on seed, vegetative parts, and tubers respectively [130]. Copper is a well-known biocide that has been used for centuries, as well as an important nutrient for living things like plants and bacteria. As chemicals to kill bacteria and fungi on veins, plants, and reproductive parts, some copper-based insecticides are currently permitted in organic farming. Originally employed as lime, copper sulfate was neutralized in the Bordeaux mixture to treat downy mildew-infected grapes [131]. Products to kill pests that are formed of copper include those with the chemical compositions copper hydroxide, cuprous oxide copper ammonium carbonate, copper oxychloride, and copper octanoate. Since copper sulfate's solubility encourages phytotoxicity and reduces the persistence on tree leaves and fruits as well as the fungicide action, fixed copper-less soluble variations have really been developed. These particles are known as fixed-coppers, and their size affects how well they cover and cling to plant leaves as well as how much copper ions they discharge. Copper nanoparticles have been developed and marketed for their ability to improve the growth of plants by preventing the release of copper ions. These nanoparticles were originally sold as micronized particles. There are two brands of nanosized copper formulations available: NANOCU and Kocide®3000, both produced by the same company [131].

The authors of the reference [132] examined the effect of nano-sized CuO vs copper ions (CuSO4) in five various agricultural soils with pH ranging from 6.4 to 8.21 in order to understand the complexes in biology and physio-chemical diversity of the soil. After the soil water content had been maintained to a holding capacity of H_2O particular to every soil, microbes present in soil were incubated for 90 days without light at a temperature of twenty-eight centigrade. When maximum dosage was applied, 100 $mg{\cdot}kg^{-1}$, CuONPs dramatically inhibit microbial activity involved in C and N cycles, respiration, denitrification, and nitrification in the five soils under study. These results worsen overtime. The lowest doses usually have little to no impact, with respiration in sandy loam soils reducing at 1 $mg{\cdot}kg^{-1}$ and denitrification in loamy soils decreasing at 1 $mg{\cdot}kg^{-1}$ after 90 days. Copper oxide nanoparticles have a different impact on soil microbial activity in relation to the carbon and nitrogen cycles compared to ionic copper. They are most effective at promoting denitrification in maximum soils, while nitrification and soil respiration are more influenced in coarse soils. However, when used at agriculturally suitable levels, copper oxide nanoparticles have little impact on soil microbes. Soils with a coarse texture and low levels of organic matter or clay may be more susceptible to the effects of copper oxide nanoparticles. In a study, plants cultivated for 50 days in climatic chambers showed that copper oxide nanoparticles had a negative impact on microbial respiration and denitrification when applied at a dose of 1 $mg{\cdot}kg^{-1}$. Copper-based nanoparticles have also been found to be harmful to nitrifiers, with copper oxide nanoparticles and copper ions exhibiting distinct behaviors such as the release and uptake of copper ions and their effects on microbial activity [133].

In a study, copper nanoparticles between 40 and 60 nm in size were incubated for 30 days in soil with high levels of organic matter at concentrations of 0.05% and 0.15% by weight and 3 $mg{\cdot}kg^{-1}$ ATZ. The bacterial, fungal, and nitrifying bacterial population profiles, as identified by PCR denaturing Gradient Gel Electrophoresis, remained relatively stable throughout the experiment. However, the 0.15% weight concentration of copper nanoparticles significantly reduced the dissipation of ATZ, indicating increased persistence of ATZ in soil. Most of the interactions between this persistence and soil particles were physical-chemical in nature. Paddy soils, which are commonly used for agriculture in China and are exposed to cycles of flood and dry conditions and frequent changes in soil oxidation-reduction environments, are the most common type of agricultural soils in the country [116,133].

Cu and AgNPs' behavior and fate in soil are influenced by factors that are intrinsic to the NPs, such as their size, charge on the surface, and pH, as well as factors that are extrinsically linked to the characteristics of the intricate soil matrix. Nanoparticle dissolution and interactions with cells are significantly influenced by their form. Exposure time also influences the characteristics of common nanoparticles and of how silver can change the makeup of communities of microorganisms [134]. Nanoparticles may dissolve, undergo changes due to redox reactions, form aggregations with particles of soil and adsorb, particularly to clays, on a global scale [135,136]. Acidity tends to source solubilization of Silver nanoparticles, while high pH of soil facilitates Ag-sorption. According to the authors of the reference [120,133], across a range of soils, AgNP toxicity towards microbial processes including substrates persuade breakdown of glusose and to bacteria that oxidize NH_3 decreased as clay content and pH increased [120,133]. The similar line of reasoning is reached in the conclusion by the authors of the reference [132] regarding the sporadic effects of copper oxide nanoparticles at agriculturally related concentrations on coarse soil texture having lower content of organic matter or clay [133–135]. Almost 4.5% of the world's soils with low pH are used for cultivation purposes. This is despite the fact that acidic soils make up roughly 30% of the ice-free terrain of the world [137]. Short-term effects of the nanoparticles can be increased by using acidic soil, which encourages the breakdown of copper- and Silver nanoparticles releasing free ions. An intriguing finding in the literature is that the pesticide's ionic or nanoform can have distinct effects, perhaps due to the ion release percentage. Ionic and nanoforms of metals may exhibit parallels and variations in the mode of antibacterial action or the influence on a microbial population when exposed in vitro to AgNPs, according to certain scientists [134,138]. The toxicity of nanoparticles is kinetic in long-term investigations and appears to be connected to soil breakdown or transformation processes that result in momentary adjustments and adaptations of the microbial population. As demonstrated by the authors of the reference [139], adjusting the surface features of NPs could aid in controlling the dissolution and phase shifts and probably lessen the toxicity towards microbial cells.

5.3. *Zinc Oxide*

The third most common metal-based NP is zinc oxide nanoparticles (ZnO NPs) [140]. Wastewater from factories and waste material from homes that are used in agricultural systems to enhance fertility is the basic source of entry of Zinc oxide nanoparticles into the environment [141]. Zinc oxide nanoparticles create a network of soil particles that preserve NPs' emulsion characteristics, but they can also form bigger aggregates or become soluble and excrete zinc ions [142]. ZnO-nanoparticles have a regionally variable PEC, similar to Ag-NPs. In Europe, PECs of 0.085–0.661 $g{\cdot}kg^{-1}$ were projected, while in the United States, PECs of 0.041–0.271$g{\cdot}kg^{-1}$ were predicted [39]. The values viz., 0.018–0.9 and 0.008–0.35 $g{\cdot}kg^{-1}$ were anticipated by a Danish study comparing uncultivated and agricultural soil [39]. Among the various varieties of NPs, ZnO NPs are becoming more common and are now the third most widely used nanomaterial. As a result, the direct influence of ZnO NPs discharged into the environment on soil microbial communities and processes must be carefully assessed. Despite many potential benefits, various experimental research studies have found that ZnO NPs can affect soil productivity by modulating microbial community features [143]. ZnO-NPs have an ecotoxicological impact on soil microorganisms wherein according to the authors of the reference [144], the microbial respiration, dehydrogenase, ammonification, and fluorescent diacetate hydrolase activity decrease in the soil.

Another study demonstrated that ZnO NPs inhibit enzymatic activities such as diacetatehydrolysis, urease, and catalase. They also reduced thermogenic metabolism, lowered Azotobacter, Phosphate-solubilizing, and potassium solubilizing bacterial colonies [145]. According to reports ammonification was suppressed by up to 37.8% in soil that had received three months' worth of treatment with 1 mg of ZnO NPs per gram of soil. Respiration inhibition was 14.2% during first the month. In soil treated with 1–10 mg ZnO NPs a similar pattern was observed for dehydrogenase and fluorescent diacetate hydrolase activities. A study by the authors of the reference [146] was conducted on Staphylococcus aureus and Escherichia coli which indicate that ZnO NPs are hazardous for both of these gram-positive and gram-negative bacteria. ZnO-NPs inhibited completely the growth of *S. aureus* at doses of 1 mmol/L and *E. coli* at doses of 3.4 mmol/L [146]. These findings validated ZnO-NPs' toxicity to many bacterial systems, paving the way for biomedical and antimicrobial applications. It has been demonstrated that ZnO-NPs are antibacterial against a variety of bacteria, Including *B. subtilis*, *E. coli*, *P. flourescens*, *S. aureus*, and *S. typhimurium* as well as fungus including *A. flavus* and *A. fumigates* [147]. According to the authors of the reference [148], exposure to ZnO-NPs results in bacterial morphological abnormalities and ultimately mortality for food-borne and water-borne diseases such *E. coli*, *C. jejuni*, and *V. cholera*. Although there is little research using environmental strains, there is a lot of data concerning the environmental influence of ZnO-NPs [149].

It also has been proven that NPs have an antimicrobial effect against the helpful soil bacterium *P. putida*. ZnO-NPs were bacteriostatic, but CuO and AgO-NPs were largely bactericidal [150]. An intriguing set of findings were produced by *P. chlororaphis*, a member of the same genus. These plant-growth-promoting rhizobacteria produced the phytohormones more quickly thanks to CuO-NPs than ZnO-NPs [151]. It appears that ZnO-NPs have a species-specific impact. For instance, *P. putida* is bacteriostatically affected by these NPs [150]. These intricate mechanisms must surely be further investigated because both species are gram-negative rods with the same general cell layout. ZnO NPs have been widely used in environmental cleanup and as an antibacterial agent [152].

In an alternative study, the effects of ZnO NPs on the marine alga *Chlorella vulgaris* were studied by the authors of the reference [153], who discovered that cell viability was inversely correlated with nano-ZnO concentration and exposure time. The authors of the reference [154] investigated the capability of ZnO-NPs to eradicate clinical isolates of *B. subtilis*, *E. coli*, *K. pneumoniae*, *P. aeuginosa*, *S. typhi*, and S. *aureus* clinical isolates of *B. subtilis*, *E. coli*, *K. pneumoniae*, *P. aeruginosa*, *S. typhi*, and *S. aureus* [139]. ZnO NPs were found to have a greater impact than TiO_2 NPs at the same exposure dose, as shown by decreased DNA content and more pronounced changes in the genetic makeup of the bacterial population (Figure 5) [155,156].

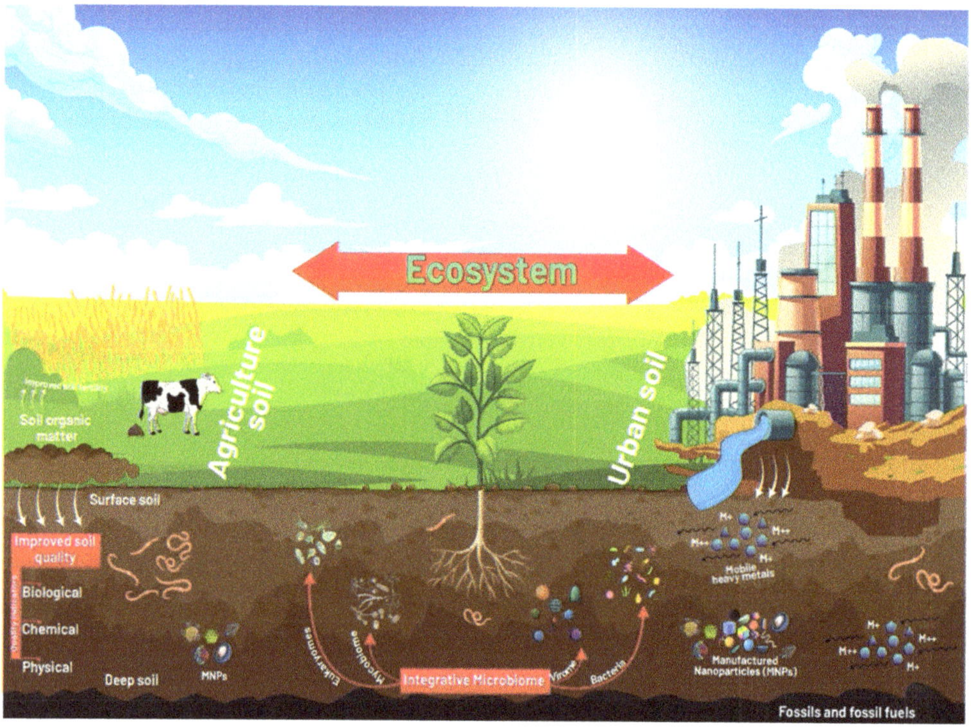

Figure 5. How manufactured nano-objects are deteriorating the soil microbiome.

5.4. Titanium Oxide

There are many applications for TiO_2-NPs, the bulk of which are commercial [157]. TiO_2 can potentially be employed in agriculture [158]. In soil PEC value of TiO_2 NPs has been determined to be 1.01 to 4.45 g per kilogram, in soil that is treated from wastes PEC value of these NPs was determined as 70.6–310 $g{\cdot}kg^{-1}$ while in sewage sludge its PEC value is 100–433 $mg{\cdot}kg^{-1}$ [39,117] and the most hazardous factor of titanium dioxide nanoparticles for soil microorganisms is sewage sludge. Importantly, even the 1000 $mg{\cdot}kg^{-1}$ concentrations claimed in several papers are not found in soils exposed to TiO_2 NPs polluted sewage sludge [157].

5.5. Cerium Oxide

Cerium (Ce) is frequently utilized as a catalyst and a fuel additive as CeO_2-NPs in optics [159] which has increased nanoscale form manufacturing levels in the worldwide market level up to 1000 $NT{\cdot}yr^{-1}$. In terms of weight cerium makes up about 0.0046% of the planet earth [160], Ce can share its electrons in three ways (1) gaining two electrons

(2) gaining three electrons (3) gaining four electrons. The PEC of CeO_2-NPs is unknown; the PECs of uncultivated and agricultural soils in Denmark is estimated to be 24 to 1500 ng·kg^{-1} and 10 to 530 ng·kg^{-1} [137].

5.6. Silica

Silica nanoparticles (SiNPs) are crystalline solids [143] where farming and trading items are the most common applications [15]. For instance, hollow porous sand nanoparticles are responsible as agents for the managed release of neurotoxins or medicines while also shielding them from UV light [161]. The annual output of Si-NPs was around 93,300 tons in 2016, which was the third highest production after $nTiO_2$ and nFeOx and rose yearly [162]. The PEC value of sand nanoparticles in different types of soils may range from 86 to 150,000 g·kg^{-1} [28]. Negative impacts on soil organic communities have not been fully recognized.

5.7. Quantum Dots

The prospective applications of semiconductor quantum dots (QDs) in electronics, solar cells, and healthcare have generated a lot of attention. These luminescent nanocrystals range in size depending on the materials employed and the synthesis process, but most are between 1 and 10 nanometers in size [70]. Global QD production peaked at around 135 tones in 2012 and has steadily climbed since then [14]. In nanosafety research, QDs are utilized as fluorescent indicators at the laboratory scale [140]. They are not yet used as nano-agrochemicals. Less is known about their possible negative effect on human health and the environment. The PEC value for QDs in soil in Europe was determined lower and that was about 9 pg·kg^{-1} [28], nevertheless, due to a lack of data on the QD production ratio, published exposure model findings are currently quite erroneous.

6. Carbon Nanotubs (CNTs)

CNTs are a cylindrical nanostructure and allotrope of carbon. Single-wall carbon nanotubes (SWCNTs) and multi-wall carbon nanotubes (MWCNTs) are the two most popular forms of CNTs [163]. CNTs used in the electronic industry and have other applications too [158]. The worldwide requirement for carbon nanotubes is estimated to be over US$ 4.5 billion in 2026, with a rise of more than US$ 10–15 billion [14]. CNTs could be used as a nano-agrochemical to stop viral multiplication and spread, for example [14]. However, nanosafety issues continue to limit their use in agriculture. The soil PEC value in Europe is 35 ng·kg^{-1} in urban and natural soil and 12 g per kg in soils treated from wastes projected for CNTs [26]. It is also critical to understand their behavior and potential environmental implications. CNTs may behave differently in natural environments depending on their length, diameter, functionalization, and environmental conditions [164].

CNTs' impact on the microbiological activity of soil is debatable and has gotten little study. However, the majority of these studies seem to indicate that CNTs lessen soil microbial activity [165–167]. Both MWCNTs and SWCNTs prevented soil bacteria from producing enzymes at 500 mg·kg^{-1}, and MWCNTs decreased the enzyme production of two natural soils [165]. Similarly, the authors of the reference [166] discovered that SWCNTs at concentrations of 30 to 300 mg·kg^{-1} dramatically reduced enzyme activity in a natural sandy loam soil. Another study found that after 3 days, SWCNTs had a substantial effect on the bacterial soil community, but after 14 days, the bacteria had entirely recovered [168,169]. The authors of the reference [162] discovered that MWCNTs at low concentrations (0.2 mg·kg^{-1}) promoted microorganisms to mineralize an agricultural soil. The authors of the reference [170] used long-term dry soil research to examine the effects of MWCNTs in contrast to natural or manufactured carbonaceous materials on soil microbial populations. After one year of exposure, they discovered that MWCNTs reduced soil DNA diversity and altered bacterial populations. These results are equivalent to those of carbonaceous materials, both man-made and natural. As of now, there aren't enough studies to determine if functionalized and unfunctionalized CNTs affect soil microbial activity differently [170]. There have been few investigations on the effects of CNTs on soil microorganisms to date. They were all interested in earthworms that were uncovered in the soil [170–172].

7. Nanoplastics

Nanoplastics, which are closely connected to microplastics, are an increasing pollutant of concern. Microplastics are most commonly found in macroplastic objects that are mistakenly discharged into the environment and breakdown into secondary microplastics [86,173]. Due to the analytical equipment's resolution limits in identifying and quantifying these materials, the majority of research has focused on microplastics with particles larger than 10 micrometres [147]. As a result, there aren't many articles that examine the relationship between earthworm species and nanoplastics. In particular, the inclusion of traceable materials like fluorescently labeled polystyrene (PS) beads enables higher-resolution identification using fluorescence microscopy [28]. The mobility and bioavailability of metallic MNOs in the soil is influenced by the material's physico-chemical characteristics and a number of soil variables, the most important of which are pH and organic matter. Other soil parameters are pore size, texture and other surface properties of soil are also important [28].

8. Nano-Agrochemicals/Nanopesticides

These are active compounds that have been improved by nanotechnology and nanoformulation. Nanopesticides, nanofertilizers, and nanosensors are all examples of nanoenabled agrochemical [127]. Nanoparticles can interface with the soil community when they are applied to soils, plants or used for coating seeds potentially influencing soil microbiome and soil fertility. The microorganisms that dwell in the soil make up the soil microbiota. Cycles of C, N, P, S, and other elements are significantly influenced by microbiota, as well as soil formation, pollution, degradation, and water and microbial modification of rock is also affected by microorganisms. Soil organisms are vital to the ecosystem services provided by the agricultural landscape, including pest control, biodiversity, soil structure, and nutrient cycling. Significantly, for the cultivation of land, soil offers a collection of microbes out of which plants select a community of microbes to support its production and wellbeing. As a result, the plant responds more quickly to stress, whether it be biotic (pathogens) or abiotic (drought, floods, chemical toxins) (plant pathogens). The plant microbiome and the gut microbiome in humans are frequently compared. The second genome of the plant and its potential for agriculture are believed to reside in the soil microbiome. Some soil microbiome can naturally help to prevent plant illnesses by acting as pathogen suppressors. The creation of smart nano-agrochemicals that combine efficacy and eco-compatibility while protecting soil microbial diversity depends on an understanding of how nanopesticides interact with soil and plant microbiome [21].

9. Conclusions

It is undeniable that scientific breakthroughs in nanotechnology have become extremely important. However, it has had some negative consequences for the ecosystem. The abundant production of nano-goods, as well as their discharge and permanence within the soil ecosystem, have damaged beneficial microorganisms and soil composites. Positive interactions between soil, plant, and bacteria are prevented by surface charges, area, size, and responsiveness. Some of them, for instance, cling to or enter microbial cells and cause significant harm. Our comprehensive review of the literature reveals that MNOs may have adverse, beneficial, or even neutral impacts on soil microbiota. Extensive literature assessment indicates that MNOs may have negative, neutral, or even positive effects on soil microbiota. To maintain ecosystem functioning and resilience, the study of interactions between MNOs and these vital root microbial symbioses must be increased. Procedures for safe disposal in soil agro ecosystems are widespread and safe, and they should be devised to avoid contact with soil microflora given the structural and functional toxicity of NPs to the environment. To put it another way, the items should be tailored to their intended use. Many illusive outcomes could be established to open new roads in this discipline by undertaking microcosm study based on this. These studies are critical for a thorough understanding of this topic and the protection of the ecosystem.

Author Contributions: Conceptualization, B.M., A.H., S.A.H.N., and A.u.R.; methodology, B.M. and A.H.; validation, S.A.H.N. and M.M.; formal analysis, M.A.S.; investigation, B.M. and A.H.; resources, A.H. and S.A.H.N.; data curation, B.M.; writing—original draft preparation, B.M., A.H., S.A.H.N., M.A.S., and M.M.; writing—review and editing, J.W., H.Z., M.A.K., R.P., M.F., and M.Z.H.; visualization, R.P.; supervision, S.A.H.N.; project administration, M.A.S.; funding acquisition, M.A.S. All authors have read and agreed to the published version of the manuscript.

Funding: Central Public-Interest Scientific Institution Basal Research Fund (No. 1610232022002) and Science and Technology Cooperation Project of Shandong and Gansu (YDZX2021098, YDZX2022162).

Institutional Review Board Statement: Not applicable.

Informed Consent Statement: Not applicable.

Data Availability Statement: Not applicable.

Acknowledgments: Authors extend their sincere thanks to Central Public-Interest Scientific Institution Basal Research Fund (No. 1610232022002) and Science and Technology Cooperation Project of Shandong and Gansu (YDZX2021098, YDZX2022162).

Conflicts of Interest: The authors declare no conflict of interest.

References

1. Dipak, S.C.; Srirama, D. A review of stabilization of expansive soils by using nanomaterials. In Proceedings of the 50th Indian Geotech. Conference, Maharashtra, India, 17–19 December 2015; p. 8.
2. Grolimund, D.; Barmettler, K.; Borkovec, M. *Colloid Facilitated Transport in Natural Porous Media: Fundamental Phenomena and Modelling, Colloidal Transport in Porous Media*; Springer: Berlin/Heidelberg, Germany, 2007; pp. 3–27.
3. Maria, E.; Crançon, P.; Le Coustumer, P.; Bridoux, M.; Lespes, G. Comparison of preconcentration methods of the colloidal phase of a uranium-containing soil suspension. *Talanta* **2020**, *208*, 120383. [CrossRef] [PubMed]
4. Bayda, S.; Adeel, M.; Tuccinardi, T.; Cordani, M.; Rizzolio, F. The History of Nanoscience and Nanotechnology: From Chemical–Physical Applications to Nanomedicine. *Molecules* **2020**, *25*, 112. [CrossRef] [PubMed]
5. Shakib, K.; Tan, A.; Soskic, V.; Seifalian, A.M. Regenerative nanotechnology in oral and maxillofacial surgery. *Br. J. Oral Maxillofac. Surg.* **2014**, *52*, 884–893. [CrossRef] [PubMed]
6. Alvarez, P.J.; Chan, C.K.; Elimelech, M.; Halas, N.J.; Villagrán, D. Emerging opportunities for nanotechnology to enhance water security. *Nat. Nanotechnol.* **2018**, *13*, 634–641. [CrossRef] [PubMed]
7. Duhan, J.S.; Kumar, R.; Kumar, N.; Kaur, P.; Nehra, K.; Duhan, S. Nanotechnology: The new perspective in precision agriculture. *Biotechnol. Rep.* **2017**, *15*, 11–23. [CrossRef]
8. Hussein, A.K. Applications of nanotechnology to improve the performance of solar collectors—Recent advances and overview. *Renew. Sustain. Energy Rev.* **2016**, *62*, 767–792. [CrossRef]
9. Mathew, J.; Joy, J.; George, S.C. Potential applications of nanotechnology in transportation: A review. *J. King Saud Univ.—Sci.* **2019**, *31*, 586–594. [CrossRef]
10. Froggett, S.J.; Clancy, S.F.; Boverhof, D.R.; Canady, R.A. A review and perspective of existing research on the release of nanomaterials from solid nanocomposites. *Part. FibreToxicol.* **2014**, *11*, 1–28. [CrossRef]
11. Part, F.; Berge, N.; Baran, P.; Stringfellow, A.; Sun, W.; Bartelt-Hunt, S.; Mitrano, D.; Li, L.; Hennebert, P.; Quicker, P.; et al. A review of the fate of engineered nanomaterials in municipal solid waste streams. *Waste Manag.* **2018**, *75*, 427–449. [CrossRef]
12. Sun, X.-D.; Yuan, X.-Z.; Jia, Y.; Feng, L.-J.; Zhu, F.-P.; Dong, S.-S.; Liu, J.; Kong, X.; Tian, H.; Duan, J.-L.; et al. Differentially charged nanoplastics demonstrate distinct accumulation in Arabidopsis thaliana. *Nat. Nanotechnol.* **2020**, *15*, 755–760. [CrossRef]
13. Vighi, M.; de Voogt, P.; Rizzo, L.; Krätke, R.; Linders, J.; Scott, M. Proposed EU minimum quality requirements for water reuse in agricultural irrigation and aquifer recharge: SCHEER scientific advice. *Curr. Opin. Environ. Sci. Health* **2018**, *2*, 7–11.
14. Adeel, M.; Shakoor, N.; Shafiq, M.; Pavlicek, A.; Part, F.; Zafiu, C.; Raza, A.; Ahmad, M.A.; Jilani, G.; White, J.C.; et al. A critical review of the environmental impacts of manufactured nano-objects on earthworm species. *Environ. Pollut.* **2021**, *290*, 118041. [CrossRef]
15. Kah, M.; Beulke, S.; Tiede, K.; Hofmann, T. Nanopesticides: State of knowledge, environmental fate, and exposure modeling. *Crit. Rev. Environ. Sci. Technol.* **2013**, *43*, 1823–1867. [CrossRef]
16. Farooq, T.; Adeel, M.; He, Z.; Umar, M.; Shakoor, N.; da Silva, W.; Elmer, W.; White, J.C.; Rui, Y. Nanotechnology and Plant Viruses: An Emerging Disease Management Approach for Resistant Pathogens. *ACS Nano* **2021**, *15*, 6030–6037. [CrossRef]

17. Markiewicz, M.; Kumirska, J.; Lynch, I.; Matzke, M.; Köser, J.; Bemowsky, S.; Docter, D.; Stauber, R.; Westmeier, D.; Stolte, S. Changing environments and biomolecule coronas: Consequences and challenges for the design of environmentally acceptable engineered nanoparticles. *Green Chem.* **2018**, *20*, 4133–4168. [CrossRef]
18. Adeel, M.; Shakoor, N.; Ahmad, M.A.; White, J.C.; Jilani, G.; Rui, Y. Bioavailability and toxicity of nanoscale/bulk rare earth oxides in soil: Physiological and ultrastructural alterations in *Eisenia fetida*. *Environ. Sci. Nano* **2021**, *8*, 1654–1666. [CrossRef]
19. Rocha, T.L.; Mestre, N.C.; Sabóia-Morais, S.M.T.; Bebianno, M.J. Environmental behaviour and ecotoxicity of quantum dots at various trophic levels: A review. *Environ. Int.* **2017**, *98*, 1–17. [CrossRef]
20. Maynard, A.D.; Aitken, R.J.; Butz, T.; Colvin, V.; Donaldson, K.; Oberdörster, G.; Warheit, D.B. Safe handling of nanotechnology. *Nature* **2006**, *444*, 267. [CrossRef]
21. Santaella, C.; Plancot, B. *Interactions of Nanoenabled Agrochemicals with Soil Microbiome, Nanopesticides*; Springer: Berlin/Heidelberg, Germany, 2020; pp. 137–163.
22. Vance, M.E.; Kuiken, T.; Vejerano, E.P.; McGinnis, S.P.; Hochella, M.F., Jr.; Rejeski, D.; Hull, M.S. Nanotechnology in the real world: Redeveloping the nanomaterial consumer products inventory. *Beilstein J. Nanotechnol.* **2015**, *6*, 1769–1780. [CrossRef]
23. Jean, J. Getting high with quantum dot solar cells. *Nat. Energy* **2020**, *5*, 10–11. [CrossRef]
24. Janković, N.Z.; Plata, D.L. Engineered nanomaterials in the context of global element cycles. *Environ. Sci. Nano* **2019**, *6*, 2697–2711. [CrossRef]
25. Song, R.; Qin, Y.; Suh, S.; Keller, A.A. Dynamic model for the stocks and release flows of engineered nanomaterials. *Environ. Sci. Technol.* **2017**, *51*, 12424–12433. [CrossRef] [PubMed]
26. Sun, T.Y.; Mitrano, D.M.; Bornhöft, N.A.; Scheringer, M.; Hungerbühler, K.; Nowack, B. Envisioning nano release dynamics in a changing world: Using dynamic probabilistic modeling to assess future environmental emissions of engineered nanomaterials. *Environ. Sci. Technol.* **2017**, *51*, 2854–2863. [CrossRef] [PubMed]
27. Sun, T.Y.; Bornhöft, N.A.; Hungerbühler, K.; Nowack, B. Dynamic probabilistic modeling of environmental emissions of engineered nanomaterials. *Environ. Sci. Technol.* **2016**, *50*, 4701–4711. [CrossRef] [PubMed]
28. Wang, Y.; Nowack, B. Dynamic probabilistic material flow analysis of nano-SiO_2, nano iron oxides, nano-CeO_2, nano-Al_2O_3, and quantum dots in seven European regions. *Environ. Pollut.* **2018**, *235*, 589–601. [CrossRef]
29. Garner, K.L.; Suh, S.; Keller, A.A. Assessing the risk of engineered nanomaterials in the environment: Development and application of the nanoFate model. *Environ. Sci. Technol.* **2017**, *51*, 5541–5551. [CrossRef]
30. Adam, V.; Caballero-Guzman, A.; Nowack, B. Considering the forms of released engineered nanomaterials in probabilistic material flow analysis. *Environ. Pollut.* **2018**, *243*, 17–27. [CrossRef]
31. Nel, A.; Xia, T.; Meng, H.; Wang, X.; Lin, S.; Ji, Z.; Zhang, H. Nanomaterial Toxicity Testing in the 21st Century: Use of a Predictive Toxicological Approach and High-Throughput Screening. *Acc. Chem. Res.* **2013**, *46*, 607–621. [CrossRef]
32. Hund-Rinke, K.; Schlich, K.; Kühnel, D.; Hellack, B.; Kaminski, H.; Nickel, C. Grouping concept for metal and metal oxide nanomaterials with regard to their ecotoxicological effects on algae, daphnids and fish embryos. *Nanoimpact* **2018**, *9*, 52–60. [CrossRef]
33. Lamon, L.; Aschberger, K.; Asturiol, D.; Richarz, A.; Worth, A. Grouping of nanomaterials to read-across hazard endpoints: A review. *Nanotoxicology* **2019**, *13*, 100–118. [CrossRef]
34. Lynch, I.; Weiss, C.; Valsami-Jones, E. A strategy for grouping of nanomaterials based on key physico-chemical descriptors as a basis for safer-by-design NMs. *Nano Today* **2014**, *9*, 266–270. [CrossRef]
35. Zhang, P.; Guo, Z.; Zhang, Z.; Fu, H.; White, J.C.; Lynch, I. Nanomaterial transformation in the soil–plant system: Implications for food safety and application in agriculture. *Small* **2020**, *16*, 2000705. [CrossRef]
36. Tian, H.; Kah, M.; Kariman, K. Are Nanoparticles a Threat to Mycorrhizal and Rhizobial Symbioses? A Critical Review. *Front. Microbiol.* **2019**, *10*, 1660. [CrossRef]
37. Hou, W.-C.; Westerhoff, P.; Posner, J.D. Biological accumulation of engineered nanomaterials: A review of current knowledge. *Environ. Sci. Process. Impacts* **2013**, *15*, 103–122. [CrossRef]
38. Asadishad, B.; Chahal, S.; Akbari, A.; Cianciarelli, V.; Azodi, M.; Ghoshal, S.; Tufenkji, N. Amendment of agricultural soil with metal nanoparticles: Effects on soil enzyme activity and microbial community composition. *Environ. Sci. Technol.* **2018**, *52*, 1908–1918. [CrossRef]
39. Gottschalk, F.; Lassen, C.; Kjoelholt, J.; Christensen, F.; Nowack, B. Modeling flows and concentrations of nine engineered nanomaterials in the Danish environment. *Int. J. Environ. Res. Public Health* **2015**, *12*, 5581–5602. [CrossRef]
40. Kumar, N.; Shah, V.; Walker, V.K. Perturbation of an arctic soil microbial community by metal nanoparticles. *J. Hazard. Mater.* **2011**, *190*, 816–822. [CrossRef]
41. Shinde, S.S. Antimicrobial activity of ZnO nanoparticles against pathogenic bacteria and fungi. *Sci. Med. Central* **2015**, *3*, 1033.
42. Johansen, A.; Pedersen, A.L.; Jensen, K.A.; Karlson, U.; Hansen, B.M.; Scott-Fordsmand, J.J.; Winding, A. Effects of c60 fullerene nanoparticles on soil bacteria and protozoans. *Environ. Toxicol. Chem.* **2008**, *27*, 1895–1903. [CrossRef]
43. Tong, Z.; Bischoff, M.; Nies, L.; Applegate, B.; Turco, R.F. Impact of fullerene (C60) on a soil microbial community. *Environ. Sci. Technol.* **2007**, *41*, 2985–2991. [CrossRef]
44. He, L.; Liu, Y.; Mustapha, A.; Lin, M. Antifungal activity of zinc oxide nanoparticles against Botrytis cinerea and Penicillium expansum. *Microbiol. Res.* **2011**, *166*, 207–215. [CrossRef] [PubMed]

45. Karunakaran, G.; Suriyaprabha, R.; Manivasakan, P.; Yuvakkumar, R.; Rajendran, V.; Prabu, P.; Kannan, N. Effect of nanosilica and silicon sources on plant growth promoting rhizobacteria, soil nutrients and maize seed germination. *IET Nanobiotechnol.* **2013**, *7*, 70–77. [CrossRef] [PubMed]
46. Raffi, M.M.; Husen, A. *Impact of Fabricated Nanoparticles on the Rhizospheric Microorganisms and Soil Environment, Nanomaterials and Plant Potential*; Springer: Berlin/Heidelberg, Germany, 2019; pp. 529–552.
47. Saleem, M.; Pervaiz, Z.H.; Traw, M.B. *Theories, Mechanisms and Patterns of Microbiome Species Coexistence in an Era of Climate Change, Microbiome Community Ecology*; Springer: Berlin/Heidelberg, Germany, 2015; pp. 13–53.
48. Abd-Alla, M.H.; Nafady, N.A.; Khalaf, D.M. Assessment of silver nanoparticles contamination on faba bean-Rhizobium leguminosarumbv. viciae-Glomus aggregatumsymbiosis: Implications for induction of autophagy process in root nodule. *Agric. Ecosyst. Environ.* **2016**, *218*, 163–177. [CrossRef]
49. Huang, Y.C.; Fan, R.; Grusak, M.A.; Sherrier, J.D.; Huang, C. Effects of nano-ZnO on the agronomically relevant Rhizobium–legume symbiosis. *Sci. Total Environ.* **2014**, *497*, 78–90. [CrossRef] [PubMed]
50. Jing, X.-X.; Su, Z.-Z.; Xing, H.-E.; Wang, F.-Y.; Shi, Z.-Y.; Liu, X.-Q. Biological Effects of ZnO Nanoparticles as Influenced by Arbuscular Mycorrhizal Inoculation and Phosphorus Fertilization. *Huan Jing KeXue = HuanjingKexue* **2016**, *37*, 3208–3215.
51. Noori, A.; White, J.C.; Newman, L.A. Mycorrhizal fungi influence on silver uptake and membrane protein gene expression following silver nanoparticle exposure. *J. Nanoparticle Res.* **2017**, *19*, 1–13. [CrossRef]
52. Chen, C.; Tsyusko, O.V.; McNear, D.H., Jr.; Judy, J.; Lewis, R.W.; Unrine, J.M. Effects of biosolids from a wastewater treatment plant receiving manufactured nanomaterials on Medicago truncatula and associated soil microbial communities at low nanomaterial concentrations. *Sci. Total Environ.* **2017**, *609*, 799–806. [CrossRef]
53. Feng, Y.; Cui, X.; He, S.; Dong, G.; Chen, M.; Wang, J.; Lin, X. The role of metal nanoparticles in influencing arbuscular mycorrhizal fungi effects on plant growth. *Environ. Sci. Technol.* **2013**, *47*, 9496–9504. [CrossRef]
54. Van Der Heijden, M.G.; Bardgett, R.D.; Van Straalen, N.M. The unseen majority: Soil microbes as drivers of plant diversity and productivity in terrestrial ecosystems. *Ecol. Lett.* **2008**, *11*, 296–310. [CrossRef]
55. Kariman, K.; Barker, S.; Tibbett, M. Structural plasticity in root-fungal symbioses: Diverse interactions lead to improved plant fitness. *Peer J.* **2018**, *6*, e6030. [CrossRef]
56. Brundrett, M.C.; Tedersoo, L. Evolutionary history of mycorrhizal symbioses and global host plant diversity. *New Phytol.* **2018**, *220*, 1108–1115. [CrossRef]
57. Ramirez, M.D.A.; Silva, J.D.; Ohkama-Ohtsu, N.; Yokoyama, T. In vitro rhizobia response and symbiosis process under aluminum stress. *Can. J. Microbiol.* **2018**, *64*, 511–526. [CrossRef]
58. Timoshenko, A.; Kolesnikov, S.; Rajput, V.D.; Minkina, T. *Effects of Zinc-Oxide Nanoparticles on Soil Microbial Community and Their Functionality, Zinc-Based Nanostructures for Environmental and Agricultural Applications*; Elsevier: Amsterdam, The Netherlands, 2021; pp. 267–284.
59. Li, S.; Liu, X.; Wang, F.; Miao, Y. Effects of ZnO Nanoparticles, $ZnSO_4$ and Arbuscular Mycorrhizal Fungus on the Growth of Maize. *Huan Jing keXue = HuanjingKexue* **2015**, *36*, 4615–4622.
60. Wu, J.; Zhai, Y.; Liu, G.; Bosker, T.; Vijver, M.G.; Peijnenburg, W.J. Dissolution Dynamics and Accumulation of Ag Nanoparticles in a Microcosm Consisting of a Soil–Lettuce–Rhizosphere Bacterial Community. *ACS Sustain. Chem. Eng.* **2021**, *9*, 16172–16181. [CrossRef]
61. Sarabia-Castillo, C.; Fernández-Luqueño, F. TiO_2, ZnO, and Fe_2O_3 nanoparticles effect on Rhizobium leguminosarum-Pisum sativum L. symbiosis. In Proceedings of the 3rd Biotechnology Summit 2016, Ciudad Obregón, Sonora, Mexico, 24–28 October 2016; pp. 144–149.
62. Priester, J.H.; Ge, Y.; Mielke, R.E.; Horst, A.M.; Moritz, S.C.; Espinosa, K.; Gelb, J.; Walker, S.L.; Nisbet, R.M.; An, Y.-J. Soybean susceptibility to manufactured nanomaterials with evidence for food quality and soil fertility interruption. *Proc. Natl. Acad. Sci. USA* **2012**, *109*, E2451–E2456. [CrossRef]
63. Judy, J.D.; Kirby, J.K.; Creamer, C.; McLaughlin, M.J.; Fiebiger, C.; Wright, C.; Cavagnaro, T.R.; Bertsch, P.M. Effects of silver sulfide nanomaterials on mycorrhizal colonization of tomato plants and soil microbial communities in biosolid-amended soil. *Environ. Pollut.* **2015**, *206*, 256–263. [CrossRef]
64. Burke, D.J.; Pietrasiak, N.; Situ, S.F.; Abenojar, E.C.; Porche, M.; Kraj, P.; Lakliang, Y.; Samia, A.C.S. Iron Oxide and Titanium Dioxide Nanoparticle Effects on Plant Performance and Root Associated Microbes. *Int. J. Mol. Sci.* **2015**, *16*, 23630–23650. [CrossRef]
65. Klingenfuss, F. Testing of TiO_2 Nanoparticles on Wheat and Microorganisms in a Soil Microcosm. Ph.D. Dissertation, University of Gothenburg, Gothenburg, Sweden, 2014.
66. Li, Y.; Zhang, W.; Niu, J.; Chen, Y. Mechanism of Photogenerated Reactive Oxygen Species and Correlation with the Antibacterial Properties of Engineered Metal-Oxide Nanoparticles. *ACS Nano* **2012**, *6*, 5164–5173. [CrossRef]
67. Reinsch, B.; Levard, C.; Li, Z.; Ma, R.; Wise, A.; Gregory, K.; Brown, G., Jr.; Lowry, G. Sulfidation of silver nanoparticles decreases Escherichia coli growth inhibition. *Environ. Sci. Technol.* **2012**, *46*, 6992–7000. [CrossRef]
68. Watts-Williams, S.J.; Turney, T.; Patti, A.; Cavagnaro, T. Uptake of zinc and phosphorus by plants is affected by zinc fertiliser material and arbuscular mycorrhizas. *Plant Soil* **2014**, *376*, 165–175. [CrossRef]
69. Wang, F.; Liu, X.; Shi, Z.; Tong, R.; Adams, C.A.; Shi, X. Arbuscular mycorrhizae alleviate negative effects of zinc oxide nanoparticle and zinc accumulation in maize plants—A soil microcosm experiment. *Chemosphere* **2016**, *147*, 88–97. [CrossRef] [PubMed]

70. Liu, J.; Wang, Z.; Liu, F.D.; Kane, A.B.; Hurt, R.H. Chemical Transformations of Nanosilver in Biological Environments. *ACS Nano* **2012**, *6*, 9887–9899. [CrossRef] [PubMed]
71. Adeleye, A.S.; Conway, J.R.; Perez, T.; Rutten, P.; Keller, A.A. Influence of Extracellular Polymeric Substances on the Long-Term Fate, Dissolution, and Speciation of Copper-Based Nanoparticles. *Environ. Sci. Technol.* **2014**, *48*, 12561–12568. [CrossRef] [PubMed]
72. Baijukya, F.; Semu, E. Effects of Kocide 101® on the bean (*Phaseolus vulgaris* L.)-Rhizobium symbiosis. *Acta Agric. Scand. B—Plant Soil Sci.* **1998**, *48*, 175–183.
73. Moghaddam, M.N.; Sabzevar, A.H.; Mortazaei, Z. Impact of ZnO and silver nanoparticles on legume-Sinorhizobium symbiosis. *Adv. Stud. Biol.* **2017**, *9*, 83–90. [CrossRef]
74. Wang, X.; Liu, X.; Chen, J.; Han, H.; Yuan, Z. Evaluation and mechanism of antifungal effects of carbon nanomaterials in controlling plant fungal pathogen. *Carbon* **2014**, *68*, 798–806. [CrossRef]
75. Djurišić, A.B.; Leung, Y.H.; Ng, A.M.C.; Xu, X.Y.; Lee, P.K.H.; Degger, N.; Wu, R.S.S. Toxicity of Metal Oxide Nanoparticles: Mechanisms, Characterization, and Avoiding Experimental Artefacts. *Small* **2015**, *11*, 26–44. [CrossRef]
76. Khanna, K.; Kohli, S.K.; Handa, N.; Kaur, H.; Ohri, P.; Bhardwaj, R.; Yousaf, B.; Rinklebe, J.; Ahmad, P. Enthralling the impact of engineered nanoparticles on soil microbiome: A concentric approach towards environmental risks and cogitation. *Ecotoxicol. Environ. Saf.* **2021**, *222*, 112459. [CrossRef]
77. Dickson, J.S.; Koohmaraie, M. Cell surface charge characteristics and their relationship to bacterial attachment to meat surfaces. *Appl. Environ. Microbiol.* **1989**, *55*, 832–836. [CrossRef]
78. Fabrega, J.; Luoma, S.N.; Tyler, C.R.; Galloway, T.S.; Lead, J.R. Silver nanoparticles: Behaviour and effects in the aquatic environment. *Environ. Int.* **2011**, *37*, 517–531. [CrossRef]
79. Mensch, A.C.; Hernandez, R.T.; Kuether, J.E.; Torelli, M.D.; Feng, Z.V.; Hamers, R.J.; Pedersen, J.A. Natural organic matter concentration impacts the interaction of functionalized diamond nanoparticles with model and actual bacterial membranes. *Environ. Sci. Technol.* **2017**, *51*, 11075–11084. [CrossRef]
80. Hussain, S.; Garantziotis, S.; Rodrigues-Lima, F.; Dupret, J.-M.; Baeza-Squiban, A.; Boland, S. Intracellular signal modulation by nanomaterials. *Nanomateria l* **2014**, *811*, 111–134.
81. Ameen, F.; Alsamhary, K.; Alabdullatif, J.A.; Alnadhari, S. A review on metal-based nanoparticles and their toxicity to beneficial soil bacteria and fungi. *Ecotoxicol. Environ. Saf.* **2021**, *213*, 112027. [CrossRef]
82. Jacobson, K.H.; Gunsolus, I.L.; Kuech, T.R.; Troiano, J.M.; Melby, E.S.; Lohse, S.E.; Hu, D.; Chrisler, W.B.; Murphy, C.J.; Orr, G.; et al. Lipopolysaccharide Density and Structure Govern the Extent and Distance of Nanoparticle Interaction with Actual and Model Bacterial Outer Membranes. *Environ. Sci. Technol.* **2015**, *49*, 10642–10650. [CrossRef]
83. Mensch, A.C.; Buchman, J.T.; Haynes, C.L.; Pedersen, J.A.; Hamers, R.J. Quaternary amine-terminated quantum dots induce structural changes to supported lipid bilayers. *Langmuir* **2018**, *34*, 12369–12378. [CrossRef]
84. Abbas, Q.; Yousaf, B.; Amina; Ali, M.U.; Munir, M.A.M.; El-Naggar, A.; Rinklebe, J.; Naushad, M. Transformation pathways and fate of engineered nanoparticles (ENPs) in distinct interactive environmental compartments: A review. *Environ. Int.* **2020**, *138*, 105646. [CrossRef]
85. Abbas, Q.; Yousaf, B.; Ullah, H.; Ali, M.U.; Ok, Y.S.; Rinklebe, J. Environmental transformation and nano-toxicity of engineered nano-particles (ENPs) in aquatic and terrestrial organisms. *Crit. Rev. Environ. Sci. Technol.* **2020**, *50*, 2523–2581. [CrossRef]
86. Amorim, M.J.B.; Scott-Fordsmand, J.J. Toxicity of copper nanoparticles and CuCl2 salt to Enchytraeusalbidus worms: Survival, reproduction and avoidance responses. *Environ. Pollut.* **2012**, *164*, 164–168. [CrossRef]
87. Williams, D.N.; Pramanik, S.; Brown, R.P.; Zhi, B.; McIntire, E.; Hudson-Smith, N.V.; Haynes, C.L.; Rosenzweig, Z. Adverse interactions of luminescent semiconductor quantum dots with liposomes and *Shewanella oneidensis*. *ACS Appl. Nano Mater.* **2018**, *1*, 4788–4800. [CrossRef]
88. Dupont, C.L.; Grass, G.; Rensing, C. Copper toxicity and the origin of bacterial resistance—New insights and applications. *Metallomics* **2011**, *3*, 1109–1118. [CrossRef]
89. Ahmed, S.; Chaudhry, S.A.; Ikram, S. A review on biogenic synthesis of ZnO nanoparticles using plant extracts and microbes: A prospect towards green chemistry. *J. Photochem. Photobiol. B Biol.* **2017**, *166*, 272–284. [CrossRef] [PubMed]
90. Hudson-Smith, N.V.; Clement, P.L.; Brown, R.P.; Krause, M.O.; Pedersen, J.A.; Haynes, C.L. Research highlights: Speciation and transformations of silver released from Ag NPs in three species. *Environmental Science: Nano* **2016**, *3*, 1236–1240. [CrossRef]
91. Polívková, M.; Hubáček, T.; Staszek, M.; Švorčík, V.; Siegel, J. Antimicrobial treatment of polymeric medical devices by silver nanomaterials and related technology. *Int. J. Mol. Sci.* **2017**, *18*, 419. [CrossRef] [PubMed]
92. Zhang, H.; Lv, X.; Li, Y.; Wang, Y.; Li, J. P25-Graphene Composite as a High Performance Photocatalyst. *ACS Nano* **2010**, *4*, 380–386. [CrossRef]
93. Hang, M.N.; Gunsolus, I.L.; Wayland, H.; Melby, E.S.; Mensch, A.C.; Hurley, K.R.; Pedersen, J.A.; Haynes, C.L.; Hamers, R.J. Impact of nanoscale lithium nickel manganese cobalt oxide (NMC) on the bacterium *Shewanella oneidensis* MR-1. *Chem. Mater.* **2016**, *28*, 1092–1100. [CrossRef]
94. Hang, M.N.; Hudson-Smith, N.V.; Clement, P.L.; Zhang, Y.; Wang, C.; Haynes, C.L.; Hamers, R.J. Influence of nanoparticle morphology on ion release and biological impact of nickel manganese cobalt oxide (NMC) complex oxide nanomaterials. *ACS Appl. Nano Mater.* **2018**, *1*, 1721–1730. [CrossRef]

95. Imlay, J.A. The molecular mechanisms and physiological consequences of oxidative stress: Lessons from a model bacterium. *Nat. Rev. Microbiol.* **2013**, *11*, 443–454. [CrossRef]
96. Jiang, Y.; Dong, Y.; Luo, Q.; Li, N.; Wu, G.; Gao, H. Protection from oxidative stress relies mainly on derepression of OxyR-dependent KatB and Dps in *Shewanella oneidensis*. *J. Bacteriol.* **2014**, *196*, 445–458. [CrossRef]
97. Symonds, D.A.; Merchenthaler, I.; Flaws, J.A. Methoxychlor and Estradiol Induce Oxidative Stress DNA Damage in the Mouse Ovarian Surface Epithelium. *Toxicol. Sci.* **2008**, *105*, 182–187. [CrossRef]
98. Peng, Z.; Ni, J.; Zheng, K.; Shen, Y.; Wang, X.; He, G.; Jin, S.; Tang, T. Dual effects and mechanism of TiO_2 nanotube arrays in reducing bacterial colonization and enhancing C3H10T1/2 cell adhesion. *Int. J. Nanomed.* **2013**, *8*, 3093.
99. Anjem, A.; Imlay, J.A. Mononuclear Iron Enzymes Are Primary Targets of Hydrogen Peroxide Stress. *J. Biol. Chem.* **2012**, *287*, 15544–15556. [CrossRef]
100. Liu, Z.; Lin, C.-H.; Hyun, B.-R.; Sher, C.-W.; Lv, Z.; Luo, B.; Jiang, F.; Wu, T.; Ho, C.-H.; Kuo, H.-C. Micro-light-emitting diodes with quantum dots in display technology. *Light Sci. Appl.* **2020**, *9*, 1–23. [CrossRef]
101. Domínguez, G.A.; Torelli, M.D.; Buchman, J.T.; Haynes, C.L.; Hamers, R.J.; Klaper, R.D. Size dependent oxidative stress response of the gut of Daphnia magna to functionalized nanodiamond particles. *Environ. Res.* **2018**, *167*, 267–275. [CrossRef]
102. Wang, F.Y.; Lin, X.G.; Yin, R. Effect of Arbuscular Mycorrhizal Fungal Inoculation on Heavy Metal Accumulation of Maize Grown in a Naturally Contaminated Soil. *Int. J. Phytoremediation* **2007**, *9*, 345–353. [CrossRef]
103. Xie, J.; Ming, Z.; Li, H.; Yang, H.; Yu, B.; Wu, R.; Liu, X.; Bai, Y.; Yang, S.-T. Toxicity of graphene oxide to white rot fungus *Phanerochaete chrysosporium*. *Chemosphere* **2016**, *151*, 324–331. [CrossRef]
104. Akhavan, O.; Ghaderi, E. Escherichia coli bacteria reduce graphene oxide to bactericidal graphene in a self-limiting manner. *Carbon* **2012**, *50*, 1853–1860. [CrossRef]
105. Zarzuela, R.; Carbú, M.; Gil, M.A.; Cantoral, J.M.; Mosquera, M.J. CuO/SiO_2 nanocomposites: A multifunctional coating for application on building stone. *Mater. Des.* **2017**, *114*, 364–372. [CrossRef]
106. De Filpo, G.; Palermo, A.M.; Rachiele, F.; Nicoletta, F.P. Preventing fungal growth in wood by titanium dioxide nanoparticles. *Int. Biodeterior. Biodegrad.* **2013**, *85*, 217–222. [CrossRef]
107. Vogel-Mikuš, K.; Pongrac, P.; Kump, P.; Nečemer, M.; Regvar, M. Colonisation of a Zn, Cd and Pb hyperaccumulator Thlaspi praecox Wulfen with indigenous arbuscular mycorrhizal fungal mixture induces changes in heavy metal and nutrient uptake. *Environ. Pollut.* **2006**, *139*, 362–371. [CrossRef]
108. Gurunathan, S.; Han, J.W.; Dayem, A.A.; Eppakayala, V.; Kim, J.-H. Oxidative stress-mediated antibacterial activity of graphene oxide and reduced graphene oxide in Pseudomonas aeruginosa. *Int. J. Nanomed.* **2012**, *7*, 5901–5914. [CrossRef]
109. Imlay, J.A. Cellular Defenses against Superoxide and Hydrogen Peroxide. *Annu. Rev. Biochem.* **2008**, *77*, 755–776. [CrossRef] [PubMed]
110. Zhao, Y.; Ran, W.; He, J.; Huang, Y.; Liu, Z.; Liu, W.; Tang, Y.; Zhang, L.; Gao, D.; Gao, F. High-performance asymmetric supercapacitors based on multilayer MnO_2/graphene oxide nanoflakes and hierarchical porous carbon with enhanced cycling stability. *Small* **2015**, *11*, 1310–1319. [CrossRef] [PubMed]
111. McKee, M.S.; Filser, J. Impacts of metal-based engineered nanomaterials on soil communities. *Environ. Sci. Nano* **2016**, *3*, 506–533. [CrossRef]
112. Lin, D.; Tian, X.; Wu, F.; Xing, B. Fate and Transport of Engineered Nanomaterials in the Environment. *J. Environ. Qual.* **2010**, *39*, 1896–1908. [CrossRef] [PubMed]
113. Wightwick, A.; Walters, R.; Allinson, G.; Reichman, S.; Menzies, N. Environmental risks of fungicides used in horticultural production systems. *Fungicides* **2010**, *1*, 273–304.
114. Durán, N.; Durán, M.; de Jesus, M.B.; Seabra, A.B.; Fávaro, W.J.; Nakazato, G. Silver nanoparticles: A new view on mechanistic aspects on antimicrobial activity. *Nanomedicine* **2016**, *12*, 789–799. [CrossRef]
115. Jung, J.-H.; Kim, S.-W.; Min, J.-S.; Kim, Y.-J.; Lamsal, K.; Kim, K.S.; Lee, Y.S. The Effect of Nano-Silver Liquid against the White Rot of the Green Onion Caused by *Sclerotiumcepivorum*. *Mycobiology* **2010**, *38*, 39–45. [CrossRef]
116. Parada, J.; Rubilar, O.; Sousa, D.; Martínez, M.; Fernández-Baldo, M.A.; Tortella, G. Short term changes in the abundance of nitrifying microorganisms in a soil-plant system simultaneously exposed to copper nanoparticles and atrazine. *Sci. Total Environ.* **2019**, *670*, 1068–1074. [CrossRef]
117. Gottschalk, F.; Sonderer, T.; Scholz, R.W.; Nowack, B. Modeled environmental concentrations of engineered nanomaterials (TiO_2, ZnO, Ag, CNT, fullerenes) for different regions. *Environ. Sci. Technol.* **2009**, *43*, 9216–9222. [CrossRef]
118. Maqueda, C.; Villaverde, J.; Sopena, F.; Undabeytia, T.; Morillo, E. Effects of soil characteristics on metribuzin dissipation using clay-gel-based formulations. *J. Agric. Food Chem.* **2009**, *57*, 3273–3278. [CrossRef]
119. Hund-Rinke, K.; Hümmler, A.; Schlinkert, R.; Wege, F.; Broll, G. Evaluation of microbial shifts caused by a silver nanomaterial: Comparison of four test systems. *Environ. Sci. Eur.* **2019**, *31*, 1–13. [CrossRef]
120. Schlich, K.; Hoppe, M.; Kraas, M.; Schubert, J.; Chanana, M.; Hund-Rinke, K. Long-term effects of three different silver sulfide nanomaterials, silver nitrate and bulk silver sulfide on soil microorganisms and plants. *Environ. Pollut.* **2018**, *242*, 1850–1859. [CrossRef]
121. Schlich, K.; Hund-Rinke, K. Influence of soil properties on the effect of silver nanomaterials on microbial activity in five soils. *Environ. Pollut.* **2015**, *196*, 321–330. [CrossRef]

122. Vitali, F.; Raio, A.; Sebastiani, F.; Cherubini, P.; Cavalieri, D.; Cocozza, C. Environmental pollution effects on plant microbiota: The case study of poplar bacterial-fungal response to silver nanoparticles. *Appl. Microbiol. Biotechnol.* **2019**, *103*, 8215–8227. [CrossRef]
123. Grün, A.-L.; Straskraba, S.; Schulz, S.; Schloter, M.; Emmerling, C. Long-term effects of environmentally relevant concentrations of silver nanoparticles on microbial biomass, enzyme activity, and functional genes involved in the nitrogen cycle of loamy soil. *J. Environ. Sci.* **2018**, *69*, 12–22. [CrossRef]
124. Marshall, B.M.; Levy, S.B. Food Animals and Antimicrobials: Impacts on Human Health. *Clin. Microbiol. Rev.* **2011**, *24*, 718–733. [CrossRef]
125. Vandevoort, A.R.; Arai, Y. Effect of Silver Nanoparticles on Soil Denitrification Kinetics. *Ind. Biotechnol.* **2012**, *8*, 358–364. [CrossRef]
126. Dugal, S.; Mascarenhas, S. Chemical synthesis of copper nanoparticles and its antibacterial effect against gram negative pathogens. *J. Adv. Sci. Res.* **2015**, *6*, 1–4.
127. Adisa, I.O.; Pullagurala, V.L.R.; Peralta-Videa, J.R.; Dimkpa, C.O.; Elmer, W.H.; Gardea-Torresdey, J.L.; White, J.C. Recent advances in nano-enabled fertilizers and pesticides: A critical review of mechanisms of action. *Environ. Sci. Nano* **2019**, *6*, 2002–2030. [CrossRef]
128. Rajput, V.; Minkina, T.; Ahmed, B.; Sushkova, S.; Singh, R.; Soldatov, M.; Laratte, B.; Fedorenko, A.; Mandzhieva, S.; Blicharska, E.; et al. Interaction of Copper-Based Nanoparticles to Soil, Terrestrial, and Aquatic Systems: Critical Review of the State of the Science and Future Perspectives. *Rev. Environ. Contam. Toxicol.* **2019**, *252*, 51–96.
129. Rajput, V.D.; Minkina, T.M.; Behal, A.; Sushkova, S.N.; Mandzhieva, S.; Singh, R.; Gorovtsov, A.; Tsitsuashvili, V.S.; Purvis, W.O.; Ghazaryan, K.A.; et al. Effects of zinc-oxide nanoparticles on soil, plants, animals and soil organisms: A review. *Environ. Nanotechnol. Monit. Manag.* **2018**, *9*, 76–84. [CrossRef]
130. He, X.; Deng, H.; Hwang, H.-M. Nanosensors for Heavy Metal Detection in Environmental Media: Recent Advances and Future Trends. *NanosensorsEnviron. Food Agric.* **2021**, *1*, 23–51.
131. Millardet, A.; Gayon, U. *The Discovery of Bordeaux Mixture: Three Papers: I. Treatment of Mildew and Rot. II. Treatment of Mildew with Copper Sulphate and Lime Mixture. III. Concerning the History of the Treatment of Mildew with Copper Sulphate (No. 3)*; American Phytopathological Society: St. Paul, MN, USA, 1933.
132. Simonin, M.; Colman, B.P.; Tang, W.; Judy, J.D.; Anderson, S.M.; Bergemann, C.M.; Rocca, J.D.; Unrine, J.M.; Cassar, N.; Bernhardt, E.S. Plant and microbial responses to repeated Cu (OH) 2 nanopesticide exposures under different fertilization levels in an agro-ecosystem. *Front. Microbiol.* **2018**, *9*, 1769. [CrossRef] [PubMed]
133. VandeVoort, A.R.; Skipper, H.; Arai, Y. Macroscopic Assessment of Nanosilver Toxicity to Soil Denitrification Kinetics. *J. Environ. Qual.* **2014**, *43*, 1424–1430. [CrossRef] [PubMed]
134. Zhai, Y.; Hunting, E.R.; Wouters, M.; Peijnenburg, W.J.G.M.; Vijver, M.G. Silver Nanoparticles, Ions, and Shape Governing Soil Microbial Functional Diversity: Nano Shapes Micro. *Front. Microbiol.* **2016**, *7*, 1123. [CrossRef]
135. Ottoni, C.A.; Neto, M.L.; Léo, P.; Ortolan, B.D.; Barbieri, E.; De Souza, A.O. Environmental impact of biogenic silver nanoparticles in soil and aquatic organisms. *Chemosphere* **2020**, *239*, 124698. [CrossRef]
136. Cornelis, G.; Hund-Rinke, K.; Kuhlbusch, T.; Van den Brink, N.; Nickel, C. Fate and bioavailability of engineered nanoparticles in soils: A review. *Crit. Rev. Environ. Sci. Technol.* **2014**, *44*, 2720–2764. [CrossRef]
137. Von Uexküll, H.R.; Mutert, E. Global extent, development and economic impact of acid soils. *Plant Soil* **1995**, *171*, 1–15. [CrossRef]
138. Kędziora, A.; Speruda, M.; Krzyżewska, E.; Rybka, J.; Łukowiak, A.; Bugla-Płoskońska, G. Similarities and differences between silver ions and silver in nanoforms as antibacterial agents. *Int. J. Mol. Sci.* **2018**, *19*, 444. [CrossRef]
139. Venkataraju, J.L.; Sharath, R.; Chandraprabha, M.; Neelufar, E.; Hazra, A.; Patra, M. Synthesis, characterization and evaluation of antimicrobial activity of zinc oxide nanoparticles. *J. Biochem. Technol.* **2014**, *3*, 151–154.
140. Murray, R.A.; Escobar, A.; Bastús, N.G.; Andreozzi, P.; Puntes, V.; Moya, S.E. Fluorescently labelled nanomaterials in nanosafety research: Practical advice to avoid artefacts and trace unbound dye. *Nanoimpact* **2018**, *9*, 102–113. [CrossRef]
141. Dempsey, M.A.; Fisk, M.C.; Yavitt, J.B.; Fahey, T.J.; Balser, T.C. Exotic earthworms alter soil microbial community composition and function. *Soil Biol. Biochem.* **2013**, *67*, 263–270. [CrossRef]
142. Tourinho, P.S.; Van Gestel, C.A.; Lofts, S.; Svendsen, C.; Soares, A.M.; Loureiro, S. Metal-based nanoparticles in soil: Fate, behavior, and effects on soil invertebrates. *Environ. Toxicol. Chem.* **2012**, *31*, 1679–1692. [CrossRef]
143. Tchalala, M.R.; Kara, A.; Lachgar, A.; Yagoubi, S.; Foy, E.; Vega, E.; Nitsche, S.; Chaudanson, D.; Aufray, B.; EL Firdoussi, L.; et al. Silicon nanoparticles synthesis from calcium disilicide by redox assisted chemical exfoliation. *Mater. Today Commun.* **2018**, *16*, 281–284. [CrossRef]
144. Shen, Z.; Chen, Z.; Hou, Z.; Li, T.; Lu, X. Ecotoxicological effect of zinc oxide nanoparticles on soil microorganisms. *Front. Environ. Sci. Eng.* **2015**, *9*, 912–918. [CrossRef]
145. Chai, H.; Yao, J.; Sun, J.; Zhang, C.; Liu, W.; Zhu, M.; Ceccanti, B. The Effect of Metal Oxide Nanoparticles on Functional Bacteria and Metabolic Profiles in Agricultural Soil. *Bull. Environ. Contam. Toxicol.* **2015**, *94*, 490–495. [CrossRef]
146. Reddy, K.M.; Feris, K.; Bell, J.; Wingett, D.G.; Hanley, C.; Punnoose, A. Selective toxicity of zinc oxide nanoparticles to prokaryotic and eukaryotic systems. *Appl. Phys. Lett.* **2007**, *90*, 213902–2139023. [CrossRef]
147. Shim, W.J.; Hong, S.H.; Eo, S.E. Identification methods in microplastic analysis: A review. *Anal. Methods* **2016**, *9*, 1384–1391. [CrossRef]

148. Manzoor, U.; Siddique, S.; Ahmed, R.; Noreen, Z.; Bokhari, H.; Ahmad, I. Antibacterial, structural and optical characterization of mechano-chemically prepared ZnO nanoparticles. *PLoS ONE* **2016**, *11*, e0154704. [CrossRef]
149. Suman, T.; Rajasree, S.R.; Kirubagaran, R. Evaluation of zinc oxide nanoparticles toxicity on marine algae chlorella vulgaris through flow cytometric, cytotoxicity and oxidative stress analysis. *Ecotoxicol. Environ. Saf.* **2015**, *113*, 23–30. [CrossRef]
150. Gajjar, P.; Pettee, B.; Britt, D.W.; Huang, W.; Johnson, W.P.; Anderson, A.J. Antimicrobial activities of commercial nanoparticles against an environmental soil microbe, Pseudomonas putida KT2440. *J. Biol. Eng.* **2009**, *3*, 1–13. [CrossRef] [PubMed]
151. Dimkpa, C.O.; Zeng, J.; McLean, J.E.; Britt, D.W.; Zhan, J.; Anderson, A.J. Production of indole-3-acetic acid via the indole-3-acetamide pathway in the plant-beneficial bacterium Pseudomonas chlororaphis O_6 is inhibited by ZnO nanoparticles but enhanced by CuO nanoparticles. *Appl. Environ. Microbiol.* **2012**, *78*, 1404–1410. [CrossRef]
152. Anjum, N.A.; Gill, S.S.; Duarte, A.C.; Pereira, E.; Ahmad, I. Silver nanoparticles in soil–plant systems. *J. Nanoparticle Res.* **2013**, *15*, 1–26. [CrossRef]
153. Simonin, M.; Cantarel, A.A.; Crouzet, A.; Gervaix, J.; Martins, J.M.; Richaume, A. Negative effects of copper oxide nanoparticles on carbon and nitrogen cycle microbial activities in contrasting agricultural soils and in presence of plants. *Front. Microbiol.* **2018**, *9*, 3102. [CrossRef] [PubMed]
154. Svendsen, C.; Walker, L.A.; Matzke, M.; Lahive, E.; Harrison, S.; Crossley, A.; Park, B.; Lofts, S.; Lynch, I.; Vázquez-Campos, S.; et al. Key principles and operational practices for improved nanotechnology environmental exposure assessment. *Nat. Nanotechnol.* **2020**, *15*, 731–742. [CrossRef]
155. Ge, Y.; Schimel, J.P.; Holden, P.A. Evidence for negative effects of TiO_2 and ZnO nanoparticles on soil bacterial communities. *Environ. Sci. Technol.* **2011**, *45*, 1659–1664. [CrossRef]
156. Ge, Y.; Priester, J.H.; Mortimer, M.; Chang, C.H.; Ji, Z.; Schimel, J.P.; Holden, P.A. Long-term effects of multiwalled carbon nanotubes and graphene on microbial communities in dry soil. *Environ. Sci. Technol.* **2016**, *50*, 3965–3974. [CrossRef]
157. Hu, C.; Li, M.; Cui, Y.; Li, D.; Chen, J.; Yang, L. Toxicological effects of TiO_2 and ZnO nanoparticles in soil on earthworm Eisenia fetida. *Soil Biol. Biochem.* **2010**, *42*, 586–591. [CrossRef]
158. Kim, B.; Park, C.-S.; Murayama, M.; Hochella, M.F., Jr. Discovery and characterization of silver sulfide nanoparticles in final sewage sludge products. *Environ. Sci. Technol.* **2010**, *44*, 7509–7514. [CrossRef]
159. Baker, S.; Volova, T.; Prudnikova, S.V.; Satish, S.; Prasad, N. Nanoagroparticles emerging trends and future prospect in modern agriculture system. *Environ. Toxicol. Pharmacol.* **2017**, *53*, 10–17. [CrossRef]
160. Johnson, A.C.; Park, B. Predicting contamination by the fuel additive cerium oxide engineered nanoparticles within the United Kingdom and the associated risks. *Environ. Toxicol. Chem.* **2012**, *31*, 2582–2587. [CrossRef]
161. Li, Z.-Z.; Chen, J.-F.; Liu, F.; Liu, A.-Q.; Wang, Q.; Sun, H.-Y.; Wen, L.-X. Study of UV-shielding properties of novel porous hollow silica nanoparticle carriers for avermectin. *Pest Manag. Sci.* **2007**, *63*, 241–246. [CrossRef]
162. Shan, J.; Ji, R.; Yu, Y.; Xie, Z.; Yan, X. Biochar, activated carbon and carbon nanotubes have different effects on fate of 14C-catechol and microbial community in soil. *Sci. Rep.* **2015**, *5*, 16000. [CrossRef]
163. Liné, C.; Larue, C.; Flahaut, E. Carbon nanotubes: Impacts and behaviour in the terrestrial ecosystem—A review. *Carbon* **2017**, *123*, 767–785. [CrossRef]
164. Jackson, P.; Jacobsen, N.R.; Baun, A.; Birkedal, R.; Kühnel, D.; Jensen, K.A.; Vogel, U.; Wallin, H. Bioaccumulation and ecotoxicity of carbon nanotubes. *Chem. Central J.* **2013**, *7*, 154. [CrossRef]
165. Chung, H.; Son, Y.; Yoon, T.K.; Kim, S.; Kim, W. The effect of multi-walled carbon nanotubes on soil microbial activity. *Ecotoxicol. Environ. Saf.* **2011**, *74*, 569–575. [CrossRef]
166. Jin, L.; Son, Y.; DeForest, J.L.; Kang, Y.J.; Kim, W.; Chung, H. Single-walled carbon nanotubes alter soil microbial community composition. *Sci. Total. Environ.* **2014**, *466*, 533–538. [CrossRef]
167. Jin, L.; Son, Y.; Yoon, T.K.; Kang, Y.J.; Kim, W.; Chung, H. High concentrations of single-walled carbon nanotubes lower soil enzyme activity and microbial biomass. *Ecotoxicol. Environ. Saf.* **2013**, *88*, 9–15. [CrossRef]
168. Gigault, J.; Halle, A.T.; Baudrimont, M.; Pascal, P.Y.; Gauffre, F.; Phi, T.L.; El Hadri, H.; Grassl, B.; Reynaud, S. Current opinion: What is a nanoplastic? *Environ. Pollut.* **2018**, *235*, 1030–1034. [CrossRef]
169. Rodrigues, D.F.; Jaisi, D.P.; Elimelech, M. Toxicity of Functionalized Single-Walled Carbon Nanotubes on Soil Microbial Communities: Implications for Nutrient Cycling in Soil. *Environ. Sci. Technol.* **2013**, *47*, 625–633. [CrossRef]
170. Petersen, E.J.; Pinto, R.A.; Landrum, P.F.; Weber, J.; Walter, J. Influence of carbon nanotubes on pyrene bioaccumulation from contaminated soils by earthworms. *Environ. Sci. Technol.* **2009**, *43*, 4181–4187. [CrossRef] [PubMed]
171. Petersen, E.J.; Huang, Q.; Weber, J.W.J. Bioaccumulation of Radio-Labeled Carbon Nanotubes by *Eisenia foetida*. *Environ. Sci. Technol.* **2008**, *42*, 3090–3095. [CrossRef] [PubMed]

172. Petersen, E.J.; Pinto, R.A.; Zhang, L.; Huang, Q.; Landrum, P.F.; Weber, W.J., Jr. Effects of polyethyleneimine-mediated functionalization of multi-walled carbon nanotubes on earthworm bioaccumulation and sorption by soils. *Environ. Sci. Technol.* **2011**, *45*, 3718–3724. [CrossRef] [PubMed]
173. Qi, R.; Jones, D.L.; Li, Z.; Liu, Q.; Yan, C. Behavior of microplastics and plastic film residues in the soil environment: A critical review. *Sci. Total. Environ.* **2019**, *703*, 134722. [CrossRef]

coatings

Article

The Role of Biochar Nanoparticles Performing as Nanocarriers for Fertilizers on the Growth Promotion of Chinese Cabbage (*Brassica rapa* (Pekinensis Group))

Ruiping Yang [1,2], Jiamin Shen [1], Yuhan Zhang [1], Lin Jiang [1], Xiaoping Sun [1], Zhengyang Wang [3], Boping Tang [1,*] and Yu Shen [2,*]

1 Jiangsu Key Laboratory for Bioresources of Saline Soils, Jiangsu Synthetic Innovation Center for Coastal Bio-Agriculture, School of Wetlands, Yancheng Teachers University, Yancheng 224007, China
2 Co-Innovation Center for the Sustainable Forestry in Southern China, College of Biology and the Environment, Nanjing Forestry University, Nanjing 210037, China
3 Department of Environmental Sciences, The Connecticut Agricultural Experiment Station, New Haven, CT 06504, USA
* Correspondence: boptang@163.com (B.T.); sheyttmax@hotmail.com or yushen@njfu.edu.cn (Y.S.)

Abstract: Chinese cabbage (*Brassica rapa*) belongs to the Pekinensis Group and is grown annually as a salad crop. It is one of the most important food crops in Eastern Asia and the most widely grown vegetable in China, accounting for more one-quarter of the total annual vegetable consumption in northern parts of the country. It is reported that nitrogen (N), phosphorus (P), and potassium (K) fertilizations play important roles in the physio-morphological traits and yields of Chinese cabbage. However, N, P, and K use in agriculture continues to increase. Excessive application of fertilizers has a harmful impact on the environment. Yet how to improve the irrigation effects on Chinese cabbage growth is still limited. In this study, we firstly selected biochar nanoparticles (BNPs) prepared from corn straw, which had been air-dried and heated in a muffle furnace at 350 °C for 120 min, with K (potassium sulfate), N (calcium nitrate tetrahydrate), and P (sodium dihydrogen phosphate dihydrate) fertilizers. Then, a screening experiment (Experiment I) was performed via the response model to find the best solution for Chinese cabbage growth. Treatment with 2 g/kg of N and 2 g/kg of K for 4 weeks was the optimum application to promote Chinese cabbage growth. Then, a comparison experiment (Experiment II) was carried out to test the best formula for Chinese cabbage growth with or without BNPs. After co-irrigation with N and K for 4 weeks, treatment with a combination of 2 g/kg of BNPs, 2 g/kg of N, and 2 g/kg of K was the optimum formula for Chinese cabbage growth. Plant biomass increased by more than 1796.86% and 32.80%, respectively, in two combined treatments of BNPs and fertilizers as compared to the control treatment. After the addition of BNPs, Chinese cabbage height (aboveground) and the dry weight of belowground biomass in the N + K treatment increased to 10.97% and 20.48%, respectively. These results suggest that BNPs have great potential as a nanocarrier for fertilization as they are highly efficient (over 50% increase), reducing fertilizer use while promoting plant growth. The use of BNPs as a nanocarrier for fertilizers represents a step toward more environmentally friendly agriculture.

Keywords: biochar nanoparticle; Chinese cabbage; conventional fertilizers; plant growth; nanocarrier

Citation: Yang, R.; Shen, J.; Zhang, Y.; Jiang, L.; Sun, X.; Wang, Z.; Tang, B.; Shen, Y. The Role of Biochar Nanoparticles Performing as Nanocarriers for Fertilizers on the Growth Promotion of Chinese Cabbage (*Brassica rapa* (Pekinensis Group)). *Coatings* **2022**, *12*, 1984. https://doi.org/10.3390/coatings12121984

Academic Editor: Zivile Luksiene

Received: 31 October 2022
Accepted: 15 December 2022
Published: 18 December 2022

1. Introduction

Chinese cabbage (*Brassica rapa*) is known as napa, napa cabbage, petsai, wongbok, and chihli in Asian countries; and it is also called Chinese leaves or celery cabbage, which belongs to the Pekinensis group. The vegetable has a long history in China and is of major importance, with over 300,000 ha grown in China. Chinese cabbage is an important food in Korea, Taiwan, and Japan, where it is grown as an annual crop. Most Chinese cabbage cultivars are biennial and produce tight, compact, cylindrical heads [1]. Similar to other cruciferous vegetables, Chinese cabbage has a shallow root system, which limits its ability to absorb water and nutrients from deeper soil [2]. The plant's nutritional demands are significantly higher during the growing period when the leaf mass is at its highest [3].

Chinese cabbage is a fast-growing vegetable with high nutritional value. For optimum growth, Chinese cabbage requires an adequate supply of both soil nutrients and soil water [4]. In recent decades, crop production has depended largely on the use of chemical fertilizers, with nutrient fertilization playing an important role in improving crop productivity and maintaining soil fertility. Farmers use large quantities of chemical fertilizers to increase yields. The average cost of fertilizers is a minimum of USD 23.6 per m^2 and 0.4 kg per m^2 of co-fertilizers (organic fertilizers and N-P-K fertilizers). Over 50–70% of fertilizers applied to crops/fields are not absorbed by plants but lost to surface runoff [5]. This surface runoff leads to groundwater pollution [6]. Thus, new methods are needed to improve the effectiveness of fertilizers and reduce loss lost to runoff.

In terms of nutrition, plants require a correct proportion of nitrogen (N), phosphorus (P), and potassium (K), which have a synergistic effect on plant health, plant growth, and final plant yield [7]. Fertilizers are fundamental in the development of plants and crops. Fertilizers help plants grow faster. This goal can be achieved in two ways. The first is through the use of nutrient-rich additives. The second mechanism by which certain fertilizers work is to improve the soil's efficacy by altering water retention and aeration [8]. However, it is known that fertilizers are one of the main pollution sources of soil and water, with nitrates leaking into groundwater and soil, and N and P runoff into water bodies [9]. The use of large quantities of fertilizers also contributes to greenhouse gas emissions, leading to soil pollution, bioaccumulation of pollutants in the soil, and transfer to the food chain [10,11]. Thus, how to reduce fertilizer use is important for environmental protection and sustainable agricultural development.

According to a previous study, 40–70%, 80–90%, and 50–90% of N, P, and K, respectively, applied to the soil is lost, representing a considerable cost in terms of key macronutrient resources [12]. Previous research suggested that nanocarrier-bound fertilizers exhibit higher delivery efficiencies than fertilizers applied using traditional irrigation [13]. Carbon nanomaterials, including nanotubes and graphene oxide, have been proposed as nanocarriers for micronutrients [14,15]. Although promising, thus far, most proposed nanotechnologies for micronutrient delivery have been tested only in the laboratory.

There is much interest in agricultural research in the idea of nanocarriers for fertilizers [16,17]. However, the problem of how to improve the delivery efficiency of nanomacronutrient elements remains [18]. In this respect, the current best strategy is to find a kind of material to slow nutrient release. Biochar pores are divided into micropores (<2 nm), mesopores (2–50 nm), and macropores (>50 nm), according to pore size [19]. Larger biochar pores, such as those of BNPs, offer a significantly greater specific surface area and pore volume than smaller biochar pores [20]. Biochar is known for its high potential absorption of heavy metals [21] and organic contaminants [22,23]. Nanosized materials have markedly high surface areas. Previous research proposed that biochar NPs (BNPs) have great potential as carriers for fertilizers and that they can slow down the release of fertilizers in the soil. Zein-based NPs have been shown to be a safe, biocompatible, and effective nanocarriers for botanical pest repellents [24]. Mesoporous silica nanoparticles were delivered through soil-free nutrient media to wheat translocated from roots to shoots and localized to chloroplasts, which promoted photosynthesis and seedling growth [25]. Compared with chemical fertilizers used in agriculture, BNPs are more cost effective, eco-friendly, nontoxic, and stable. In this study, it is hypothesized that BNPs are a kind of material that could reduce fertilizer use for plant growth. BNP- made nanocarriers would be effective materials for fertilizer use in sustainable agriculture in the future.

2. Material and Methods

2.1. BNPs Preparation

Fresh corn stalks were collected from Jiangning District, Nanjing, Jiangsu Province, China, and aid-dried. The waste biomass materials were then cut into 30–50 mm pieces and heated in a muffle furnace at 350 °C for 120 min. The biochar preparation method followed that of Shen et al. [26]. The aim was to convert 35% of the biomass into biochar.

After preparation of the biochar, a Planetary Ball Mill (Shunchi Tech, PMQW2; Nanjing, China) was used to process the BNPs. Ethanol was used as a grinding aid, and the weight ratio of ZrO_2 balls to powder was 15:1 in the vials in the system. The vials were spun at 400 rpm for 6 h. The biochar was then removed and placed in sealed bags and stored in desiccators until use. Characterization of the prepared nano biochar using transmission electron microscopy (TEM) was conducted, and X-ray diffractometer (XRD) and X-ray photoelectron spectroscopy (XPS) were performed via Da 8 Venture Single Crystal X-ray Diffractometer (Bruker, Germany), Ultima IV X-ray diffractometer (Rigaku, Japan), and Surface Area and Micropore Size Analyzer (V-Sorb 2800P, Gold APP Instruments Corporation, Xi'an, China), respectively.

In the experimental treatments, the BNPs were added to the N-P-K solution in a 1:1 (*w*/*w*) ratio. Before the root application, the BNPs and N-P-K solution were mixed together and then placed in a shaker at 140 rpm for 12 h. In terms of the fertilizer additions, N was derived from calcium nitrate tetrahydrate, P was derived from sodium dihydrogen phosphate dihydrate, and K was derived from potassium sulfate. All the chemicals were of analytical pure grade and were purchased from Nanjing Chemical Reagent Co., Ltd. (Nanjing, China). Leaf quality and stem diameter determined the Chinese cabbage grade.

The experimental soil was an air-dried commercial seedling medium purchased from Shaanxi Yangling Yufeng Seed Industry Co., Ltd. (Xianyang, China). The cabbage seeds used were commercial Suzhou Qing seeds (Xingyun Vegetable Seed Breeding Center, Qing County, China). The seeds were germinated in a seedling box at a temperature of 20/25 °C (day/night), with humidity of 50% (ZLC-100D; Shuolian Equipment Co., Ltd., Hangzhou, China). Four weeks later, when the seedlings had three true leaves, uniform seedlings were selected for subsequent experiments. Every three seedlings were placed in a plastic pot (10 × 7 × 8.5 cm) with a drainage hole of 1 cm diameter at the bottom. A 20-mesh insect-proof net of 5 × 5 cm was placed on the bottom of the pot to prevent soil leakage.

2.1.1. Experiment I: Screening Experiment

In experiment I, we used a response surface model (Design Expert 9; Stat-Ease, Minneapolis, MN, USA) to determine the optimum N-P-K solution for Chinese cabbage growth (Table S1). In the model, the variates were as follows: unite 0 = 0 g/L, unite 1 = 0.5 g/L, unite 2 = 1 g/L, and unite 3 = 1.5 g/L. There were five replicates of each treatment, and the fertilizer treatment was applied to the seedlings every 7 days, with a total of four fertilizer applications and one control one. During the experiment, plant heights and stem diameters were recorded once a week. After four weeks, the plants were harvested. Nine seedlings were randomly selected from each treatment. The above- and belowground parts were then separated, and the stem diameters, root lengths, and above- and belowground fresh weights were measured. The root length was measured from the base of the stem to the tip of the main root. The plants were then transferred to a 105 °C oven for 30 min and then 80 °C for 24 h. The dry weights of the plants were then measured.

2.1.2. Experiment II

Based on the results of experiment I, the fertilizer treatments were compared to determine the best ratio of N-P-K solution with or without BNPs. In the protocol design (Table S2), the variates were as follows: unite 0 = 0 g/L, unite 1 = 0.5 g/L, unite 2 = 1 g/L, and unite 3 = 1.5 g/L. There were five replicates of each treatment. In experiment II, the fertilizer treatments (Table S2) were added to the plants every 2 weeks. Four weeks later, the Chinese cabbage plants were harvested, and the plant heights, root lengths, above- and belowground fresh weights, stem diameters, and dry weights were recorded for further analysis.

2.2. Content of Elements

The dried above- and belowground tissues were ground and passed through a 1 mm sieve for elemental analysis. Each plant tissue sample (0.5 g) was digested in 50 mL polypropylene digestion tubes with 5 mL of nitric acid for 45 min on a heat block at 115 °C. The K, P, calcium (Ca), magnesium (Mg), sodium (Na), and sulfur (S) contents of the plant tissues were determined by inductively coupled plasma emission spectroscopy (Agilent 710 Series; Santa Clara, CA, USA). The elemental content was expressed in mg kg^{-1} (dry weight) plant tissue.

2.3. Statistical Analysis

Statistical analysis of the data was performed using SPSS 21.0 statistical software (IBM Crop., Armonk, NY, USA). The data with the replicates was performed for analysis with variance. The figures were constructed using PAST 4.03 (University of Oslo, Norway). Means were compared using the least significant difference test and Duncan's new multiple range test. In this study, statistical significance was set at the level of $p < 0.05$.

3. Results

3.1. The Properties of BNPs

TEM characterization of the prepared BNPs revealed that they were of uniform size (approximately 85 nm) (Figure 1c,d). In an XRD analysis, the BNPs showed higher peaks from 18° to 30°, pointing to higher carbon contents in the BNPs. Peaks observed from 15° to 36° revealed stacking of monoatomic carbon layers as crystallites in the BNPs (Figure S1). Furthermore, the XPS result presents that the BNPs are rich in oxygen-containing groups on the surface (Figure S2). Based on the surface-area analysis, it is suggested that the BNPs are rich in pores on the surface, and it can perform the absorption model with unrestricted monolayer-multiplayer absorption (Table S1).

3.2. Experiment I: The Optimum N-P-K Solution for Chinese Cabbage Growth

After 4 weeks of growth and four fertilizer applications, based on the treatments of response surface model, Chinese cabbage growth was best in the T4, T13, and T15 treatments with N-P-K ratios of 2-0-2, 1-1-2, and 2-1-1, respectively. Chinese cabbage growth was the worst in the T7 and T11 treatments with higher N-P-K ratios of 2-3-2 and 3-2-2, respectively, (Figure 1a).

In the control (T1), the plant size and leaf area of Chinese cabbage were the smallest in all of the treatments (Figure 1a). Furthermore, the Chinese cabbage above-ground part fresh biomass and dry biomass was the lowest, at 4.90 and 0.51 in the control compared with other treatments, respectively (Figure 1d,f). However, the Chinese cabbage under-ground part presented the highest fresh biomass and dry biomass, which reached 1.30 and 0.22 in the control and other treatments, respectively (Figure 1e,g).

Plant height and root length reached 12.05 and 16.23 cm, respectively, in the T4 treatment. The maximum increase of 0.05 cm in stem dimeter was found in the T5 treatment. The T13 and T15 treatments also showed trends toward increases in plant heights, root lengths, and stem dimeters (11.56, 14.18, and 3.04 cm, respectively) as compared with the control after 4 weeks. The plants in the T7, T8, T9, T10, and T11 treatments showed a decreasing trend in plant heights, root lengths, and stem dimeters. In addition, root lengths in the T3, T6, and T7 treatments exhibited a decreasing trend (Figure 2a–c).

Figure 1. The phenotype of Chinese cabbage in the screening experiment after four weeks (**a**); and the phenotype of Chinese cabbage in the comparison experiment after four weeks (**b**). The BNPs TEM image (**c**). (Note, in Figure a, T1: N-P-K=0-0-0, T2: N-P-K=0-2-2, T3: N-P-K=1-2-2, T4: N-P-K=2-0-2, T5: N-P-K=2-1-2, T6: N-P-K=2-2-2, T7: N-P-K=2-3-2, T8: N-P-K=2-2-3, T9: N-P-K=2-2-0, T10: N-P-K=2-2-1, T11: N-P-K=3-2-2, T12: N-P-K=1-3-2, T13: N-P-K=1-1-2, T14: N-P-K=1-2-1, T15: N-P-K=2-1-1; in Figure 2, T1: BNPs-N-K=1:2:2, T2: BNPs-N-K=2:2:2, T3: BNPs -N-K=2:1:1, T4: BNPs-N-K=2:0.5:0.5, T5: BNPs-N-K=2:0:0, T6: BNPs-N-K=0:2:2; BNPs, biochar nanoparticles; 1 unit means 0.5 g/L nutrient).

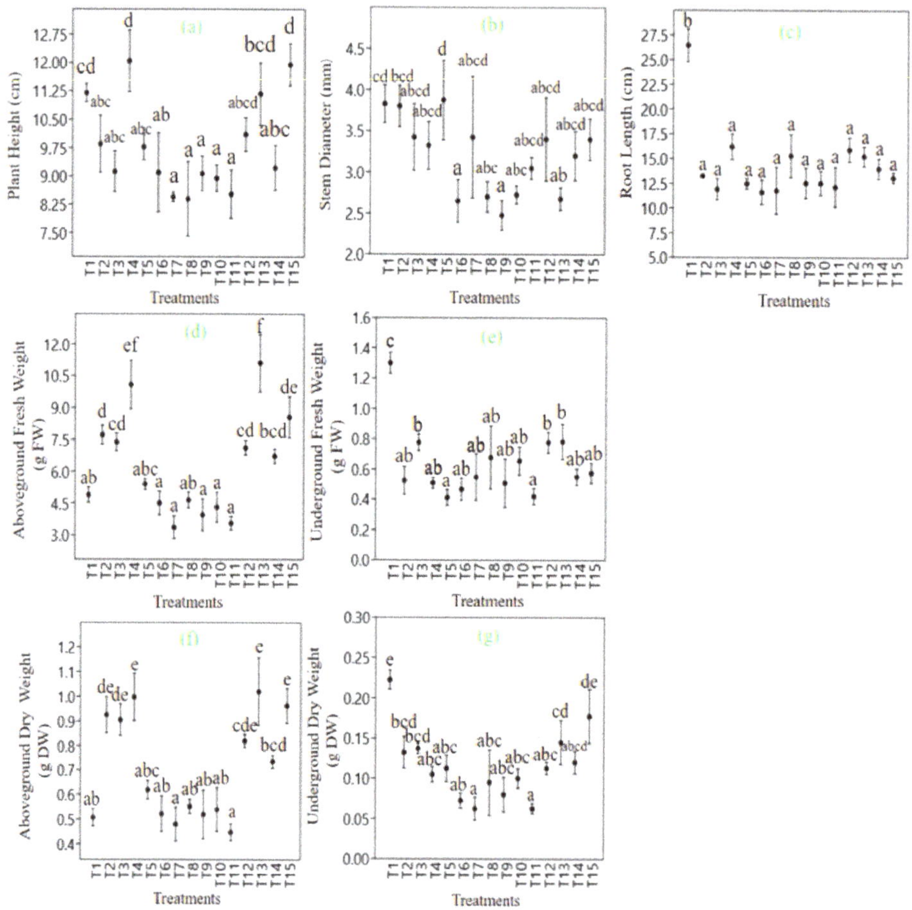

Figure 2. The ration effects of nitrogen, phosphorus, and potassium on plant height (**a**), stem diameter (**b**), root length (**c**), fresh weight (**d**,**e**) and dry weight (**f**,**g**) after four weeks treatments. (Note, T1: N-P-K=0-0-0, T2: N-P-K=0-2-2, T3: N-P-K=1-2-2, T4: N-P-K=2-0-2, T5: N-P-K=2-1-2, T6: N-P-K=2-2-2, T7: N-P-K=2-3-2, T8: N-P-K=2-2-3, T9: N-P-K=2-2-0, T10: N-P-K=2-2-1, T11: N-P-K=3-2-2, T12: N-P-K=1-3-2, T13: N-P-K=1-1-2, T14: N-P-K=1-2-1, T15: N-P-K=2-1-1; 1 unit means 0.5 g/L nutrient; Error bars indicate standard error of the mean; Different letters indicate the significant differences with $p < 0.05$.).

Among all the treatments, the fresh weights of the aboveground parts of Chinese cabbage were the best in the T4 and T13 treatments, with weights of 10.09 and 11.11 g, respectively. However, in terms of dry weights, the belowground parts of the control treatment were the highest (0.22 g). The average of fresh weight and dry weight in the T7, T9, T10, and T11 treatments was approximately 3.80 and 0.50 g in the Chinese cabbage, respectively, and the fresh and dry weights in these treatments were significantly lower than those in the other treatments ($p < 0.05$) (Figure 2c,e). Based on the plant phenotype, height, root length, and biomass, the optimum solution of N-P-K was 1.0-0-1.0 g/L (T4, 2-0-2).

3.3. Experiment II: The Application of BNPs with N-P-K for Chinese Cabbage Growth

After milling, corn BNPs with a diameter of 85 nm were obtained (Figure 1c,d). These BNPs were applied in experiment II. Based on the results of experiment I, 2 g/L of N, and

2 g/L of K were mixed with BNPs in experiment II to find out the best formula of BNPs and fertilizers. After 4 weeks treatments, Chinese cabbage showed the best phenotype in the T2 treatment (2 g/L of BNPs, 2 g/L of N, and 2 g/L of K), with a brighter green color and better growth than the other treatments. Plant growth in the experimental control (BNPs only) was poor (T5). In the T5 treatment, all the leaves had chlorosis. Plant size was smaller in the N-P-K-only treatment (T4 in Experiment I) when compared with that in the N-P-K plus BNP treatment (T6).

After the addition of the BNPs, Chinese cabbage height and stem diameter were 1.25 and 0.63 times greater, respectively, than the height and stem diameter of the control (T5). The treatment (T2) to which the BNPs and fertilizer were added resulted in an increase of 2.23 cm and 1.31 cm in plant height and stem diameter, respectively, as compared to that of the T5 (Figure 3a,b). However, there was no significant difference in the root length among these treatments ($p < 0.05$) (Figure 3c).

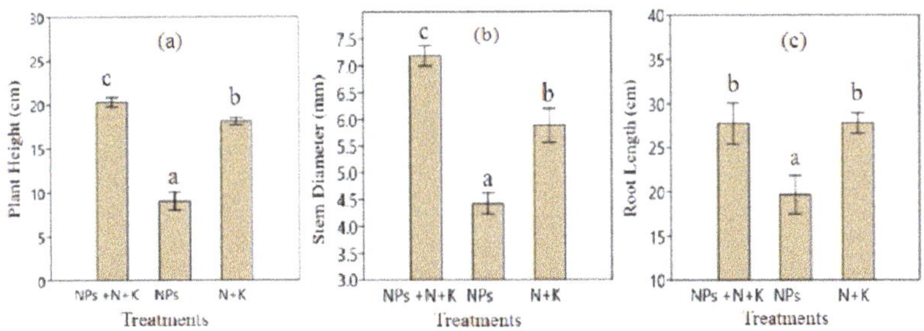

Figure 3. The ration effects of biochar nanoparticles, nitrogen, and potassium on Chinese cabbage of height (**a**), stem diameter (**b**) and root length (**c**) after four weeks treatments. (Note, NPs+N+K: BNPs-N-K=1:2:2 (T1), NPs: BNPs-N-K=2:0:0 (T5), N+K: BNPs-N-K=0:2:2 (T6); BNPs, biochar nanoparticles; 1 unit means 0.5 g/L nutrient; Error bars indicate standard error of the mean; Different letters indicate the significant differences with $p < 0.05$).

The fresh weight of the aboveground plant parts significantly increased after the addition of BNPs ($p < 0.05$). The aboveground weight reached 52.60 g after 4 weeks in the T2, with a significant increase in the dry weights of the above- and belowground parts (3.66 g and 0.52 g, respectively) ($p < 0.05$) as compared with that of the control treatment (Figure 4a,c,d). However, Chinese cabbage showed no significant increase, with or without the addition of BNPs (Figure 4b).

In the elemental analysis, there was no significant difference in the K content in the treatments with or without the addition of BNPs or in the treatments with or without the addition of fertilizers ($p < 0.05$). The average K content of above- and belowground parts was 28.45 g/kg and 46.93 g/kg, respectively, (Figure 5a,d). The P content of the above- and belowground parts of Chinese cabbage were significantly higher in the BNPs-only treatments than in T1 treatment ($p < 0.05$); however, the fertilized treatments were at the lower level, with contents of 2.13 and 1.81 g/kg, respectively (Figure 5b,e). The Ca content of Chinese cabbage was higher in the fertilizer treatments than in the T5 (Figure 5c,f). There was no significant difference in the Mg contents of Chinese cabbage leaves compared with the plants in T2 and the control ($p < 0.05$). The Mg content of underground parts was significantly lower in T6 (1.92 g/kg) than those treated in T5 (Figure 5g,j). The Na content of the aboveground parts was significantly higher in the BNP-only treatments than in the BNP plus fertilizer treatments. The Na content of the belowground parts was significantly higher in the BNPs plus fertilizer treatments ($p < 0.05$) (Figure 5h,k). The S contents were significantly lower only in the BNP treatments (T5), with contents of 4.44 and 3.56 g/kg, respectively (Figure 5i,l).

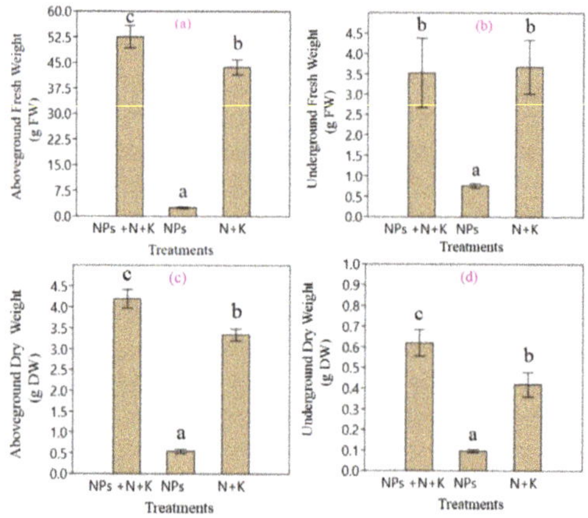

Figure 4. The ration effects of biochar nanoparticles, nitrogen, and potassium on Chinese cabbage of fresh (**a**,**b**) and dry (**c**,**d**) weight of plants after four weeks treatments. (Note, NPs+N+K: BNPs-N-K=1:2:2 (T1), NPs: BNPs-N-K=2:0:0 (T5), N+K: BNPs-N-K=0:2:2 (T6); BNPs, biochar nanoparticles; 1 unit means 0.5 g/L nutrient; Error bars indicate standard error of the mean; Different letters indicate the significant differences with $p < 0.05$.).

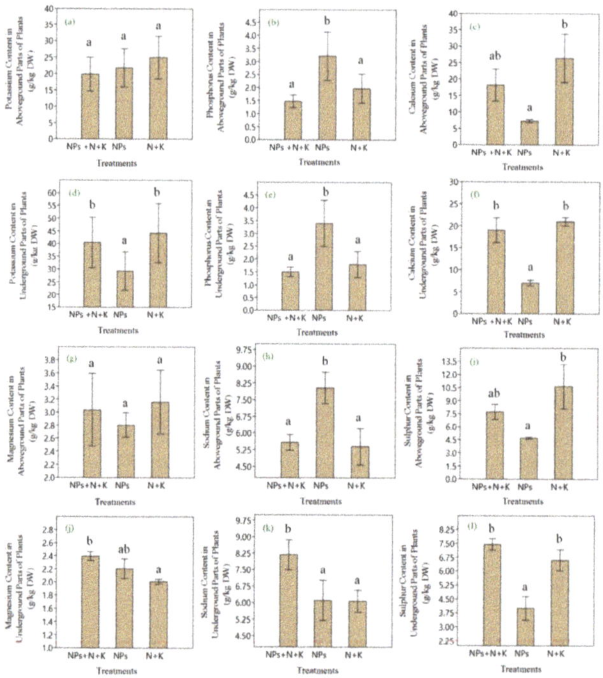

Figure 5. The ration effects of biochar nanoparticles, nitrogen, and potassium on Chinese cabbage of K (**a**,**d**), P (**b**,**e**), Ca (**c**,**f**), Mg (**g**,**j**), Na (**h**,**k**), and S (**i**,**l**) content of plants. (Note, NPs+N+K: BNPs-N-K=1:2:2 (T?), NPs: BNPs-N-K=2:0:0 (T5), N+K: BNPs-N-K=0:2:2 (T6); BNPs, biochar nanoparticles; 1 unit means 0.5 g/L nutrient; Error bars indicate standard error of the mean; Different letters indicate the significant differences with $p < 0.05$.)

4. Discussion

4.1. N-P-K Nutrient Management for Chinese Cabbage Growth

It is reported that N and P are important for plant growth [27]. In experiment I, we found that there is no positive correlation between plant growth and an increase in fertilizer applications. In this study, P was not a significant element for Chinese cabbage growth (Figures 1 and 2). Previous research reported that rhizosphere colonization by phosphate-solubilizing bacteria enhanced Chinese cabbage growth by 7.21%, without the addition of P fertilizer [28]. It was reported that current agricultural production systems require high quantities of P fertilizers, which have led to a build-up of legacy-P in soils [29]. The left P has become another P source for plant use; the rhizobacterium *Proteus vulgaris* JBLS202 stimulated Chinese cabbage growth, with an increase of 32.6% [30]. In a study that used green fluorescent protein to study YL6 colonization of Chinese cabbage roots, the biomass of the colonized plants increased 400% as compared with that of noncolonized plants [31]. It was reported that N additions did not significantly improve Chinese cabbage or maize growth [32], and we found that N has no significant relationship with the Chinese cabbage growth (Figure 1a). Our findings are in accordance with those of this previous study.

As reported previously, the application of phosphate-solubilizing bacteria can reduce the consumption of fertilizer and aid sustainable agricultural development. Exogenous additions of N and K are necessary for Chinese cabbage growth, but P is not. This is the intended reason that P is missing from experiment II.

4.2. The Effects of BNPs on Chinese Cabbage Growth

In experiment II, the BNPs significantly improved Chinese cabbage growth, especially the aboveground parts (heights, fresh and dry weights, and stem diameters) (Figures 1b and 2). Chinese cabbage was significantly smaller and lighter in the N-K-only fertilizer treatments as compared with the N-K with BNPs treatments. These results indicate that BNPs and N-K fertilizers applied in a two-time fertilizer application regime led to slow release of fertilizers and therefore increase the efficiency of the fertilizer treatment as compared with traditional applications of fertilizers. Compared to the fresh weights and heights of Chinese cabbage in T2, those in T5 and T6 were reduced 718.10 and 107.45%, respectively, (Figure 2). In terms of the efficiency of BNPs as a nanocarrier for fertilizers, BNPs improved the biomass harvest by 18.49% as compared to the treatments without BNPs. As compared with the other treatments, the N-K treatment combined with BNPs increased the biomass of the aboveground parts of Chinese cabbage. As reported previously, using nanocarriers for fertilizers can increase plant height because they have greater potential to provide plant nutrients continuously than conventional fertilizers [33]. Compared with the fertilizer-only treatments, the fertilizer plus BNP treatments significantly increased the dry matter biomass. The results suggest that BNPs are an efficient nanocarrier for fertilizers.

In addition, we found the BNPs on their own did not have positive effects on Chinese cabbage growth (Figure 1b), with Chinese cabbage treated with BNPs showing dwarfism and chlorosis. This result confirmed that BNPs do not provide nutrients for plant growth. Thus, we suggest that BNPs function only as a nanocarrier for fertilizers.

As we know, Mg contributes to plant growth and photosynthesis, and it is important for leaf development [34]. Higher Mg contents lead to higher fresh and dry weights of vegetables [35]. Na is a macronutrient required for plant growth. Na has been reported to increase WUE and stomatal diffusion and augment carbon dioxide uptake efficiency, thereby resulting in significant gains in nutritional status and positive plant physiological responses [36]. In the present study, in the treatments containing BNPs, more Mg and Na were transferred from the soil and the plant roots to the aboveground parts, leading to an increase in fresh biomass (Figure 5). In addition, Chinese cabbage quality increased in accordance with an increase of Mg and Na in the N-K-loaded BNPs treatments. As the fourth major plant nutrient, S, a constituent of three S-containing amino acids, participates in the formation of chlorophyll for photosynthesis [37]. In this study, the addition of BNPs reduced the uptake of S from the roots. However, the accumulation of S in the roots had no effects on the aboveground parts of the plant. Therefore, we suggest that BNPs can increase Chinese cabbage absorption of Mg and Na while not limiting aspects of other elements. Additionally, due to the properties of biochar, P contents are around 60 to 85 ppm in the biochar [38], and this is the reason that Chinese cabbage can have more P in the above- and under-ground parts. In addition, the added biochar can help plants to absorb P [39], and it makes P more effective for plant absorption.

5. Conclusions

BNPs have been applied in agriculture. In a previous study, we applied BNPs to alleviate the allelopathic effects caused from *Imperata cylindrica* on rice growth [26], and it was the first time that we applied the BNPs working as an immunity promoter to induce the disease resistance in *Nicotiana benthamiana* [40]. Furthermore, extend research studied the application of BNPs in agriculture. Using nano-enabled fertilizers and nanocarriers, such as BNPs, fertilizers can be formulated or "tuned" to release nutrients in a controlled manner [41]; Figure 6 presents the potential mechanism of BNPs working as nanocarrier for Chinese cabbage. Using this "smart" technology enables a slower release of fertilizers, which can help plants absorb nutrients more efficiently than traditional fertilizers, improving nutrient usage at the same time. The present study demonstrates that using BNPs as a nanocarrier can improve plant nutrient absorption and fertilizer efficiency and improve the phenotype and quality of the plant in terms of biomass, height, and root length. In this study, the application of the BNP treatment to the roots increased the transfer of Mg and Na for plant growth while not affecting the transfer of other elements. Thus, BNPs can be considered a nanocarrier for the application of fertilizers in agriculture.

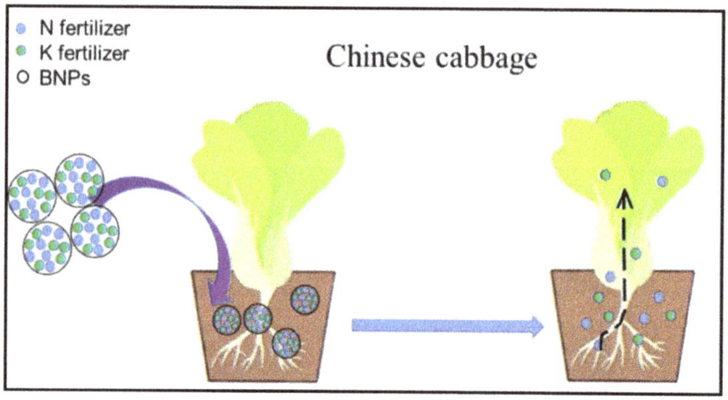

Figure 6. The potential mechanism of BNPs working as nanocarriers to regulate Chinese cabbage growth.

This study provides the use of BNPs as a nanocarrier that can better to manage the fertilizer release than traditional fertilizer treatments, thereby benefiting Chinese cabbage growth. Soil application of BNPs has benefits for Chinese cabbage in terms of plant growth, plant height, stem diameter, dry mass, and fresh biomass. Using only half the amount of fertilizers (2 g/L of BNPs and N-P-K) could achieve the same target in terms of Chinese cabbage growth as twice the amount of fertilizers applied using traditional methods. In this study, compared with the fertilizer treatments alone, Chinese cabbage height increased more than 18–25% and the fresh biomass increased 44% in the BNP plus fertilizer treatments. We conclude that BNPs provide a novel and efficient nutrient delivery method to improve plant growth, which is essential in agriculture to achieve more sustainable crop systems. This study demonstrates the potential of nanocarriers in agricultural fertilizer applications.

Supplementary Materials: The following supporting information can be downloaded at: https://www.mdpi.com/article/10.3390/coatings12121984/s1, The BNPs physical properties of The X-ray diffractometer (XRD), X-ray photoelectron spectroscopy (XPS), and the surface-area and the pore size were present in the Figures S1, S2 and Table S1, respectively. Response Surface Model information was presented in Table S2. The Treatments for Experiment of Comparison was presented in Table S3. Plant weight and Nutrient contents for Experiment of Comparison were shown in Tables S4 and S5.

Author Contributions: Conceptualization, R.Y. and Y.S.; methodology, Y.S.; formal analysis, R.Y.; investigation, J.S., Y.Z. and L.J.; data curation, R.Y., Z.W. and X.S.; writing—original draft preparation, R.Y.; writing—review and editing, Y.S.; visualization, R.Y. and J.S.; supervision, B.T. and Y.S.; project administration, B.T.; funding acquisition, R.Y. and Y.S. All authors have read and agreed to the published version of the manuscript.

Funding: This research was supported by the Primary Research & Development Plan of Jiangsu Province [BE2020673]; the Innovation Project of Jiangsu Academy of Agricultural Science and Technology [SCX(22)3985], and National Key R&D Program of China [2019YFD0900404-05].

Institutional Review Board Statement: Not applicable.

Informed Consent Statement: Not applicable.

Data Availability Statement: Not applicable.

Conflicts of Interest: The authors declare no competing financial interests.

References

1. Fahey, J.W. Brassica: Characteristics and Properties. In *Encyclopedia of Food and Health*; Elsevier: Amsterdam, The Netherlands, 2015; pp. 469–477.
2. Pasakdee, S.; Banuelos, G.; Shennan, C.; Cheng, W. Organic N fertilizers and irrigation influence organic broccoli production in two regions of California. *J. Veg. Sci.* **2007**, *12*, 27–46. [CrossRef]
3. Krezel, J.; Kolota, E. The effects of nitrogen fertilization on yielding and biological value of Chinese cabbage grown from seedlings for autumn harvest. *J. Elementol.* **2008**, *13*, 255–260.
4. Okorogbona, A.O.M.; Van Averbeke, W.; Ramusandiwa, T.D. Growth and yield response of Chinese cabbage (*Brassica rapa* L. ssp. *campestris*) as affected by nutrient availability in air-dried and pulverized different types of animal manure using low biological activity soil. *World J. Agric. Sci.* **2011**, *7*, 1–12.
5. Ha, N.M.C.; Nguyen, T.H.; Wang, S.; Nguyen, A.D. Preparation of NPK nanofertilizer based on chitosan nanoparticles and its effect on biophysical characteristics and growth of coffee in green house. *Res. Chem. Intermediat.* **2019**, *45*, 51–63. [CrossRef]
6. Marchiol, L. An outlook of crop nutrition in the fourth agricultural revolution. *Ital. J. Agron.* **2019**, *14*, 1367. [CrossRef]
7. Agyin-Birikorang, S.; Tindjina, I.; Adu-Gyamfi, R.; Dauda, H.W.; Fugice, J.; Sanabria, J. Managing essential plant nutrients to improve maize productivity in the savanna agroecological zones of northern Ghana: The role of secondary and micronutrients. *J. Plant. Nutr.* **2022**, *46*, 38–57. [CrossRef]
8. Roberts, T.L. The role of fertilizer in growing the world's food. *Better Crops Plant Food* **2009**, *93*, 12–15.
9. Chien, S.H.; Prochnow, L.I.; Cantarella, H. Chapter 8 recent developments of fertilizer production and use to improve nutrient efficiency and minimize environmental impacts. *Adv. Agron.* **2009**, *102*, 267–322.
10. Xiao, Y.; Peng, F.; Zhang, Y.; Wang, J.; Zhu, Y.; Zhang, S.; Gao, H. Effect of bag-controlled release fertilizer on nitrogen loss, greenhouse gas emissions, and nitrogen applied amount in peach production. *J. Clean. Prod.* **2019**, *234*, 258–274. [CrossRef]
11. O'Connor, J.; Hoang, S.A.; Bradney, L.; Dutta, S.; Xiong, X.; Tsang, D.C.W.; Ramadass, K.; Vinu, A.; Kirkham, M.B.; Bolan, N.S. A review on the valorisation of food waste as a nutrient source and soil amendment. *Environ. Pollut.* **2021**, *272*, 115985. [CrossRef]
12. Chen, J.; Wei, X. Chapter 3 controlled-release fertilizers as a means to reduce nitrogen leaching and runoff in container-grown plant production. In *Nitrogen in Agriculture*; InTech: London, UK, 2018; pp. 33–52.
13. Hofmann, T.; Lowry, G.V.; Ghoshal, S.; Tufenkji, N.; Wilkinson, K.J. Technology readiness and overcoming barriers to sustainably implement nanotechnology-enabled plant agriculture. *Nat. Food* **2020**, *1*, 416–425. [CrossRef]
14. Kumar, R.; Ashfaq, M.; Verma, N. Synthesis of novel PVA-starch formulation-supported Cu-Zn nanoparticle carrying carbon nanofibers as a nanofertilizer: Controlled release of micronutrients. *J. Mater. Sci.* **2018**, *53*, 7150–7164. [CrossRef]
15. Meurer, R.A.; Kemper, S.; Knopp, S.; Eichert, T.; Jakob, F.; Goldbach, H.E.; Schwaneberg, U.; Pich, A. Biofunctional microgel-based fertilizers for controlled foliar delivery of nutrients to plants. *Angew. Chem. Int. Edit.* **2017**, *56*, 7380–7386. [CrossRef] [PubMed]
16. Jakhar, A.M.; Aziz, I.; Kaleri, A.R.; Hasnain, M.; Haider, G.; Ma, J.; Abideen, Z. Nano-fertilizers: A sustainable technology for improving crop nutrition and food security. *NanoImpact* **2022**, *27*, 100411. [CrossRef] [PubMed]
17. Astaneh, N.; Bazrafshan, F.; Zare, M.; Amiri, B.; Bahrani, A. Nano-fertilizer prevents environmental pollution and improves physiological traits of wheat grown under drought stress conditions. *Sci. Agropecu.* **2021**, *12*, 41–47. [CrossRef]

18. Das, S.K.; Ghosh, G.K. Development and evaluation of biochar-based secondary and micronutrient enriched slow-release nano-fertilizer for reduced nutrient losses. *Biomass Convers. Biorefin.* **2021**, 1–12. [CrossRef]
19. Yuan, H.; Lu, T.; Huang, H.; Zhao, D.; Kobayashi, N.; Chen, Y. Influence of pyrolysis temperature on physical and chemical properties of biochar made from sewage sludge. *J. Anal. Appl. Pyrol.* **2015**, *112*, 284–289. [CrossRef]
20. Tomczyk, A.; Sokolowska, Z.; Boguta, P. Biochar physicochemical properties: Pyrolysis temperature and feedstock kind effects. *Rev. Environ. Sci. Bio/Technol.* **2020**, *19*, 191–215. [CrossRef]
21. Mansoor, S.; Kour, N.; Manhas, S.; Zahid, S.; Wani, O.A.; Sharma, V.; Wijaya, L.; Alyemeni, M.N.; Alsahli, A.A.; El-Serehy, H.A.; et al. Biochar as a tool for effective management of drought and heavy metal toxicity. *Chemosphere* **2021**, *271*, 129458. [CrossRef]
22. Sun, Y.; Xiong, X.; He, M.; Xu, Z.; Tsang, D. Roles of biochar-derived dissolved organic matter in soil amendment and environmental remediation: A critical review. *Chem. Eng. J.* **2021**, *424*, 130387. [CrossRef]
23. Krasucka, P.; Pan, B.; Ok, Y.S.; Mohan, D.; Sarkar, M.; Oleszczuk, P. Engineered biochar—A sustainable solution for the removal of antibiotics from water. *Chem. Eng. J.* **2020**, *405*, 126926. [CrossRef]
24. Oliveira, J.L.D.; Campos, E.V.R.; Pereira, A.E.S.; Pasquoto, T.; Lima, R.; Grillo, R.; Andrade, D.J.; Santos, F.A.D.; Fraceto, L.F. Zein nanoparticles as eco-friendly carrier systems for botanical repellents aiming sustainable agriculture. *J. Agric. Food Chem.* **2018**, *66*, 1330–1340. [CrossRef]
25. Sun, D.; Hussain, H.I.; Yi, Z.; Rookes, J.E.; Kong, L.; Cahill, D.M. Mesoporous silica nanoparticles enhance seedling growth and photosynthesis in wheat and lupin. *Chemosphere* **2016**, *152*, 81–91. [CrossRef]
26. Shen, Y.; Tang, H.; Wu, W.; Shang, H.; Zhang, D.; Zhan, X.; Xing, B. Role of nano-biochar in attenuating the allelopathic effect from Imperata cylindrica on rice seedlings. *Environ. Sci.-Nano* **2020**, *7*, 116–126. [CrossRef]
27. Prasad, P.H.; Bhunia, P.; Naik, A.; Thapa, U. Response of nitrogen and phosphorus levels on the growth and yield of Chinese cabbage (*Brassica campestris* ssp. *chinensis*) in the gangetic plains of West Bengal. *J. Crop Weed* **2009**, *5*, 75–77.
28. Ku, Y.; Xu, G.; Tian, X.; Xie, H.; Yang, X.; Cao, C.; Chen, Y. Root colonization and growth promotion of soybean, wheat and Chinese cabbage by bacillus cereus YL6. *PLoS ONE* **2018**, *13*, e0210035.
29. El Attar, I.; Hnini, M.; Taha, K.; Aurag, J. Phosphorus availability and its sustainable use. *J. Soil Sci. Plant Nut.* **2022**, *22*, 5036–5048. [CrossRef]
30. Yu, S.M.; Lee, Y.H. Plant growth promoting rhizobacterium proteus vulgaris JBLS202 stimulates the seedling growth of Chinese cabbage through indole emission. *Plant Soil* **2013**, *370*, 485–495. [CrossRef]
31. Wang, Z.; Xu, G.; Ma, P.; Lin, Y.; Cao, C. Isolation and characterization of a phosphorus-solubilizing bacterium from rhizosphere soils and its colonization of Chinese cabbage (*Brassica campestris* ssp. *chinensis*). *Front. Microbiol.* **2017**, *8*, 1270. [CrossRef]
32. Zhao, H.; Xie, T.; Xiao, H.; Gao, M. Biochar-based fertilizer improved crop yields and N utilization efficiency in a maize-Chinese cabbage rotation system. *Agriculture* **2022**, *12*, 1030. [CrossRef]
33. Liu, R.; Lal, R. Potentials of engineered nanoparticles as fertilizers for increasing agronomic productions. *Sci. Total Environ.* **2015**, *514*, 131–139. [CrossRef] [PubMed]
34. Hauer-Jákli, M.; Tränkner, M. Critical leaf magnesium thresholds and the impact of magnesium on plant growth and photo-oxidative defense: A systematic review and meta-analysis from 70 years of research. *Front. Plant Sci.* **2019**, *10*, 766. [CrossRef] [PubMed]
35. Shebl, A.; Hassan, A.A.; Salama, D.M.; Abd El-Aziz, M.E.; Abd Elwahed, M.S.A. Template-free microwave-assisted hydrothermal synthesis of manganese zinc ferrite as a nanofertilizer for squash plant (*Cucurbita pepo* L.). *Heliyon* **2020**, *6*, e03596. [CrossRef] [PubMed]
36. Nikolas, D.S.M.; Eric, V.D.O.F.; Junior, J.A.; Domec, J.C.; Jordan-Meille, L.; José, L.D.M.G.; Lavres, J. The ideal percentage of k substitution by Na in eucalyptus seedlings: Evidences from leaf carbon isotopic composition, leaf gas exchanges and plant growth. *Plant Physiol. Biochem.* **2019**, *137*, 102–112.
37. de Bang, T.C.; Husted, S.; Laursen, K.H.; Persson, D.P.; Schjoerring, J.K. The molecular physiological functions of mineral macronutrients and their consequences for deficiency symptoms in plants. *New Phytol.* **2021**, *229*, 2446–2469. [CrossRef]
38. Kong, M.; Liang, J.; White, J.C.; Elmer, W.H.; Wang, Y.; Xu, H.; He, W.; Shen, Y.; Gao, X. Biochar nanoparticle-induced plant immunity and its application with the elicitor methoxyindole in *Nicotiana benthamiana*. *Environ. Sci. Nano* **2022**, *9*, 3514–3524. [CrossRef]
39. Glaser, B.; Lehr, V.I. Biochar effects on phosphorus availability in agricultural soils: A meta-analysis. *Sci. Rep.* **2019**, *9*, 9338. [CrossRef]
40. Li, H.; Li, Y.; Xu, Y.; Lu, X. Biochar phosphorus fertilizer effects on soil phosphorus availability. *Chemosphere* **2020**, *244*, 125471. [CrossRef]
41. Guo, H.; White, J.C.; Wang, Z.; Xing, B. Nano-enabled fertilizers to control the release and use efficiency of nutrients. *Curr. Opin. Environ. Sci. Health* **2018**, *6*, 77–83. [CrossRef]

coatings

MDPI

Article

Synergic Effect of Microorganism and Colloidal Biochar-Based Organic Fertilizer on the Growth and Fruit Quality of Tomato

Shiguo Gu [1], Fei Lian [2,*], Hanyue Yang [1], Yaru Han [1], Wei Zhang [1], Fan Yang [1] and Jie Gao [3]

[1] Institute of Environmental Processes and Pollution Control, School of Environmental and Civil Engineering, Jiangnan University, Wuxi 214122, China; ychgushiguo@163.com (S.G.); yang18861831936@gmail.com (H.Y.); xxhanyaru@163.com (Y.H.); zw19952730051@163.com (W.Z.); fanyang1126@163.com (F.Y.)
[2] School of Energy and Environmental Engineering, Hebei University of Technology, Tianjin 300401, China
[3] Inner Mongolia Sunture Environmental Technology, Co., Hohhot 010020, China; XC4000400065@163.com
* Correspondence: lianfei2000@126.com; Tel.: +86-022-60435775

Abstract: It is well known that carbon-based organic fertilizer can effectively promote crop growth and improve nutrient utilization efficiency. However, little is known about the effect of microorganisms on the nutrient availability of carbon-based organic fertilizer. To elucidate the contribution of microorganisms to the agricultural benefit of colloidal biochar-based fertilizer, a 5-month pot experiment was conducted to study the effect of different combinations of Methyltrophic bacillus, colloidal biochar, and organic fertilizer on physical–chemical properties of soil, plant growth, physiological-biochemical reactions, yield, and quality of tomato. The results show that the addition of Methyltrophic bacillus effectively promoted the availability of soil nutrients (such as nitrate nitrogen and available potassium) and increased soil cation exchange capacity; meanwhile, it significantly increased the content of chlorophyll-a (9.42–27.41%) and promoted the net photosynthetic rate (10.86–13.73%) and biomass of tomato fruit (17.84–26.33%). The contents of lycopene, vitamin C, total sugar, and soluble sugar in the fruits treated by the ternary combination of Methyltrophic bacillus, colloidal biochar, and organic fertilizer increased by 58.40%, 46.53%, 29.45%, and 26.65%, respectively. The above results demonstrate that the addition of beneficial microorganisms could further improve the performance of biochar-based fertilizer on plant growth, yield, and fruit quality of tomato. This information provides evidence for the promising performance of microorganism-supported biochar organic fertilizer in agricultural applications.

Keywords: microorganism; colloidal biochar; organic fertilizer; tomato; fruit quality

Citation: Gu, S.; Lian, F.; Yang, H.; Han, Y.; Zhang, W.; Yang, F.; Gao, J. Synergic Effect of Microorganism and Colloidal Biochar-Based Organic Fertilizer on the Growth and Fruit Quality of Tomato. *Coatings* **2021**, *11*, 1453. https://doi.org/10.3390/coatings11121453

Academic Editor: Raffaele Porta

Received: 2 November 2021
Accepted: 24 November 2021
Published: 26 November 2021

1. Introduction

Biochar (BC), a man-made form of black carbon, is produced by pyrolysis or gasification of organic matter without or with limited oxygen and is rich in organic carbon, mineral elements, and inorganic carbonates [1]. BC mainly consists of aromatic hydrocarbons and alkyl structures, which are highly inert and difficult to be oxidized or decomposed by microorganisms and is extremely stable in soil and can persist for hundreds or even thousands of years [2,3]. BC usually has a large number of surface functional groups and a porous structure, which can improve the pH and ion exchange capacity of acidic soil [4,5] and exhibit high capacity of binding and retaining soil nutrients, so as to reduce the loss of nutrients and promote the uptake of nutrients by plants [6–8]. Therefore, a synergic effect is likely to be achieved by loading organic fertilizer into the pores of BC (as a carrier) to prepare BC-based compound fertilizers, which could increase the nutrient content of BC and more effectively regulate the release of nutrient elements into the soil.

Carbon-based compound fertilizer mainly includes three typical types: carbon-based organic fertilizer, carbon-based inorganic fertilizer, and carbon-based organic–inorganic compound fertilizer. Environmentally friendly carbon-based compound fertilizer is capable of saving energy and improving nutrient utilization efficiency, which is an important development direction for improving the quality and efficiency of chemical fertilizers. Glaser et al. showed that adding BC to compost in a field experiment increased corn yield by 26% compared to pure compost [9]. Combined addition of BC and inorganic fertilizer significantly improved the maize grain yield and biomass yield compared to the unmanned plot in all irrigation treatments [10]. It is thus clear that the combination of organic fertilizer and BC has evident advantages in crop growth and soil improvement than the individual application [11–13]. In recent years, microbial agents have been widely used to improve the crop growth and can also improve soil properties. Methylotrophic bacteria represent an important group of multi-functional plant growth-promoting bacteria. Methylotrophic bacteria utilize the plant waste to product methanol as the source of carbon and energy and enhances plant growth by producing growth hormones, such as indole-3-acetic acid [14] and cytokinins [15]. Simultaneously, it can produce beneficial metabolites and promote plant growth through antagonism, competition, and induction [16,17]. Pot experiments indicated that the plant height and dry weight of tomato treated with T. afroharzianum TM2-4 microbial agent were increased by 36.1% and 32.3%, respectively, compared with the control [18]. Moreover, some reports showed that microbial-organic fertilizers could replace 23–52% of N fertilizer without any loss of yield [19]. However, there are few studies investigating the synergic effects of microorganisms and BC-based organic fertilizer on crop growth and quality. Therefore, we put forward a scientific hypothesis that microbial agents can further accelerate the nutrient transformation in BC-based organic fertilizer, effectively promote the absorption and utilization of nutrients by plants, and improve the fertilizer utilization efficiency.

Therefore, the purpose of this study was to examine the synergic effect (if any) of microorganisms and BC organic fertilizer on the growth and development of plants. Previous studies found that colloidal BC possesses more abundant oxyl groups, higher surface area, and more negative zeta potential than bulk-BC due to its plenty of submicron and even nano-sized BC fractions [20]. Thus, colloidal BC derived from sawdust was selected to prepare the BC-based fertilizer in the present study. Methyltrophic bacillus and tomato were used as typical plant growth-promoting bacteria and plant, respectively. The growth, physiological and biochemical reactions, and fruit quality of tomato were determined based on a 5-month pot experiment. The results can not only provide new information for the development of promising BC-based organic fertilizer but also provide theoretical support for the agricultural sustainable development.

2. Materials and Experimental Design

2.1. Materials

The air-dried tree branches were used as the raw material to prepare BC. After ground and sieved (20 mesh), the wood powder was fed into a lab-scale tubular reactor within a muffle furnace (Tianjin Zhonghuan Electric Furnace Co., Ltd., Tianjin, China) and slowly pyrolyzed (10 °C/min) in a N_2 atmosphere at 550 °C for 120 min. The obtained BC was ground and sieved through a 100-mesh sieve. Colloidal BC was obtained by wet sieving technique [20–22]. Briefly, a certain amount of the BC powder was added to 1 L of deionized water and stirred (150 rpm) for 24 h. After sonication (1 h at 120 W), the suspension was slowly poured through a 1-micron sieve and the remaining particles were carefully washed with deionized water to ensure that most of the desired fractions (<1-micron) can pass through the sieve and the BC in the filtrate was obtained after it was centrifuged (referred to as colloidal BC).

The organic fertilizer was purchased from a local environmental technology company, which was mainly composed of starter, rice bran, and pig manure. Methyltrophic bacillus was obtained from China microbial strain collection center (Serial number, CCTCC-AB2014337). A 0.1 mL volume of raw bacillus was spread on Luria Bertani (LB) liquid medium (4.9 mL), which contained (g/L) tryptone, 10; yeast extract, 5; NaCl, 10; and agar, 18, and the solutions were then incubated in a constant temperature shaking incubator (IGS100, Thermo, Waltham, MA, USA) at 180 rpm and 30 °C for 48 h. The colony was extracted from LB liquid medium and dissolved in sterilized phosphate-buffered saline (PBS, pH 7.4) buffer. The concentration of bacterial solution (1.0×10^6 CFU/mL) was determined by measuring the light absorption value at OD600 with a spectrophotometer (Shimadzu, Kyoto, Japan). Tomato seeds, "Jinpeng No. 1", were purchased from Xi'an Jinpeng seeds Co., Ltd. (Xi'an, China) The used soil is brown soil and its basic properties are as follows: pH = 6.34 ± 0.16, cation exchange capacity (CEC) 14.12 ± 1.16 cmol/kg, organic matter content 35.76 ± 2.04 g/kg, nitrate nitrogen (N) 12.74 ± 0.36 mg/kg, available phosphorus (P) 12.64 ± 2.37 mg/kg, and available potassium (K) 126.5 ± 9.08 mg/kg. Basic properties of the colloidal BC and organic fertilizer are shown in Table 1.

Table 1. Basic properties of colloidal BC and organic fertilizer.

Category	CEC (cmoL/kg)	Available K (g/kg)	Available P (g/kg)	Nitrate N (mg/kg)	Organic Matter Content (g/kg)	pH
Colloidal BC	65.63 ± 4.08	31.56 ± 1.63	83.47 ± 16.02	2.26 ± 0.08	395.33 ± 27.19	8.98 ± 0.90
Organic fertilizer	73.33 ± 2.45	112.54 ± 2.73	185.70 ± 8.17	15.60 ± 0.86	380.82 ± 23.03	9.98 ± 0.87

2.2. Experimental Design

The tomato seeds were sorted and disinfected by washing with 5% (v/v) NaClO for 10 min and then germinated in a deionized water irrigated vermiculite medium. The seedlings were incubated in a growth chamber, with 12 h photoperiod, 25/20 °C day/night temperature, and 60% relative humidity. Seedlings with consistent growth (the third leaflet) were transplanted into a pot containing 3 kg of dry soil.

A pot experiment was conducted with 6 treatments, including soil control (CK), soil amended by colloidal BC (3% of soil weight) (T1), soil amended by organic fertilizer (3% of soil weight) (T2), soil amended by colloidal BC (1.5% of soil weight) and organic fertilizer (1.5% of soil weight) (T3), soil amended by colloidal BC (3% of soil weight) and microbial bacterial solution (30 mL, 1.0×10^6 CFU/mL) (T4), soil amended by organic fertilizer (3% of soil weight) and microbial bacterial solution (30 mL, 1.0×10^6 CFU/mL) (T5), soil amended by colloidal BC (1.5% of soil weight), organic fertilizer (1.5% of soil weight), and microbial bacterial solution (30 mL, 1.0×10^6 CFU/mL) (T6). For the treatments containing both colloidal BC and organic fertilizer, the colloidal BC was thoroughly mixed with organic fertilizer, composted, and then fermented at a constant temperature of 30 °C for 4 weeks before use. Colloidal BC-based organic fertilizer was modified by adding Methyltrophic bacillus. Briefly, after fermenting, Methyltrophic bacillus was added to the pots with microbial bacterial treatments. The bacterial treatment group was homogenized with 3 kg of the soil at 30 rpm for 30 min.

2.3. Characterization of Soil, BC, and Organic Fertilizer

Before blending with soil, the properties of soil, colloidal BC, and organic fertilizer were measured, respectively. CEC was analyzed by sodium acetate extraction flame photometer method [23]. Available K and P were tested by modified kelowna methods [24]. Nitrate N was estimated by ultraviolet spectrophotometry [25]. Organic matter was measured by TOC analyzer [26]. The pH value was determined by potentiometric method and the water–soil ratio (W:V) was 2.5:1.

2.4. Measurements of Plant Growth and Fruit Quality

The heights of the tomato seedlings (from the base to the growth point) were measured using a ruler (centimeters). Tomato fruit was harvested after ripening and the yield was calculated. The photosynthetic indexes were measured in the morning (9:00–11:00) and the second leaf was selected from the top of each plant. Each chosen leaf was measured three times and the average of the three readouts was recorded. Various photosynthetic parameters, including net photosynthetic rate (Pn), transpiration rate (Tr), stomatal conductance (Gs), and intercellular CO_2 concentration (Ci), were measured by the photosynthetic instrument (GFS-3000 WALZ) [27].

For chlorophyll content, the middle and upper leaves of tomato were selected and quickly ground into powder with liquid nitrogen, weighed 0.5 g powder into 10 mL centrifuge tube, and then added cold acetone. The homogenate was extracted in the dark at 4 °C for 24 h and was centrifuged at 2500 rpm for 10 min. The supernatant was separated and the absorbances were read at 662 nm (Chlorophyll a) and 646 nm (Chlorophyll b) on Shimadzu UV-1800 spectrophotometer (Shimadzu, Kyoto, Japan) [28]. Tomato fruit quality was determined by selecting mature fruits (repeat picking 4 plants) with the same growth status of the first ear and the third ear and the average value was taken as the final value. The total sugar in tomato was obtained by anthrone colorimetry [29]. The soluble sugar was determined by sulfuric acid anthrone colorimetry [30]. Soluble protein content was measured using Coomassie Brilliant Blue G-250 (Shanghai Macklin Biochemical Co., Ltd., Shanghai, China) according to the method of Bradford [31]. Vitamin C content was determined using the method of Deutsch and Weeks [32]. Lycopene content was obtained from spectrophotometry at 502 nm [33]. The concentration of soluble solids in tomato was measured by means of a digital refractometer PR-32a (ATAGO Co. Ltd., Fukui, Japan) [34].

2.5. Statistical Analysis

Statistical analysis of the growth curve data was performed using standard analysis of variance (ANOVA), followed by a one-way ANOVA with Duncan's Multiple Range Test (DMRT) and were analyzed using SPSS version 21.0 (IBM, Armonk, NY, USA). All the experiments were carried out at least in triplicate and a significant difference was considered at $p < 0.05$.

3. Results and Discussion

3.1. Effect of Colloidal BC-Based Organic Fertilizer on Physical and Chemical Properties of Soil

The CEC of T4 and T5 treatments was significantly higher than that of T1, which may be enhanced by the self-metabolism of microorganisms in the processes of both compost fermentation and plant growth. Note that the treatment of T6 (i.e., 1.5% colloidal BC + 1.5% organic fertilizer + Methyltrophic bacillus) exhibited the highest CEC values among the different treatments. With the increasing incubation time of colloidal BC in soil, more oxygen-containing functional groups can be formed on its surfaces under the modifications of biotic and abiotic (chemical oxidation) processes, which increased the surface charge density and thereby effectively improved the soil CEC [35,36]. Organic fertilizer had a large number of humus and oxygen-containing functional groups and its surfaces might have a variety of binding sites for cations, which also improved the CEC of soil [37]. Figure 1b shows that the addition of BC or organic fertilizer (T1–T3) significantly ($p < 0.05$) enhanced the content of nitrate N compared to CK. However, the presence of Methyltrophic bacillus (T4, T5, and T6) did not significantly increase the content of nitrate N, relative to treatments without Methyltrophic bacillus, which may be due to the fact that it would take a longer time for microorganisms to adapt to the soil environment.

The contents of available K and P in all the treatments were increased, where the values in the T6 treatment were the highest (Figure 1c,d) compared with CK. The microorganism inoculated treatments (T4, T5, and T6) had a higher available K content relative to the non-microorganism treatments (T1, T2, and T3), which was probably attributed to the conversion of K from the insoluble species to the soluble species through the metabolism process in microorganisms (Figure 1c). However, the growth of microorganisms needs to consume energy, which may consume some available K. The opposite trend is observed for the soil available P (Figure 1d), which can be explained by the fact that microorganisms cannot significantly increase the content of available P in a short incubation time. The data demonstrated that the lack of the organic fertilizer in soil media significantly decreased available P, which was consistent with previous studies [38]. These results indicated that the microorganism amendment can greatly improve the contents of CEC, nitrate N, and available K in the soil, especially for CEC and available K.

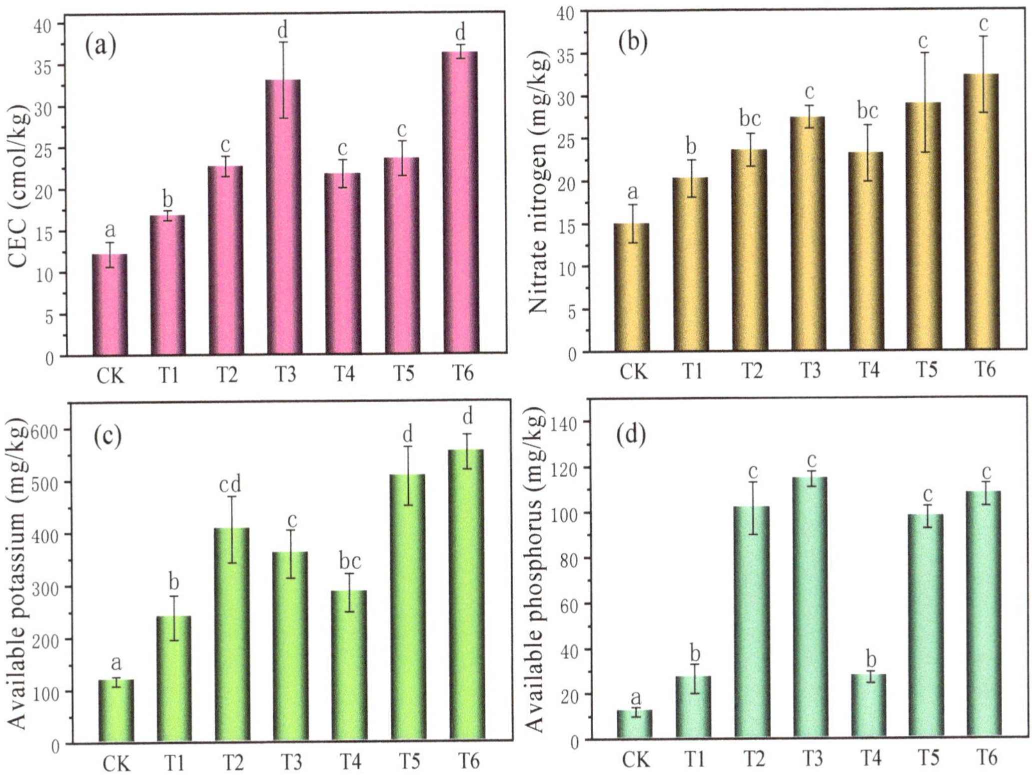

Figure 1. Effects of different treatments on soil CEC (**a**), nitrate N (**b**), available K (**c**), and available P (**d**). Soil control (CK), the amount of colloidal BC accounts for 3% of dry soil weight (T1), the amount of organic fertilizer accounts for 3% of dry soil weight (T2), 1.5% colloidal BC +1.5% organic fertilizer (T3), 3% colloidal BC + Methyltrophic bacillus (T4), 3% organic fertilizer + Methyltrophic bacillus (T5), 1.5% colloidal BC + 1.5% organic fertilizer + Methyltrophic bacillus (T6). Data with different letters are significantly different ($p < 0.05$).

3.2. *Effect of Colloidal BC-Based Organic Fertilizer on the Growth of Plants*

As shown in Figure 2a, there was little difference in the growth of plants for each treatment in the first three weeks; however, a significant growth increase occurred from the fourth week. After that, the tomato maintained a rapid growth trend, where the plant height in all the treatments was significantly higher than that in the CK. We found that the growth of tomato reached a peak in the 9th week and then decreased gradually.

For the fruit, the total biomass of tomato fruit in all treatments was significantly higher than that of CK and the fruit biomass in the microorganism inoculated treatments (T4, T5, and T6) was significantly higher than that of non-microorganism inoculated treatments (T1, T2, and T3). It can be seen in Figure 2c that the tomato yield of T6, T5, and T4 treatments were 26.33%, 18.11%, and 17.84% higher than that of T3, T2, and T1 treatments, respectively, indicating that the combined application of microorganism and colloidal BC-based organic fertilizer can significantly promote the growth of plants and improve the yield of tomato simultaneously.

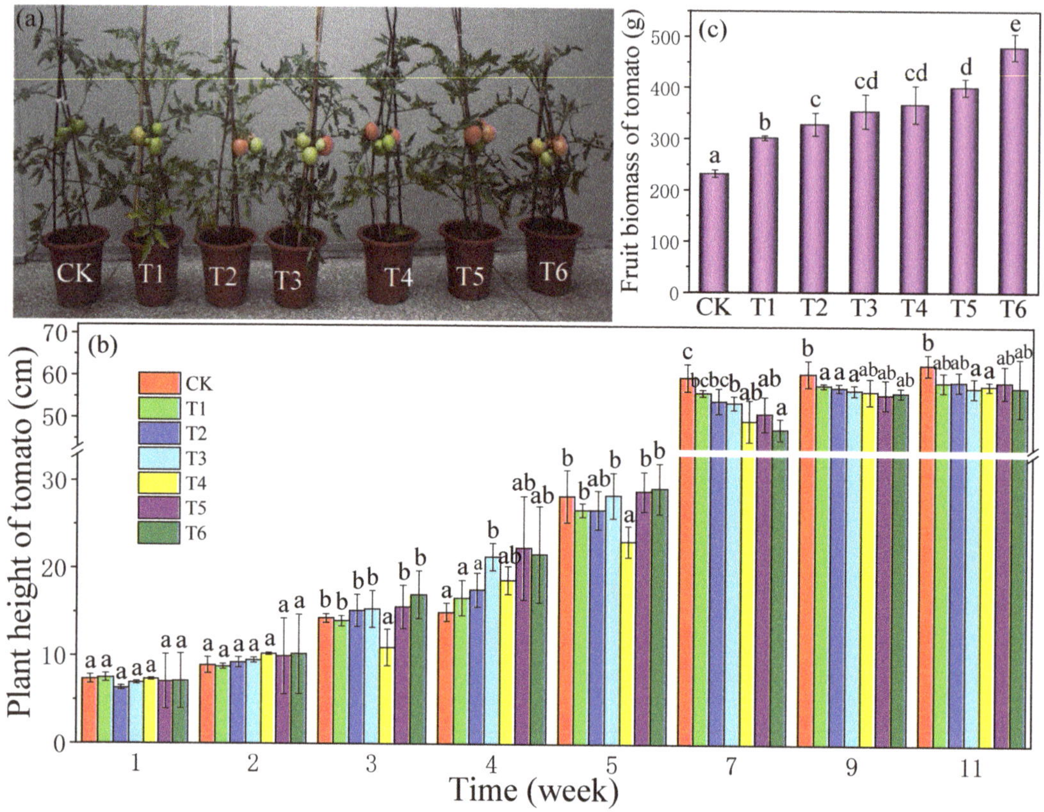

Figure 2. Effect of different treatments on growth status (**a**), plant height (**b**), and fruit biomass (**c**) of tomato. Data with different letters are significantly different ($p < 0.05$).

3.3. Effect of Colloidal BC-Based Organic Fertilizer on Photosynthesis of Plants

As shown in Figure 3a, the net photosynthetic rate of tomato leaves in all the treatments reached a significant increase (up to 22.01–54.03%) compared with that of CK where the T6 treatment had the largest value. Similarly, the leaf transpiration rate of tomato in all the treatments (except T1) was also significantly higher than that of CK (Figure 3b), among which T6 treatment was the highest (11.69 mmoL H_2O /(m^2s)). The leaf transpiration rate of T4 was significantly higher than that of single application of BC ($p < 0.05$) and the T5 treatment increased the leaf transpiration rate of tomato by 10.66%. Compared with only BC-based organic fertilizer treatment, tomato leaf transpiration rate increased by 11.21% under the T6 treatment (i.e., 1.5% colloidal BC + 1.5% organic fertilizer + Methyltrophic bacillus). Moreover, the net photosynthetic rate of tomato leaves was in general, positively correlated with the amount of available P ($y = 0.10x + 12.75$, $R^2 = 0.87$), which significantly increased tomato photosynthesis. Therefore, available P could also significantly improve the net photosynthetic rate of tomato [39].

The stomatal conductance of tomato leaves in all treatments increased significantly compared with CK (Figure 3c), where T6 was the highest and reached to 461.33 mmol H_2O /(m^2s). Compared with the non-microorganism inoculated treatments (T3, T2, and T1), the air pore conductance in the microorganism inoculated treatments (T6, T5, and T4) was significantly increased by 9.18–30.86%. It can be seen from Figure 3b,c that the trend of stomatal conductance and transpiration rate of each treatment was basically consistent. The intercellular CO_2 concentration of tomato leaves in each treatment was significantly lower ($p < 0.05$) than that of CK. Meanwhile, the intercellular CO_2 concentration of each treatment decreased by 1.87–11.46% compared with the control. The lower intercellular CO_2 concentration indicates that more CO_2 might be used in photosynthesis by leaves. Moreover, compared with T3 and T2 (without microorganisms inoculated), the intercellular CO_2 concentration of T6 and T5 increased by 5.56% and 4.96%, respectively. There was no significant difference in the intercellular CO_2 concentration between T1 and T4 treatments. The change trend of intercellular CO_2 content in these treatments was basically opposite to the net photosynthesis rate of leaves, which was in line with the basic rules of plant photosynthesis.

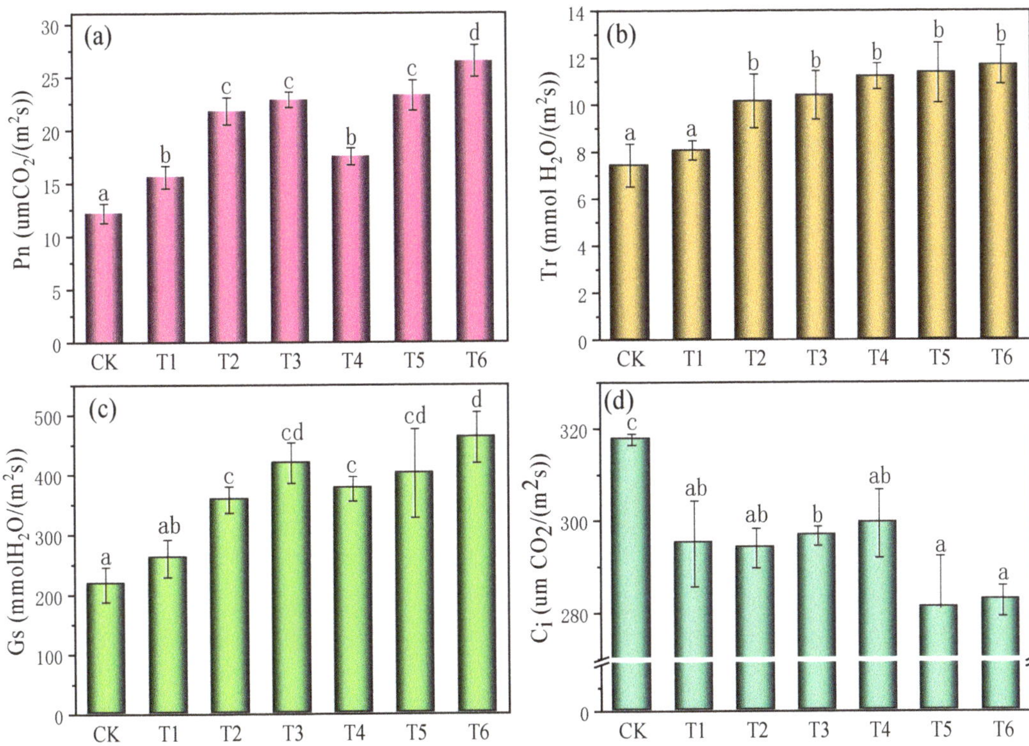

Figure 3. The gas exchange parameters: (**a**) net photosynthetic rate (Pn), (**b**) transpiration rate (Tr), (**c**) stomatal conductance (Gs), and (**d**) intercellular CO_2 concentration (Ci) of tomato. Data with different letters are significantly different ($p < 0.05$).

According to Figure 3, it is concluded that the application of colloidal BC, organic fertilizer, and microorganisms could actively promote the photosynthesis of tomato leaves. Previous studies indicated that the application of colloidal BC can significantly improve the photosynthesis of seedlings and the application of colloidal BC and organic fertilizer can also significantly improve the transpiration rate of leaves [40,41]. Microorganism inoculation (T4, T5, and T6) had an evident promoting effect on tomato photosynthesis where the photosynthesis of T6 treatment was the highest, indicating that the pores of BC might provide extra living space for the microorganisms and reduce their competition with each other [42]. Comparing the Pn of T4 and T1 with T5 and T2, it was found that the Pn of microbial solution containing colloidal BC is nearly 0.5 μm $CO_2/(m^2s)$ higher, which further indicated that colloidal BC (porous) provided growth space for microorganisms. In addition, microorganisms can help the BC-based fertilizer to provide more nutrient elements for plant growth.

3.4. Effects of Colloidal BC-Based Organic Fertilizer on Chlorophyll Content of Plants

It is found from Figures 3a and 4 that the content of chlorophyll was highly dependent on the net photosynthetic rate of tomato. The chlorophyll-a contents in the microorganism inoculated treatments (T6, T5, and T4) were 27.41%, 12.74%, and 9.42% higher than those of non-microorganism inoculated treatments (T3, T2, and T1), respectively. Some studies have reported that the application of BC and organic fertilizer can effectively improve the content of chlorophyll in plants and the increase of chlorophyll content can effectively promote photosynthesis [43,44]. In addition, the inoculation of microorganisms also promoted the synthesis of chlorophyll in tomato, where the molecular mechanism still needs to be further studied. Figure 4b shows that the chlorophyll a/b value of tomato was higher than that of the CK, indicating that the microorganisms cooperated with BC-based organic fertilizer to promote the synthesis of chlorophyll in tomato leaves.

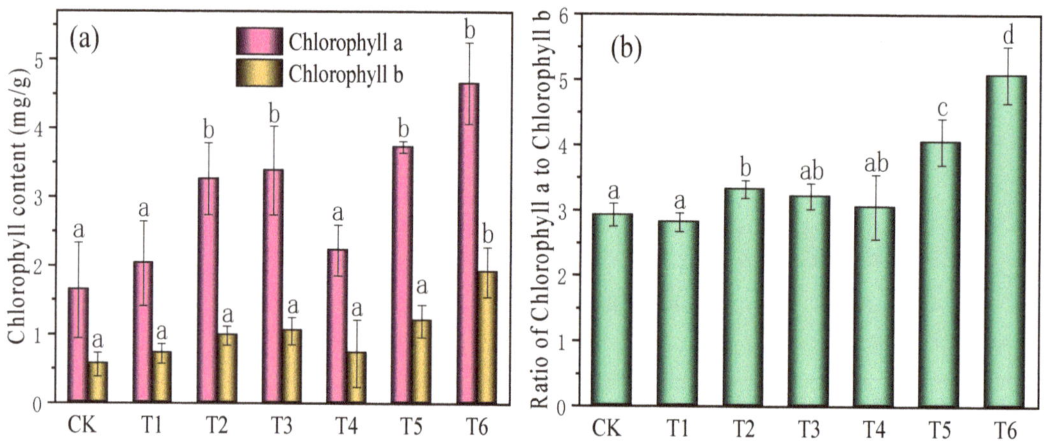

Figure 4. Effects of different treatments on chlorophyll content (**a**) and ratio of chlorophyll a to chlorophyll b (**b**) of tomato. Data with different letters are significantly different ($p < 0.05$).

3.5. Effects of Colloidal BC-Based Organic Fertilizer on Tomato Quality

Data on the quality of tomato fruits is presented in Table 1. The total sugar content of fruits in the microbial treatments (T5 and T6) was significantly higher ($p < 0.05$) than that of the non-microbial treatments (T2 and T3), for which the total sugar content of tomatoes increased by 39.95% and 30.87%, respectively (Table 1). Table 1 shows that there was no significant difference in soluble sugar of tomato fruit in T2 and T3 treatments. However, for the microorganism inoculated treatments, the soluble sugar content of fruit was 26.65% higher than that in only BC-based fertilizer treatment. It was found that there were no significant changes in soluble protein and soluble sugar contents of tomato in different treatments (Table 1). BC-based fertilizer can provide N for tomato growth and Methyltrophic bacillus can further improve the tomato plants to absorb N or improve the availability of N in BC-based fertilizer for plant growth. It has confirmed that the appropriate increase of N fertilizer can increase the content of soluble sugar in tomato fruits [45,46].

Compared with CK (Table 1), the content of lycopene in different treatments showed an increasing trend and the order from high to low was T6 > T5 > T2 > T3 > CK > T1 > T4. Among them, the lycopene in T6 treatment was 66.39% and 46.56% higher than that of CK and T5, respectively. The vitamin C content of tomato fruit in all treatments was significantly higher than that in the CK ($p < 0.05$). In addition, T6 treatment increased by 21.95% compared with T5, indicating that Methyltrophic bacillus combined with BC-based fertilizer can increase the vitamin C content. It can be seen from Table 2 that the soluble solid content of tomato fruits in different treatment groups (T2, T3, T4, T5, and T6) showed significant differences compared with CK ($p < 0.05$). Among them, the soluble solid content of fruits treated with T4, T5, and T6 was increased by 8.99%, 12.09%, and 5.81%, respectively, indicating that the synergistic effect of inoculated microorganisms was evident and could greatly increase the soluble solid content of fruits. Maryam et al. also obtained a similar conclusion that organic-based fertilizer can increase the content of soluble solids in tomato fruit [47]. The changes of soil physical and chemical properties under different treatments showed that the CEC and available P of T6 were significantly increased, which was to an extent accounted for in the increased tomato yield, fertilizer utilization efficiency, and fruit quality.

Table 2. Effect of different treatments on tomato quality.

Treatment	Total Sugar (g/kg)	Soluble Sugar (%)	Soluble Protein (g/kg)	Lycopene (mg/100 g)	Vitamin C (%)	Soluble Solids (%)
CK	28.13 ± 2.01^{a}	5.03 ± 3.21^{a}	12.31 ± 1.71^{bc}	3.25 ± 0.22^{ab}	0.17 ± 0.03^{a}	4.58 ± 0.15^{a}
T1	34.53 ± 0.80^{ab}	7.32 ± 1.21^{b}	10.34 ± 0.38^{a}	3.47 ± 0.28^{ab}	0.34 ± 0.07^{b}	5.18 ± 0.26^{b}
T2	41.37 ± 2.47^{b}	12.03 ± 2.04^{bc}	13.54 ± 0.88^{b}	4.17 ± 0.47^{b}	0.41 ± 0.03^{b}	5.38 ± 0.78^{c}
T3	55.38 ± 1.79^{c}	12.68 ± 0.25^{bc}	13.80 ± 0.51^{b}	4.03 ± 0.14^{ab}	0.32 ± 0.07^{b}	5.67 ± 0.68^{bc}
T4	54.27 ± 1.43^{c}	14.64 ± 0.33^{b}	14.94 ± 2.28^{c}	2.48 ± 0.17^{a}	0.30 ± 0.01^{b}	6.02 ± 1.24^{d}
T5	68.89 ± 0.93^{c}	15.54 ± 0.33^{bc}	$14.71 \pm 0.76b^{c}$	5.17 ± 0.06^{c}	0.47 ± 0.11^{b}	6.12 ± 0.90^{c}
T6	78.51 ± 0.25^{d}	17.29 ± 0.09^{c}	14.72 ± 0.99^{c}	9.68 ± 0.23^{c}	0.61 ± 0.03^{c}	6.23 ± 0.58^{d}

Different letters in the same column indicate significant differences ($p < 0.05$) among treatments.

4. Conclusions

Colloidal BC-based organic fertilizer loaded with Methyltrophic bacillus effectively increased the soil CEC value (3.24–24.12 cmoL/kg), nitrate N (3.30–17.28 mg/kg), available K (45.56–435.36 g/kg), and P contents (6.04–95.58 g/kg, except T3), indicating that the addition of growth-promoting bacteria greatly contributed to the transformation and absorption of soil nutrients and then promoted the growth and development of tomato. The combined application of plant growth-promoting bacteria and colloidal BC-based organic fertilizer significantly enhanced the photosynthetic efficiency (Pn, 3.24–14.30 $umCO_2/(m^2s)$) of tomato leaves and further improved the quality of tomato fruit (such as total sugar, soluble protein, lycopene, vitamin C, and soluble solids), indicating that the growth-promoting bacteria and colloidal BC-based organic fertilizer had an obvious synergistic effect on the growth of tomato.

From the perspective of soil available N and fruit quality, these results suggest that the combined application of growth-promoting microorganisms and colloidal BC-based organic fertilizer can significantly improve the yield and quality of commercial crops, which has profound practical significance and application values. Meanwhile, it also provides a scientific basis for improving the quality and efficiency of BC-based organic fertilizer and the coupling utilization of microorganisms and BC materials.

Author Contributions: Conceptualization, S.G.; Formal analysis, S.G., H.Y. and J.G.; Investigation, W.Z. and F.Y.; Methodology, S.G., H.Y., Y.H., W.Z., F.Y. and J.G.; Project administration, F.L.; Resources, F.L.; Supervision, F.L.; Validation, Y.H.; Writing—review & editing, S.G. and F.L. All authors have read and agreed to the published version of the manuscript.

Funding: The work was supported by the National Natural Science Foundation of China (41977278, 41573127), Jiangsu Agricultural Science and Technology Innovation Project (CX (20)3080).

Institutional Review Board Statement: Not applicable.

Informed Consent Statement: Not applicable.

Data Availability Statement: Data sharing is not applicable to this article.

Conflicts of Interest: The authors declare no conflict of interest.

References

1. Roberts, K.G.; Gloy, B.A.; Joseph, S.; Scott, N.R.; Lehmann, J. Life Cycle Assessment of Biochar Systems: Estimating the Energetic, Economic, and Climate Change Potential. *Environ. Sci. Technol.* **2010**, *44*, 827–833. [CrossRef] [PubMed]
2. Schmidt, M.W.I.; Noack, A.G. Black carbon in soils and sediments: Analysis, distribution, implications, and current challenges. *Glob. Biogeochem. Cycles* **2000**, *14*, 777–793. [CrossRef]
3. Dong, W.; Ma, F.; Li, C.; Fu, Z.; Huang, Y.; Liu, J. Evaluation of Anti-Aging Performance of Biochar Modified Asphalt Binder. *Coatings* **2020**, *10*, 1037. [CrossRef]
4. Major, J.; Rondon, M.; Molina, D.; Riha, S.J.; Lehmann, J. Maize yield and nutrition during 4 years after biochar application to a Colombian savanna oxisol. *Plant Soil* **2010**, *333*, 117–128. [CrossRef]
5. Xiao, Y.; Liu, A.Z.; Chun, Y.J.; Stephen, J.; Rong, J.B.; Lian, Q.L.; Gen, X.P.; Jorge, P.F. Biochar's effect on crop productivity and the dependence on experimental conditions A meta-analysis of literature data. *Plant Soil* **2013**, *373*, 583–594.
6. Akhtar, S.S.; Li, G.; Andersen, M.N.; Liu, F. Biochar enhances yield and quality of tomato under reduced irrigation. *Agric. Water Manag.* **2014**, *138*, 37–44. [CrossRef]
7. Guo, X.-X.; Liu, H.-T.; Zhang, J. The role of biochar in organic waste composting and soil improvement: A review. *Waste Manag.* **2020**, *102*, 884–899. [CrossRef]
8. Agegnehu, G.; Srivastava, A.K.; Bird, M.I. The role of biochar and biochar-compost in improving soil quality and crop performance: A review. *Appl. Soil. Ecol.* **2017**, *119*, 156–170. [CrossRef]
9. Glaser, B.; Wiedner, K.; Seelig, S.; Schmidt, H.-P.; Gerber, H. Biochar organic fertilizers from natural resources as substitute for mineral fertilizers. *Agron. Sustain. Dev.* **2015**, *35*, 667–678. [CrossRef]
10. Faloye, O.; Alatise, M.; Ajayi, A.; Ewulo, B. Effects of biochar and inorganic fertiliser applications on growth, yield and water use efficiency of maize under deficit irrigation. *Agric. Water Manag.* **2019**, *217*, 165–178. [CrossRef]
11. Chen, J.; Lü, S.; Zhang, Z.; Zhao, X.; Li, X.; Ning, P.; Liu, M. Environmentally friendly fertilizers: A review of materials used and their effects on the environment. *Sci. Total Environ.* **2018**, *613*, 829–839. [CrossRef]
12. Shang, Q.; Ling, N.; Feng, X.; Yang, X.; Wu, P.; Zou, J.; Shen, Q.; Guo, S. Soil fertility and its significance to crop productivity and sustainability in typical agroecosystem: A summary of long-term fertilizer experiments in China. *Plant Soil* **2014**, *381*, 13–23. [CrossRef]
13. Kammann, C.I.; Schmidt, H.-P.; Messerschmidt, N.; Linsel, S.; Steffens, D.; Müller, C.; Koyro, H.-W.; Conte, P.; Joseph, S. Plant growth improvement mediated by nitrate capture in co-composted biochar. *Sci. Rep.* **2015**, *5*, 11080. [CrossRef] [PubMed]
14. Senthilkumar, M.; Madhaiyan, M.; Sundaram, S.; Kannaiyan, S. Intercellular colonization and growth promoting effects of *Methylobacterium* sp. with plant-growth regulators on rice (*Oryza sativa* L. Cv CO-43). *Microbiol. Res.* **2009**, *164*, 92–104. [CrossRef] [PubMed]
15. Kwak, M.-J.; Jeong, H.; Madhaiyan, M.; Lee, Y.; Sa, T.-M.; Oh, T.K.; Kim, J.F. Genome Information of Methylobacterium oryzae, a Plant-Probiotic Methylotroph in the Phyllosphere. *PLoS ONE* **2014**, *9*, e106704. [CrossRef]
16. Essarts, Y.R.D.; Cigna, J.; Quêtu-Laurent, A.; Caron, A.; Munier, E.; Beury-Cirou, A.; Hélias, V.; Faure, D. Biocontrol of the Potato Blackleg and Soft Rot Diseases Caused by Dickeya dianthicola. *Appl. Environ. Microbiol.* **2016**, *82*, 268–278. [CrossRef]
17. Zamioudis, C.; Pieterse, M.J. Modulation of host immunity by beneficial microbes. *Mol. Plant-Microbe Interact.* **2012**, *25*, 139–150. [CrossRef]
18. Zhao, J.; Liu, T.; Liu, W.-C.; Zhang, D.-P.; Dong, D.; Wu, H.-L.; Zhang, T.-T.; Liu, D.-W. Transcriptomic insights into growth promotion effect of Trichoderma afroharzianum TM2-4 microbial agent on tomato plants. *J. Integr. Agric.* **2021**, *20*, 1266–1276. [CrossRef]
19. Rose, M.; Phuong, T.L.; Nhan, D.K.; Cong, P.T.; Hien, N.T.; Kennedy, I.R. Up to 52% N fertilizer replaced by biofertilizer in lowland rice via farmer participatory research. *Agron. Sustain. Dev.* **2014**, *34*, 857–868. [CrossRef]
20. Lian, F.; Yu, W.; Wang, Z.; Xing, B. New Insights into Black Carbon Nanoparticle-Induced Dispersibility of Goethite Colloids and Configuration-Dependent Sorption for Phenanthrene. *Environ. Sci. Technol.* **2018**, *53*, 661–670. [CrossRef]
21. Qu, X.; Fu, H.; Mao, J.; Ran, Y.; Zhang, D.; Zhu, D. Chemical and structural properties of dissolved black carbon released from biochars. *Carbon* **2016**, *96*, 759–767. [CrossRef]
22. Lian, F.; Xing, B. Black Carbon (Biochar) In Water/Soil Environments: Molecular Structure, Sorption, Stability, and Potential Risk. *Environ. Sci. Technol.* **2017**, *51*, 13517–13532. [CrossRef]
23. Bower, C.A.; Reitemeier, R.F.; Fireman, M. Exchangeable cation analysis of saline and alkali soils. *Soil Sci.* **1952**, *73*, 251–262. [CrossRef]
24. Qian, P.; Schoenaru, J.J.; Karamanos, R.E. Simultaneous extraction of available phosphorus and potassium with a new soil test: A modification of Kelowna extraction. *Commun. Soil Sci. Plant Anal.* **1994**, *25*, 627–635. [CrossRef]
25. Cawse, P.A. The determination of nitrate in soil solutions by ultraviolet spectrophotometry. *Anayst* **1967**, *92*, 311–315. [CrossRef]
26. Sleutel, S.; De Neve, S.; Singier, B.; Hofman, G. Quantification of Organic Carbon in Soils: A Comparison of Methodologies and Assessment of the Carbon Content of Organic Matter. *Commun. Soil Sci. Plant Anal.* **2007**, *38*, 2647–2657. [CrossRef]
27. Liu, Y.; Yue, L.; Wang, C.; Zhu, X.; Wang, Z.; Xing, B. Photosynthetic response mechanisms in typical C3 and C4 plants upon La2O3 nanoparticle exposure. *Environ. Sci. Nano* **2020**, *7*, 81–92. [CrossRef]
28. Şükran, D.; Güneş, T.; Rıdvan, S. Spectrophotometric determination of chlorophyll-A, B and total carotenoid contents of some algae species using different solvents. *Turk. J. Bot.* **1998**, *22*, 13–18.

29. Laurentin, A.; Edwards, C.A. A microtiter modification of the anthrone-sulfuric acid colorimetric assay for glucose-based carbohydrates. *Anal. Biochem.* **2003**, *315*, 143–145. [CrossRef]
30. Leyva, A.; Quintana, A.; Sánchez, M.; Rodríguez, E.N.; Cremata, J.; Sánchez, J.C. Rapid and sensitive anthrone–sulfuric acid assay in microplate format to quantify carbohydrate in biopharmaceutical products: Method development and validation. *Biologicals* **2008**, *36*, 134–141. [CrossRef]
31. Bradford, M.M. A rapid and sensitive method for the quantitation of microgram quantities of protein utilizing the principle of protein-dye binding. *Anal. Biochem.* **1976**, *72*, 248–254. [CrossRef]
32. Deutsch, M.J.; Weeks, C.E. Microfluorometry Assay for Vitamin C. *J. AOAC Int.* **1965**, *48*, 1248–1256. [CrossRef]
33. Rao, A.; Waseem, Z.; Agarwal, S. Lycopene content of tomatoes and tomato products and their contribution to dietary lycopene. *Food Res. Int.* **1998**, *31*, 737–741. [CrossRef]
34. Mhemdi, H.; Almohammed, F.; Bals, O.; Grimi, N.; Vorobiev, E. Impact of pulsed electric field and preheating on the lime purification of raw sugar beet expressed juices. *Food Bioprod. Process.* **2015**, *95*, 323–331. [CrossRef]
35. Cheng, C.-H.; Lehmann, J.; Engelhard, M. Natural oxidation of black carbon in soils: Changes in molecular form and surface charge along a climosequence. *Geochim. Cosmochim. Acta* **2008**, *72*, 1598–1610. [CrossRef]
36. Wu, Y.; Xu, G.; Shao, H.B. Furfural and its biochar improve the general properties of a saline soil. *Solid Earth* **2014**, *5*, 665–671. [CrossRef]
37. Schulz, H.; Glaser, B. Effects of biochar compared to organic and inorganic fertilizers on soil quality and plant growth in a greenhouse experiment. *J. Plant Nutr. Soil Sci.* **2012**, *175*, 410–422. [CrossRef]
38. Erro, J.; Urrutia, O.; Baigorri, R.; Aparicio-Tejo, P.M.; Irigoyen, I.; Storino, F.; Mandado, M.; Yvin, J.C.; Garcia-Mina, J.M. Organic Complexed Superphosphates (CSP): Physicochemical Characterization and Agronomical Properties. *J. Agric. Food Chem.* **2012**, *60*, 2008–2017. [CrossRef] [PubMed]
39. Giovannini, C.; García-Mina, J.M.; Ciavatta, C.; Marzadori, C. Effect of organic-complexed supher phosphates on microbial biomass and microbial activity of soil. *Biol. Fertil. Soils* **2013**, *49*, 395–401. [CrossRef]
40. Saha, A.; Basak, B.; Gajbhiye, N.; Kalariya, K.; Manivel, P. Sustainable fertilization through co-application of biochar and chemical fertilizers improves yield, quality of Andrographis paniculata and soil health. *Ind. Crop. Prod.* **2019**, *140*, 111607. [CrossRef]
41. Naeem, M.A.; Khalid, M.; Aon, M.; Abbas, G.; Amjad, M.; Murtaza, B.; Khan, W.; Ahmad, N. Combined application of biochar with compost and fertilizer improves soil properties and grain yield of maize. *J. Plant Nutr.* **2018**, *41*, 112–122. [CrossRef]
42. Lehmann, J.; Rillig, M.C.; Thies, J.; Masiello, C.A.; Hockaday, W.C.; Crowley, D. Biochar effects on soil biota–A review. *Soil Biol. Biochem.* **2011**, *43*, 1812–1836. [CrossRef]
43. Zhang, P.; Yang, F.; Zhang, H.; Liu, L.; Liu, X.; Chen, J.; Wang, X.; Wang, Y.; Li, C. Beneficial Effects of Biochar-Based Organic Fertilizer on Nitrogen Assimilation, Antioxidant Capacities, and Photosynthesis of Sugar Beet (*Beta vulgaris* L.) under Saline-Alkaline Stress. *Agronomy* **2020**, *10*, 1562. [CrossRef]
44. Li, X.; Shao, X.; Ding, F.; Yuan, Y.; Li, R.; Yang, X.; Gao, C.; Miao, Q. Effects of effective microorganisms biochar-based fertilizer on photosynthetic characteristics and chlorophyll content of flue-cured tobacco under water-saving irrigation strategies. *Chil. J. Agric. Res.* **2020**, *80*, 422–432. [CrossRef]
45. Li, Y.; Sun, Y.; Liao, S.; Zou, G.; Zhao, T.; Chen, Y.; Yang, J.; Zhang, L. Effects of two slow-release nitrogen fertilizers and irrigation on yield, quality, and water-fertilizer productivity of greenhouse tomato. *Agric. Water Manag.* **2017**, *186*, 139–146. [CrossRef]
46. Rasool, G.; Guo, X.; Wang, Z.; Ali, M.U.; Chen, S.; Zhang, S.; Wu, Q.; Ullah, M.S. Coupling fertigation and buried straw layer improves fertilizer use efficiency, fruit yield, and quality of greenhouse tomato. *Agric. Water Manag.* **2020**, *239*, 106239. [CrossRef]
47. Tahamolkonan, M.; Ghahsareh, A.M.; Ashtari, M.K.; Honarjoo, N. Tomato (Solanum lycopersicum) growth and fruit quality affected by organic fertilization and ozonated water. *Protoplasma* **2021**, *258*, 1–9. [CrossRef] [PubMed]

coatings

Article

A Comparison of the Effects of Several Foliar Forms of Magnesium Fertilization on 'Superior Seedless' (*Vitis vinifera* L.) in Saline Soils

Sally F. Abo El-Ezz [1], Lo'ay A. A. [2,*], Nadi Awad Al-Harbi [3], Salem Mesfir Al-Qahtani [3], Hitham M. Allam [4], Mohamed A. Abdein [5] and Zinab A. Abdelgawad [6]

1 Soil Department, Faculty of Agriculture, Mansoura University, EL-Mansoura 35516, Egypt; sallyfady@mans.edu.eg
2 Pomology Department, Faculty of Agriculture, Mansoura University, EL-Mansoura 35516, Egypt
3 Biology Department, University College of Tayma, Tabuk University, Tabuk 47512, Saudi Arabia; nalharbi@ut.edu.sa (N.A.A.-H.); salghtani@ut.edu.sa (S.M.A.-Q.)
4 Viticulture Research Department, Horticulture Research Institute, ARC, Giza 12619, Egypt; hithamelkhateeb@yahoo.com
5 Biology Department, Faculty of Arts and Science, Northern Border University, Rafha 91911, Saudi Arabia; abdeingene@yahoo.com
6 Botany Department, Women's College, Ain Shams University, Cairo 11511, Egypt; Zinababdelgawad@women.asu.edu.eg
* Correspondence: Loay_Arafat@mans.edu.eg

Citation: El-Ezz, S.F.A.; A., L.A.; Al-Harbi, N.A.; Al-Qahtani, S.M.; Allam, H.M.; Abdein, M.A.; Abdelgawad, Z.A. A Comparison of the Effects of Several Foliar Forms of Magnesium Fertilization on 'Superior Seedless' (*Vitis vinifera* L.) in Saline Soils. *Coatings* **2022**, *12*, 201. https://doi.org/10.3390/coatings12020201

Academic Editor: Yu Shen

Received: 6 December 2021
Accepted: 25 January 2022
Published: 3 February 2022

Abstract: Magnesium (Mg) is the most essential element constituent in chlorophyll molecules that regulates photosynthesis processes. The physiological response of 'Superior Seedless' grapes was evaluated under different foliar magnesium fertilization such as sulfate magnesium ($MgSO_4 \cdot 7\,H_2O$), magnesium disodium EDTA (Mg-EDTA), and magnesium nanoparticles (Mg-NPs) during the berry development stages (flowering, fruit set, veraison, and harvest). In general, the 'Superior Seedless' vine had a higher performance in photosynthesis with Mg-NPs application than other forms. The Fy/Fm ratio declined rapidly after the fruit set stage; then, it decreased gradually up until the harvesting stage. However, both $MgSO_4$ and Mg-EDTA forms showed slight differences in Fv/Fm ratio during the berry development stages. The outcomes of this research suggest that the Fv/Fm ratio during the growth season of the 'Superior Seedless' vine may be a good tool to assess magnesium fertilization effects before visible deficiency symptoms appear. Mg-NPs are more effective at improving 'Superior Seedless' berry development than the other magnesium forms. These findings suggest that applying foliar Mg-NPs to vines grown on salinity-sandy soil alleviates the potential Mg deficiency in 'Superior Seedless' vines and improves bunches quality.

Keywords: fruit quality; nutrient concentration; chlorophyll concentration

1. Introduction

Grapes are one of the most important fruit crops on the planet. Grape is a member of the Vitis genus, which is part of the Vitaceae family, which contains more than 60 genera. Grapes (*Vitis vinifera* L.) are cultivated in more than 100 countries throughout the world, with an estimated area of 7.8 million hectares in 2016. Wine, jam, juice, grape seed extract, dried grapes, vinegar, and grape seed oil are among the many goods made from grapes. In 2016, the world produced 75.8 million tons of grapes, with 39% produced in Europe, 34% produced in Asia, 18% produced in the Americas, and 9% produced in Africa [1]. Grapes are Egypt's second most important fruit crop, after citrus. Egypt's agriculture has succeeded in increasing vineyard area by 220,665 hectares over the past decade, yielding 1,586,342 tons of grapes [2]. The grapevine is one of the most important horticultural crops in the world. The high value of table grapes is primarily attributed to bio-compounds required for human health, such as antioxidants, anthocyanins, and phenolics, which include gallic acid, catechin, anthocyanins, and resveratrol [3].

The fundamental issue with newly reclaimed and cultivated fields was that they were often sandy and calcareous soils with poor nutrient concentration, especially magnesium. Recently, research on magnesium nutrition has begun, with the goal of determining the Mg requirements of Egypt's most important crops. Magnesium deficiency has been discovered in some Egyptian soils such as clay or newly reclaimed soils [4]. Therefore, magnesium (Mg) is the most essential element constituent in chlorophyll molecules that regulates the photosynthesis processes [5,6]. The deficiency of Mg during growth seasons limits photosynthesis performance [7]. The physiological functions of Mg in plants have also been characterized for flowering induction [8]. Mg is required for the growth and development of plants [9]. It is also a cofactor in the biosynthesis of various enzymes, including those involved in respiration and photosynthesis. It is a phloem-mobile nutrient that migrates between older and younger leaves [10]. Mg is also a significant component of the chlorophyll molecule's ring structure [11]. Additionally, it alleviates abiotic stress conditions, such as dryness and heat, which can significantly enhance Mg deficit by inhibiting its absorption due to its mass flow transit [9]. Additionally, it mitigates aluminum toxicity in acid soils at micromolar concentrations, as opposed to calcium, which is required at millimolar concentrations [12]. A Mg shortage has been shown to adversely influence ribulose-1,5-bisphosphate carboxylase/oxygenase (Rubisco), which is involved in CO_2 fixation [13], resulting in a decrease in photosynthetic performance [14], which is correlated to a decrease in photosynthesis performance and stomatal mechanism [15]. Furthermore, it plays a role of metabolism nitrogen in plant [16]. The inhibitory influence of Mg loss on photosynthetic capacity and net CO_2 absorption was marked in several plant species [5,17,18]. As a result, in certain species, magnesium deprivation affects the structure and function of the PSI and PSII systems [19]. As a result, a decrease in the Fv/Fm ratio (maximum quantum efficiency of PSII) was observed in citrus seedlings [20]. Despite this, the Mg shortage had no effect on Fy/Fm and other fluorescence metrics in *Helianthus annuus* plants under Mg deficiency conditions. A rise in the chlorophyll a/chlorophyll b ratio is typically reported [21]. The decrease in light-harvesting complex II (LHC-II) abundance in Mg the absence of *Arabidopsis thaliana* leaves is caused by thylakoid membrane dysfunction [22].

Many researchers have begun to investigate magnesium nutrition and the determination of magnesium requirements for economically important crops [23] such as 'Washington navel' orange trees [24] and banana plants, and they have reported on the influence of magnesium on yield and fruit quality, stating that magnesium fertilization increased the yield and fruit quality of the aforementioned fruit species [25]. In addition, using the magnesium application can induce a state of magnesium deficiency during growing [26]. Furthermore, fertilizing "Grand Nain" bananas with 100 g/plant magnesium chelate plus a foliar spray of 2% magnesium sulfate increased growth metrics, yield, and fruit quality [27]. In addition, treating Le Conte pear plants with compost 45 kg/tree + biofertilizers 20 g/tree plus 1.5% magnesium sulfate produced the best production and fruit quality [28]. The foliar Mg (137.5 ppm) application boosted the growth characteristics and yield of Washington Navel orange trees [29]. Moreover, some studies were conducted to improve bunches of color quality of Crimson seedless by using foliar application of Mg [30].

'Superior seedless' is one of the first seedless table grapes to be produced in the Mediterranean region, and it adapts well to and performs well in Egyptian circumstances as well. It was harvested when the meat was yellow-white and the skin was green, as requested by the European market [31]. 'Superior Seedless' is also considered as one of the most important international grape variety with a good economic return [32]. Consumers value this grape selection for its excellent nutritional value, great taste, versatile application, and higher economic returns [33]. The world's vineyard area is growing as a result of a continual and unrelenting shift [34]. The purpose of this study is to determine the difference among foliar magnesium forms on 'Superior Seedless' vines grown in salinity sandy soil.

Furthermore, this study also aims to determine the optimal magnesium form for vine nutrition under soil salinity conditions.

2. Materials and Methods

2.1. Vine and Experimental Setup

A commercial vineyard in the Nobaria area of Egypt (31.23° N, 29.96° E) was studied for two growth seasons (2020 and 2021). The soil was sandy in texture (*Entisol-Typic Torripsamments*), and the soil composition is described in Table 1. The farm consists of 6-year-old vines of the 'Superior Seedless' cv. grafted on 1103 Paulsen rootstock. Three-by-three-meter vines were planted in sandy soil using a drip watering system. The pruning level was done on all vines at 70 bud vines^{-1} (7 cans × 10 buds can^{-1} each on four cardons), and all vines were trained by the Y system. Table 1 summarizes the physical and chemical examination of the field experiment with 'Superior Seedless' vines [35,36]. All vines were pruned to a height of 60 buds' vine^{-1}, with the length of the cans ranging from 6 to 8 buds per can, and each can contain 12–14 buds and were produced until mid-July in European countries. Additionally, according to the Egyptian Agriculture Ministry, all vines received the same management program as for NPK fertilizer (300, 200, and 250 Kg were afforded on three portions from growth starting until harvesting (one portion was added at the vine dormancy stage) in sandy soil. Uniform vines (48) were chosen and treated with four different types of magnesium; each treatment consisted of three duplicates with four vines per replication. All treatments receive 750 g of magnesium sulfate per 600 L of irrigation water, which was employed to avoid magnesium shortages. It is distinguished by the yellowing of older leaves and a yellow tint between the veins of the leaves.

Table 1. Soil and irrigation-water traits analysis.

Soil Analysis													
Physical Properties						Soluble Anions (meq L^{-1})			Soluble Cations(meq L^{-1})				
Sand %	Clay %	Silt %	Texture	EC dsm^{-1}	pH	HCO_3	Cl^-	SO_4	Na^+	K^+	Mg++	Ca^{++}	SAR
85.8	6.90	11.30	Sandy	4.50	7.93	2.80	14.10	13.10	25.00	3.00	3.80	12.00	8.89
Irrigation-Water Analysis													
-	-		Anions (meq L^{-1})						Cations (meq L^{-1})				
pH	EC (dS m^{-1}) 0.85		CO_3^-	HCO_3^-	Cl^-	SO_4^-	Ca^{++}	Mg^{++}	Na^+	K^+	-	-	-
7.18	567 ppm		0.20	2.45	0.90	1.18	1.73	0.67	2.60	0.16	-	-	-

2.2. Magnesium Fertilization Forms Treatment Protocol

The foliar magnesium application was laid out as control (0.5 g L^{-1}), $MgSO_4{\cdot}7H_2O$ (0.5 g L^{-1}), Mg-EDTA (Mg chelate 0.5 g L^{-1}), and Mg-NPs powder (0.5 g L^{-1}). Nanomaterials provided magnesium nanoparticles (MgO, 99%+ purity, 20 nm) in powder form at sundown. This optimal concentration was used for application. At sundown during the four stages of growth (flowers, fruit set, veraison, and harvest), foliar treatments were made (7:00 pm). In a bath of warm tap water, the magnesium compounds were melted. Using a knack-sap pump, the solutions were sprayed over the entire vine monthly until the leaves became saturated. The rest of the magnesium salts were acquired from EL-Gomhoria Co. Ltd. in Egypt from EL-Gomhoria Co. Ltd. In Mansoura city, Egypt.

2.3. Magnesium Deficiency Index

Magnesium deficiency (MD) results in interveinal yellowing or reddening on old leaves, beginning at the leaf edge and proceeding to the leaf veins' petiole-connected point. These symptoms progress to necrotic brown patches, and in severe MD, the leaves exhibit necrosis, dray, and premature fall. The Mg deficiency was inspected and scored on a scale from 0 (no injury) to 5 (very severe injury) [37].

2.4. Leaf Pigments Content and Chlorophyll Fluorescence

Total chlorophyll (Chls) and carotenoid (Car) content were determined spectrophotometrically [38] on the 7th leaf (ten leaves) from the shoot base.

Individual dark-leaf CF data were recorded. The data were acquired using a commercial fluorimeter (Mini-PAM, Walz, Effeltrich, Germany) and data gathering software (Win Control, Walz, Effeltrich, Germany). These data included F_0 (minimum fluorescence), Fm (light-saturated fluorescence), and the Fv/Fm ratio (the difference between maximum fluorescence and minimum fluorescence is Fv or variable fluorescence divided by maximum fluorescence). A fall in the Fv/Fm ratio below 0.75–0.78 suggests a decline in photosystem II photochemical transformation capability [39,40]. On the 7th leaf, CF parameters were determined.

2.5. Leaf Area, Total Carbohydrate Content, Ion Leakage Percentage, and Malondialdehyde (MDA)

On the 7th leaf, the Sokkia Planix 7 Digital Planimeter was used to quantify leaf area during four developmental stages. However, the vine canes' cumulative carbohydrate content was assessed according to [41]. The leaf petiole cell permeability was also tested. After three washes with deionized water, the rachis samples were put in 10 mL of 0.4 M mannitol at 24 °C for three hours. After measuring the EC of the aqueous phase (M1), the rachis samples were killed in a water bath at 100 °C for 20 min. This was followed by room-temperature cooling. Then, it was estimated as a percentage of the relative electrolyte loss from M1 rachis samples using the equation: ion leakage percent = (M1M2)/M1 × 100 [42,43]. However, MDA was a by-product of lipid peroxidation that accumulated during salinity stress. They used 2.5 g of leaf petiole samples for MDA extraction [44,45]. This was done by measuring 0–3 mM of TBARS (equal to 0–1 mM MDA) in 1,3,3-tetraethoxypropane (Sigma, St. Louis, MO, USA). During the assay's acid-heating halt, TEOP is stoichiometrically transformed to MDA.

2.6. Leaf Minerals Content

Leaf mineral content was measured on the 7th leaf from the base of the shoot during four vegetative growth stages. Nitrogen % [46], phosphorous [47], and potassium content [35] as well as the magnesium, calcium, chloride, and sodium content percentages were demined [48].

2.7. Yield and Berry Properties

At harvest, the number of clusters per vine, average cluster weight (Kg), and yield per vine (Kg) were determined. In addition, the pruned wood was weighted. The SSC % of berry juice was measured with a digital refractometer (PR32 ALA-GO Co., Tokyo, Japan) at lab temperature, and it was represented as a percentage. As for TA %, berry juice (20 mL) was used for titrating by NaOH (0.1N). The outcome was shown as a percentage. However, the SSC/TA-ratio was computed to judge bunch maturity [49,50].

2.8. Statistical Analysis

The experiment was designed as a randomized complete block in three-way ANOVA with three factors: seasons (2 levels), berry developmental phase (4 levels), and foliar magnesium forms (3 levels) with three replicates per treatment. The mean separations were run with Tukey's HSD test ($p \leq 0.05$). Pearson's correlation matrix among the studied parameters and principal component analysis (PCA) were applied. Tukey's HSD test was run using the JMP Pro 16 software, with $p < 0.05$ taken as indicating a statistically significant difference (SAS Institute, Cary, NC, USA).

3. Results

3.1. Magnesium Deficiency Index (MD-Index)

Figure 1 depicts the magnesium deficiency index (MD-index), which is a function of berry developmental stages (BDSs) for all magnesium types. When seasons, BDSs, and magnesium application forms are examined, the MD-index demonstrates a significant influence of $p < 0.05$. Considering the different magnesium forms, it is obvious that the Mg-NPs treatment produced fewer symptoms of magnesium deficiency than the other magnesium forms. Observably, the effect of 'Mg-NPs' was that there was no evidence of deficient symptoms prior to the veraison stage (berry change color) and that it rose somewhat until the harvesting stage was completed. For vines treated with Mg-EDTA, $MgSO_4$, and control treatments, deficit symptoms were observed prior to fruit set, increased significantly during veraison, and persisted until harvesting. However, during the vegetative growth stages, the 'Control' treatment exhibited the most deficiency symptoms. The severity of Mg was noticed on 'Control' vines that were unaffected by the Mg forms, but the control vines had more symptoms throughout the vegetative growth stages. On sandy soils, symptoms of a magnesium deficit appear on vines during the growth season, necessitating monthly spraying of vines to compensate for the shortfall and thereby avoiding deficient occurrence. Regardless of the magnesium supply to the vines, 750 g of magnesium sulfate per 600 L of irrigation water is employed to avoid magnesium shortages. It is distinguished by the yellowing of older leaves and a yellow tint between the veins of the leaves.

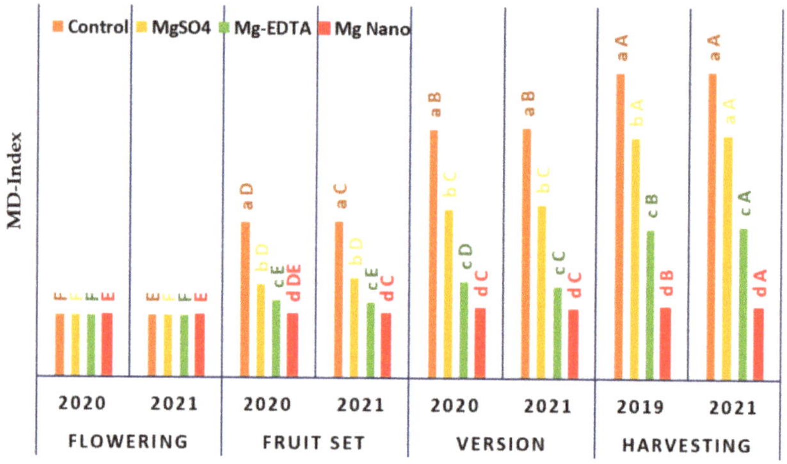

Figure 1. The influence of various magnesium fertilizer forms on 'Superior seedless' vines throughout four berry development phases (flowering, fruit set, veraison, and harvesting) under soil salinity conditions on magnesium shortage during the four stages. The values represent the mean affect levels in each application plus standard error (n = 3). Tukey's HSD test ($p \leq 0.05$) used mean severance between blocks (capital letters) to detect significant differences between growing seasons and Mg applications (capital letters) to distinguish significant differences between Mg types.

3.2. Photosynthetic Pigments: Chlorophyll (Chls) and Carotene (Car)

Photosynthetic pigments as a function of BDSs for all foliar magnesium application forms are shown in Tables 2 and 3. Leaf pigments show a significant interaction at $p \leq 0.05$ when the seasons, BDSs, and foliar magnesium treatments were considered. Generally, chlorophyll compounds (Chl A and Chl B) and carotenoid (Car) were raised gradually during BDSs until the harvest stage for all Mg treatments, whereas the untreated vines (control) treatment presented the lowest decreases in Chls and Car until the end of the experiment. Despite this, there is a significant variance between Mg treatment on pigment content that was observed during both growing seasons. The obvious outcomes are that the Mg-NPs presented the highest amount of Chl A and Chl B and Car compared to the other Mg treatments and control vines. They were marked with the highest amount at the harvest stage. Moreover, the Car exhibited the highest content at the harvest time stage compared to other foliar treatments. Regarding the Chl A:b ratio, the lowest rates at the harvesting stage of the vegetative growth period decreased progressively until grape harvesting with all Mg treatments. Nevertheless, the Chl A:b ratio of Mg-NPs had more stable outcomes than those shown with other Mg treatments throughout the growing season.

3.3. *Parameters of Chlorophyll Fluorescence (CF) (Fv/Fm, Fm, and F0)*

A significant interaction between seasons and berry developmental stages was found as well as the influence of Mg treatments on Fm and F0 ($p < 0.001$). No significant variations in Fv/Fm ratio were observed for the interaction effect of seasons, berry developmental stages, and mg treatments, but significant differences in Fm and F0 were observed, whereas a significant difference ($p < 0.01$) was noted for the magnesium effect ($p < 0.001$). The Fv/Fm ratio of 'Superior seedless' vines was proposed as a function of BDSs; when seasons, BDSs, and foliar Mg form fertilization were considered, substantial results were obtained (Table 4). On average, untreated vines exhibit a higher decline in the Fv/Fm ratio than vines treated with other Mg compounds. It is drastically reduced until the harvest stage. Except for Mg-NPs treatment, the drop in the Fv/Fm ratio appears to be more gradual and progressive, including a trend toward a more inferior Fv/Fm ratio during vegetative growth stages.

Both Fm and F0 rates increased significantly in overall Mg treatments from the initial stage (flowering) to the veraison stage (Table 4), and this increase was significant for both Fm and F0. It was discovered that the effect of Mg treatments on Fm and F0 varied according to the Mg forms. Then, both are steady until the experiment's duration expires. In comparison to other treatments, the application of Mg-NPs resulted in the greatest Fm and F0 values. Thus, when the Fv/Fm ratio of the 'Superior Seedless' vine was changed, Mg-NPs enhanced CF parameters more than other Mg treatments. As a result, this sample fluorescence parameter can detect magnesium insufficiency.

3.4. *Leaf Area, Shoot Carbohydrate, Ion Leakage, and Malondialdehyde Content*

Table 5 presents the differences in leaf area, shoot carbohydrate, ion leakage, and malondialdehyde accumulation as a function of berry developmental stages. The interaction ($p < 0.001$) was significant between the berry developmental stages and the Mg foliar fertilization forms and season. The leaf area (cm^2) and shoot carbohydrate content (%) have significantly ($p < 0.008$) higher values when vines receive the Mg-NPs form than other forms. Whereas, when considering the ion leakage percent and MDA content, there were significantly ($p < 0.0005$) lower values throughout the berry developmental stages. This implies that there is variability based on Mg type for previous variables.

3.5. *Mineral Content in Leaves*

Tables 6 and 7 exhibit the significant variances ($p > 0.001$) between seasons, BDSs, and Mg application foliar form treatments in the 7th leaf from the base of the shoot N, P, K^+, Ca^{++}, Mg^{++}, Na^+, and Cl^- content when all were considered as experimental factors. Na^+ and Cl^- content significantly decreased with Mg-NPs application compared to other Mg forms. However, the rest of the mineral increased during the growth stages.

3.6. *Yield and Berry Quality Properties*

Table 8 presents the yield and berry quality properties. The quality variables were significantly affected by foliar fertilization at harvesting time by 5%. The yield was significantly affected more by using foliar Mg-NPs (9.13 kg $vine^{-1}$) compared to other forms and control treatments.

Table 2. The influence of various magnesium fertilization types ($MgSO_4$, Mg-EDTA, and Mg-NPs) on leaf chlorophyll parameters pigment of 'Superior seedless' vines, which were used four times on various phases during berry growth (flowering, fruit set, version, and at harvest time) throughout two summers (2020 and 2021).

-	-	Berry Developmental Stages							
-	-	Flowering		Fruit Set		Veraison		At Harvesting	
-	-	Growth Seasons							
Variables	Treatment	2020	2021	2020	2021	2020	2021	2020	2021
Chl A	Control	1.77 ± 0.011 dA *	1.81 ± 0.005 d	1.62 ± 0.008 dB	1.66 ± 0.005 dB	1.33 ± 0.008 dC	1.16 ± 0.005 dC	0.86 ± 0.018 dD	0.76 ± 0.008 dE
	$MgSO_4$	1.87 ± 0.023 cC	1.91 ± 0.005 cABC	1.90 ± 0.005 cBC	1.91 ± 0.008 cABC	1.93 ± 0.008 cAB	1.96 ± 0.005 cA	1.94 ± 0.005 cAB	1.89 ± 0.005 cBC
	Mg EDTA	2.07 ± 0.024 bE	2.07 ± 0.011 bE	2.16 ± 0.011 bD	2.18 ± 0.005 bCD	2.24 ± 0.026 bABC	2.26 ± 0.005 bAB	2.30 ± 0.005 bA	2.21 ± 0.005 bBCD
	Mg-NPs	2.16 ± 0.012 aD	2.17 ± 0.005 aD	2.27 ± 0.005 aC	2.28 ± 0.008 aC	2.47 ± 0.011 aB	2.52 ± 0.014 aB	2.66 ± 0.017 aA	2.64 ± 0.005 aA
Chl b	Control	0.59 ± 0.008 cB	0.55 ± 0.005 dB	0.53 ± 0.011 dBC	0.51 ± 0.005 dCD	0.84 ± 0.005 dDE	0.45 ± 0.005 dEF	0.43 ± 0.005 dF	0.39 ± 0.005 dG
	$MgSO_4$	0.65 ± 0.011 bB	0.66 ± 0.005 cB	0.73 ± 0.028 cAB	0.71 ± 0.003 cAB	0.72 ± 0.005 cAB	0.72 ± 0.028 bAB	0.73 ± 0.008 cA	0.65 ± 0.005 cB
	Mg EDTA	0.78 ± 0.005 aD	0.79 ± 0.008 bCD	0.83 ± 0.011 bBCD	0.85 ± 0.015 bB	0.87 ± 0.005 bB	0.84 ± 0.005 cBC	0.92 ± 0.015 bA	0.81 ± 0.005 bBCD
	Mg-NPs	0.83 ± 0.012 aD	0.84 ± 0.012 aD	0.94 ± 0.005 aC	0.96 ± 0.012 aC	0.97 ± 0.005 aC	0.98 ± 0.012 aBC	1.04 ± 0.008 aA	1.03 ± 0.005 aAB
Chl A + B	Control	2.36 ± 0.020 dA	2.36 ± 0.011 dA	2.15 ± 0.020 dB	2.17 ± 0.011 dB	1.59 ± 0.014 dC	1.61 ± 0.011 dC	1.29 ± 0.024 dD	1.15 ± 0.014 dE
	$MgSO_4$	2.52 ± 0.034 cD	2.57 ± 0.011 cBCD	2.62 ± 0.033 cABCD	2.63 ± 0.012 cABCD	2.65 ± 0.014 cABC	2.68 ± 0.033 cAB	2.70 ± 0.014 cA	2.54 ± 0.011 cCD
	Mg EDTA	2.85 ± 0.029 bD	2.86 ± 0.020 dD	2.99 ± 0.017 bC	3.03 ± 0.020 bBC	3.11 ± 0.032 bBC	3.10 ± 0.011 bB	3.22 ± 0.020 bA	3.02 ± 0.011 bBC
	Mg-NPs	2.99 ± 0.023 aD	3.01 ± 0.017 aD	3.21 ± 0.032 aC	3.25 ± 0.020 aC	3.44 ± 0.017 aB	3.51 ± 0.026 aB	3.70 ± 0.026 aA	3.67 ± 0.011 aA
Chl A:B	Control	2.98 ± 0.027 aB	3.29 ± 0.023 aA	3.07 ± 0.052 aB	3.25 ± 0.026 aA	2.32 ± 0.010 aD	2.57 ± 0.020 aC	2.01 ± 0.020 aE	1.96 ± 0.006 aE
	$MgSO_4$	2.87 ± 0.014 bA	2.89 ± 0.014 bA	2.63 ± 0.092 bB	2.67 ± 0.003 bAB	2.68 ± 0.008 bAB	2.72 ± 0.095 aAB	2.54 ± 0.020 aB	2.91 ± 0.017 bA
	Mg EDTA	2.66 ± 0.020 cAB	2.61 ± 0.015 cABC	2.60 ± 0.028 bABC	2.56 ± 0.038 bBC	2.57 ± 0.013 bAB	2.69 ± 0.011 aAB	2.50 ± 0.035 aC	2.73 ± 0.011 cA
	Mg-NPs	2.63 ± 0.049 cA	2.57 ± 0.030 cA	2.42 ± 0.015 bB	2.36 ± 0.021 cB	2.54 ± 0.003 cA	2.57 ± 0.015 aA	2.55 ± 0.008 bA	2.56 ± 0.008 dA

* The mean and standard error of the mean are used to represent the data. Tukey's HSD test at $p < 0.05$ for mean separation among columns (small letters) and rows (capital letters). Data were obtained at various stages of berry growth.

Table 3. The effect of various magnesium fertilization types ($MgSO_4$, Mg-EDTA, and Mg-NPs) on leaf carotene pigment and the ratio of chlorophyll and carotenoid of 'Superior seedless' vines, which were used four times on various phases during berry growth (flowering, fruit set, version, and at harvest time) throughout two summers (2020 and 2021).

-	-	Berry Developmental Stages							
-	-	Flowering		Fruit Set		Veraison		At Harvesting	
-	-	Growth Seasons							
Variables	Treatment	2020	2021	2020	2021	2020	2021	2020	2021
Car	Control	2.18 ± 0.017 dC *	2.23 ± 0.005 dBC	2.23 ± 0.005 dBC	2.28 ± 0.005 dAB	2.31 ± 0.032 cA	2.33 ± 0.005 dA	2.20 ± 0.005 dC	2.17 ± 0.012 dC
	$MgSO_4$	2.26 ± 0.008 cE	2.28 ± 0.014 cDE	2.31 ± 0.008 cCD	2.34 ± 0.005 cBC	2.37 ± 0.008 cAB	2.39 ± 0.008 cA	2.39 ± 0.005 cA	2.37 ± 0.005 cAB
	Mg EDTA	2.60 ± 0.014 bC	2.66 ± 0.008 bBC	2.64 ± 0.008 bB	2.67 ± 0.008 bB	2.69 ± 0.005 bB	2.76 ± 0.005 bA	2.72 ± 0.008 bA	2.73 ± 0.012 bA
	Mg-NPs	2.79 ± 0.011 aE	2.81 ± 0.008 aE	2.94 ± 0.008 aD	3.07 ± 0.011 aC	3.16 ± 0.008 aB	3.19 ± 0.008 aB	3.40 ± 0.020 aA	3.36 ± 0.017 aA
Chl:Carratio	Control	1.08 ± 0.000 aA	1.05 ± 0.003 cA	0.96 ± 0.012 cB	0.95 ± 0.003 cB	0.69 ± 0.015 cC	0.69 ± 0.003 bC	0.59 ± 0.012 cD	0.53 ± 0.003 cE
	$MgSO_4$	1.11 ± 0.012 aA	1.12 ± 0.003 aA	1.13 ± 0.008 aA	1.12 ± 0.003 aA	1.11 ± 0.003 abA	1.12 ± 0.010 aA	1.13 ± 0.015 bA	1.07 ± 0.003 bB
	Mg EDTA	1.09 ± 0.017 aCD	1.07 ± 0.003 bD	1.13 ± 0.008 aB	1.13 ± 0.003 aB	1.15 ± 0.014 aB	1.12 ± 0.003 aBC	1.18 ± 0.003 aA	1.10 ± 0.003 aBCD
	Mg-NPs	1.07 ± 0.005 aAB	1.06 ± 0.003 bcAB	1.09 ± 0.010 bAB	1.06 ± 0.000 bB	1.08 ± 0.003 bAB	1.10 ± 0.005 aA	1.08 ± 0.014 bAB	1.09 ± 0.003 aAB

* The mean and standard error of the mean are used to represent the data. Tukey's HSD test at $p < 0.05$ for mean separation among columns (small letters) and rows (capital letters). Data were obtained at various stages of berry growth.

Table 4. The impact of various magnesium fertilization types ($MgSO^4$, Mg-EDTA, and Mg-NPs) on chlorophyll fluorescence parameters of 'Superior seedless' vines, which were used four times on various phases during berry growth (flowering, fruit set, version, and at harvest time) throughout two summers (2020 and 2021).

-	-	Berry Developmental Stages							
-	-	Flowering		Fruit Set		Veraison		At Harvesting	
-	-	Growth Seasons							
Variables	Treatment	2020	2021	2020	2021	2020	2021	2020	2021
Fv/Fm	Control	0.806 ± 0.00 dA *	0.800 ± 0.00 dAB	0.780 ± 0.00 dABC	0.743 ± 0.02 bC	0.756 ± 0.00 dBC	0.740 ± 0.00 dC	0.670 ± 0.00 dD	0.660 ± 0.00 dD
	$MgSO_4$	0.820 ± 0.00 cA	0.810 ± 0.00 cB	0.793 ± 0.00 cC	0.776 ± 0.00 bDE	0.780 ± 0.00 cD	0.756 ± 0.00 cF	0.770 ± 0.00 cE	0.740 ± 0.00 cG
	Mg EDTA	0.853 ± 0.00 bA	0.860 ± 0.00 bA	0.840 ± 0.00 bA	0.820 ± 0.02 abAB	0.813 ± 0.00 bAB	0.830 ± 0.00 bA	0.780 ± 0.00 bBC	0.770 ± 0.00 bC
	Mg-NPs	0.870 ± 0.00 aABC	0.860 ± 0.00 aAB	0.880 ± 0.00 aA	0.860 ± 0.01 aBC	0.880 ± 0.00 aA	0.870 ± 0.00 aABC	0.870 ± 0.00 aABC	0.856 ± 0.00 aC
Fm	Control	1697.33 ± 2.18 dA	1702.33 ± 1.45 dA	1626.67 ± 3.38 dB	1603.33 ± 1.20 dC	1591.00 ± 1.154 dD	1494.67 ± 2.60 dF	1556.33 ± 1.76 dE	1442.00 ± 0.55 dg
	$MgSO_4$	1738.33 ± 3.17 cG	1739.33 ± 0.88 cG	1886.33 ± 1.76 cD	1805.33 ± 2.60 cF	1955.00 ± 2.309 cC	1851.66 ± 1.20 cE	2020.33 ± 0.88 cA	1992.00 ± 1.52 cB
	Mg EDTA	1990.67 ± 1.20 bF	1995.00 ± 0.57 bF	2015.66 ± 1.76 bE	2105.67 ± 2.84 bD	2193.00 ± 2.309 bD	2222.66 ± 0.88 bB	2205.67 ± 1.76 bC	2314.66 ± 2.02 bA
	Mg-NPs	2137.34 ± 3.33 aH	2152.00 ± 1.52 aG	2359.00 ± 1.15 aF	2413.00 ± 1.15 aE	2585.33 ± 1.452 aD	2604.67 ± 2.02 aB	2595.33 ± 1.85 aC	2664.00 ± 1.52 aA

Table 4. *Cont.*

-	-	Berry Developmental Stages							
-	-	Flowering		Fruit Set		Veraison		At Harvesting	
-	-	Growth Seasons							
Variables	Treatment	2020	2021	2020	2021	2020	2021	2020	2021
F0	Control	364.00 ± 1.52 cB	372.00 ± 1.52 dA	357.66 ± 1.76 dB	362.00 ± 0.57 dB	304.33 ± 1.763 dD	332.00 ± 0.57 dC	296.00 ± 2.51 dE	285.66 ± 1.20 dF
	$MgSO_4$	393.00 ± 1.15 bF	403.00 ± 1.52 cE	407.33 ± 1.20 cDE	414.00 ± 1.52 cBC	420.00 ± 1.527 cAB	426.00 ± 0.57 cA	412.00 ± 0.57 cCD	417.66 ± 1.45 cBC
	Mg EDTA	422.66 ± 11.34 aD	442.33 ± 1.20 bCD	459.00 ± 1.15 bBC	457.33 ± 1.85 bBC	517.33 ± 2.333 bBC	474.00 ± 1.52 bB	532.66 ± 0.88 bA	457.00 ± 1.15 bBC
	Mg-NPs	441.00 ± 1.15 aH	461.66 ± 0.88 aG	552.66 ± 1.45 aE	526.33 ± 2.40 aF	792.33 ± 1.201 aC	693.33 ± 1.76 aD	817.66 ± 2.60 aB	827.00 ± 2.51 aA

* The mean and standard error of the mean are used to represent the data. Tukey's HSD test at $p < 0.05$ for mean separation among columns (small letters) and rows (capital letters). Data were obtained at various stages of berry growth.

Table 5. The impact of various magnesium fertilization types ($MgSO_4$, Mg-EDTA, and Mg-NPs) on leaf area (cm^2), shoot carbohydrate content percentage, ion leakage percentage, and malondialdehyde of 'Superior seedless' vines, which were used four times on various phases during berry growth (flowering, fruit set, version, and at harvest time) throughout two summers (2020 and 2021).

-	-	Berry Developmental Stages							
-	-	Flowering		Fruit Set		Veraison		At Harvesting	
-	-	Growth Seasons							
Variables	Treatment	2020	2021	2020	2021	2020	2021	2020	2021
Leaf area (cm^2)	Control	105.20 ± 0.883 dF *	113.02 ± 1.229 dE	116.11 ± 1.790 dDE	120.14 ± 0.586 dCD	125.41 ± 1.469 dBC	126.69 ± 0.904 dAB	129.94 ± 0.560 dAB	131.94 ± 0.589 dA
	$MgSO_4$	115.15 ± 1.212 bE	123.72 ± 1.212 cD	125.18 ± 0.600 cD	129.22 ± 0.586 cC	135.21 ± 0.873 cB	139.56 ± 0.583 cA	140.65 ± 0.589 cA	141.94 ± 0.335 cA
	Mg EDTA	128.32 ± 0.892 cE	132.64 ± 0.562 bD	137.85 ± 0.580 bC	139.75 ± 0.580 bC	142.00 ± 1.216 bC	148.78 ± 0.331 bB	150.79 ± 0.898 bB	154.43 ± 0.574 bA
	Mg-NPs	139.53 ± 0.881 aE	143.26 ± 0.885 aE	149.60 ± 2.623 aD	153.00 ± 1.460 aCD	154.87 ± 1.212 cBCD	159.14 ± 0.554 aBC	160.75 ± 0.586 aAB	166.40 ± 0.580 aA
Shoot carbohy-dratecontent %	Control	19.56 ± 0.591 cD	21.64 ± 0.568 cCD	21.82 ± 0.597 dCD	22.74 ± 0.568 cC	24.45 ± 0.565 cBC	26.45 ± 0.580 dB	26.62 ± 0.597 cB	29.64 ± 0.565 dA
	$MgSO_4$	23.20 ± 0.580 cD	26.34 ± 0.588 bC	26.66 ± 0.591 cC	30.96 ± 0.328 bB	30.55 ± 0.583 bB	32.66 ± 0.580 cAB	32.35 ± 0.586 bAB	34.36 ± 0.597 cA
	Mg EDTA	27.88 ± 1.208 bD	29.35 ± 0.574 bD	30.75 ± 0.566 bCD	32.66 ± 0.346 bBC	32.92 ± 0.591 bBC	35.74 ± 0.594 bAB	34.65 ± 0.571 bAB	37.67 ± 0.594 bA
	Mg-NPs	33.45 ± 1.169 aD	33.45 ± 1.169 aD	36.63 ± 0.560 aCD	39.74 ± 0.560 aBC	40.87 ± 0.586 aB	42.95 ± 0.583 aAB	41.94 ± 0.586 aAB	45.57 ± 0.583 aA
Ion leakage %	Control	12.29 ± 0.502 aC	12.67 ± 0.617 aBC	14.65 ± 0.566 aBC	15.10 ± 0.345 aAB	22.57 ± 0.580 aA	23.93 ± 0.591 aA	28.76 ± 0.673 aC	30.72 ± 0.671 aC
	$MgSO_4$	10.33 ± 0.494 abE	11.14 ± 0.447 aDE	13.58 ± 0.574 abCD	13.27 ± 0.330 bBC	19.73 ± 0.560 bAB	19.95 ± 0.332 bA	24.58 ± 0.583 bA	28.97 ± 0.377 aAB
	Mg EDTA	8.36 ± 0.565 bC	9.25 ± 0.577 bB	11.27 ± 0.586 bB	10.68 ± 0.340 cB	15.84 ± 0.600 cB	17.19 ± 1.323 bA	21.05 ± 0.600 cA	20.73 ± 0.333 bA
	Mg-NPs	5.07 ± 0.048 cE	4.99 ± 0.058 cE	6.04 ± 0.338 cD	6.10 ± 0.336 dC	7.25 ± 0.571 dB	6.84 ± 0.310 cB	10.56 ± 0.588 dA	8.64 ± 0.588 cA

Table 5. Cont.

-	-	Berry Developmental Stages							
-	-	Flowering		Fruit Set		Veraison		At Harvesting	
-	-	Growth Seasons							
Variables	Treatment	2020	2021	2020	2021	2020	2021	2020	2021
Malondialdehyde (MDA; ηM g^{-1} FW)	Control	0.15 ± 0.005 aD	0.16 ± 0.003 aD	0.20 ± 0.005 aC	0.21 ± 0.005 aC	0.24 ± 0.005 aB	0.25 ± 0.005 aB	0.29 ± 0.005 aA	0.31 ± 0.008 aA
	$MgSO_4$	0.13 ± 0.003 abE	0.14 ± 0.005 aE	0.17 ± 0.005 bD	0.18 ± 0.005 bD	0.22 ± 0.005 aC	0.23 ± 0.005 aC	0.26 ± 0.005 bB	0.28 ± 0.003 aA
	Mg EDTA	0.11 ± 0.003 bD	0.11 ± 0.005 bCD	0.13 ± 0.003 cBCD	0.14 ± 0.005 cBC	0.14 ± 0.005 bBC	0.20 ± 0.005 bA	0.15 ± 0.005 cB	0.21 ± 0.005 bA
	Mg-NPs	0.09 ± 0.005 cBC	0.08 ± 0.005 cC	0.10 ± 0.005 dBC	0.09 ± 0.005 dBC	0.11 ± 0.005 cAB	0.11 ± 0.005 cAB	0.13 ± 0.005 dA	0.13 ± 0.005 cA

* The mean and standard error of the mean are used to represent the data. Tukey's HSD test at $p < 0.05$ for mean separation among columns (small letters) and rows (capital letters). Data were obtained at various stages of berry growth.

Table 6. The effect of various magnesium fertilization types ($MgSO_4$, Mg-EDTA, and Mg-NPs) on leaf mineral compositions of 'Superior seedless' vines, which were used four times on various phases during berry growth (flowering, fruit set, version, and at harvest time) throughout two summers (2020 and 2021).

-	-	Berry Developmental Stages							
-	-	Flowering		Fruit Set		Veraison		At Harvesting	
-	-	Growth Seasons							
Variables	Treatment	2020	2021	2020	2021	2020	2021	2020	2021
N%	Control	2.57 ± 0.014 dA *	2.48 ± 0.017 cA	2.64 ± 0.012 dA	2.76 ± 0.015 dA	2.76 ± 0.005 cA	2.56 ± 0.177 bA	2.61 ± 0.008 dA	2.58 ± 0.005 dA
	$MgSO_4$	2.67 ± 0.008 cE	2.77 ± 0.017 bBCD	2.75 ± 0.011 cCD	2.87 ± 0.012 cA	2.80 ± 0.005 cBC	2.80 ± 0.008 bB	2.73 ± 0.008 cD	2.65 ± 0.008 bE
	Mg EDTA	2.79 ± 0.011 bCD	2.82 ± 0.012 bBCD	2.90 ± 0.012 bABC	2.93 ± 0.014 bAB	3.00 ± 0.014 bAB	2.98 ± 0.063 abA	2.85 ± 0.014 bBC	2.72 ± 0.012 cD
	Mg-NPs	2.85 ± 0.014 aF	3.01 ± 0.018 aD	2.98 ± 0.015 aDE	3.10 ± 0.005 aC	3.10 ± 0.008 aC	3.26 ± 0.012 aA	2.93 ± 0.008 aE	3.17 ± 0.012 aB
P%	Control	0.13 ± 0.005 dC	0.14 ± 0.005 dBC	0.16 ± 0.005 dABC	0.17 ± 0.005 dAB	0.17 ± 0.008 dA	0.19 ± 0.005 dA	0.18 ± 0.008 cA	0.17 ± 0.005 dAB
	$MgSO_4$	0.20 ± 0.005 cB	0.21 ± 0.005 cAB	0.22 ± 0.005 cAB	0.23 ± 0.005 cAB	0.24 ± 0.005 cA	0.24 ± 0.005 cA	0.21 ± 0.008 cAB	0.20 ± 0.005 cB
	Mg EDTA	0.25 ± 0.005 bCD	0.25 ± 0.005 bCD	0.27 ± 0.005 bCD	0.28 ± 0.005 bAB	0.29 ± 0.005 bAB	0.30 ± 0.005 bA	0.26 ± 0.005 bBCD	0.24 ± 0.005 bD
	Mg-NPs	0.30 ± 0.005 aD	0.32 ± 0.005 aCD	0.33 ± 0.005 aC	0.34 ± 0.005 aBC	0.37 ± 0.005 aA	0.38 ± 0.005 aA	0.33 ± 0.005 aC	0.36 ± 0.005 aAB
K%	Control	1.53 ± 0.008 dC	1.60 ± 0.008 dB	1.59 ± 0.011 dB	1.66 ± 0.005 dA	1.60 ± 0.005 dB	1.69 ± 0.008 dA	1.54 ± 0.008 dC	1.44 ± 0.008 dD
	$MgSO_4$	1.62 ± 0.005 cE	1.70 ± 0.005 cC	1.67 ± 0.005 cCD	1.74 ± 0.015 cAB	1.70 ± 0.008 cBC	1.77 ± 0.005 cA	1.64 ± 0.005 cDE	1.55 ± 0.005 cF
	Mg EDTA	1.71 ± 0.008 bB	1.74 ± 0.005 bB	1.75 ± 0.005 bB	1.81 ± 0.008 bA	1.80 ± 0.005 bA	1.84 ± 0.012 bA	1.75 ± 0.011 bB	1.63 ± 0.017 bC
	Mg-NPs	1.78 ± 0.008 aE	1.85 ± 0.012 aD	1.80 ± 0.005 aE	1.91 ± 0.005 aC	1.86 ± 0.008 aD	2.03 ± 0.014 aA	1.81 ± 0.005 aE	1.96 ± 0.008 aB

* The mean and standard error of the mean are used to represent the data. Tukey's HSD test at $p < 0.05$ for mean separation among columns (small letters) and rows (capital letters). Data were obtained at various stages of berry growth.

Table 7. The influence of various magnesium fertilization types ($MgSO_4$, Mg-EDTA, and Mg-NPs) on the leaf mineral compositions of 'Superior seedless' vines was studied for four terms in various phases during berry growth (flowering, fruit set, version, and at harvest time) throughout two summers (2020 and 2021).

-	-	Berry Developmental Stages							
-	-	Flowering		Fruit Set		Veraison		At Harvesting	
-	-	Growth Seasons							
Variables	Treatment	2020	2021	2020	2021	2020	2021	2020	2021
Mg%	Control	0.31 ± 0.008 dA *	0.32 ± 0.005 dA	0.30 ± 0.008 cAB	0.30 ± 0.005 dAB	0.25 ± 0.009 dC	0.27 ± 0.005 dBC	0.21 ± 0.007 dD	0.21 ± 0.008 dD
	$MgSO_4$	0.64 ± 0.008 cE	0.65 ± 0.005 cE	0.71 ± 0.008 cbD	0.70 ± 0.005 cD	0.77 ± 0.009 cBC	0.76 ± 0.005 cC	0.81 ± 0.007 cA	0.80 ± 0.008 cB
	Mg EDTA	0.70 ± 0.008 bD	0.71 ± 0.005 bD	0.88 ± 0.008 aB	0.79 ± 0.005 bC	0.91 ± 0.009 bC	0.89 ± 0.005 bB	0.96 ± 0.007 bA	0.97 ± 0.008 bA
	Mg-NPs	0.78 ± 0.008 aE	0.79 ± 0.005 aE	0.91 ± 0.008 aD	0.94 ± 0.005 aCD	0.97 ± 0.005 aC	1.07 ± 0.005 aB	1.04 ± 0.007 aB	1.13 ± 0.008 aA
Ca%	Control	2.27 ± 0.008 dD	2.30 ± 0.007 dCD	2.32 ± 0.005 dBC	2.33 ± 0.005 dBC	2.38 ± 0.015 dA	2.36 ± 0.005 dAB	2.32 ± 0.008 dBC	2.29 ± 0.008 cCD
	$MgSO_4$	2.35 ± 0.008 cC	2.39 ± 0.007 cB	2.44 ± 0.005 cB	2.43 ± 0.011 cB	2.51 ± 0.005 cA	2.52 ± 0.008 cA	2.49 ± 0.005 cA	2.44 ± 0.005 bcB
	Mg EDTA	2.44 ± 0.008 bA	2.49 ± 0.007 bA	2.54 ± 0.005 bA	2.52 ± 0.008 bA	2.65 ± 0.011 bA	2.59 ± 0.008 bA	2.61 ± 0.005 bA	2.60 ± 0.098 abA
	Mg-NPs	2.58 ± 0.008 aF	2.65 ± 0.007 aE	2.68 ± 0.014 aDE	2.72 ± 0.005 aCD	2.75 ± 0.017 aBC	2.81 ± 0.012 aA	2.80 ± 0.005 aAB	2.82 ± 0.005 aA
Cl%	Control	1.24 ± 0.014 aE	1.25 ± 0.005 aE	1.31 ± 0.008 aD	1.35 ± 0.005 aCD	1.39 ± 0.015 aBC	1.38 ± 0.012 aBC	1.41 ± 0.014 aAB	1.45 ± 0.005 aA
	$MgSO_4$	1.23 ± 0.008 aC	1.23 ± 0.005 aC	1.26 ± 0.005 bC	1.26 ± 0.005 bC	1.30 ± 0.005 bB	1.30 ± 0.005 bB	1.34 ± 0.005 bA	1.34 ± 0.008 bA
	Mg EDTA	1.19 ± 0.005 abD	1.20 ± 0.005 bCD	1.22 ± 0.005 cC	1.22 ± 0.005 cC	1.24 ± 0.005 cC	1.28 ± 0.005 bA	1.25 ± 0.005 cB	1.29 ± 0.005 cA
	Mg-NPs	1.13 ± 0.021 bB	1.12 ± 0.005 cB	1.19 ± 0.005 cA	1.13 ± 0.005 dB	1.22 ± 0.005 cA	1.20 ± 0.005 cA	1.23 ± 0.005 cA	1.23 ± 0.005 dA
Na%	Control	0.40 ± 0.005 aE	0.42 ± 0.005 aDE	0.43 ± 0.005 aD	0.44 ± 0.005 aCD	0.46 ± 0.005 aBC	0.46 ± 0.005 aBC	0.48 ± 0.005 aAB	0.49 ± 0.005 aA
	$MgSO_4$	0.39 ± 0.005 abD	0.39 ± 0.005 bD	0.42 ± 0.005 aBC	0.41 ± 0.005 bCD	0.44 ± 0.005 aAB	0.43 ± 0.005 bABC	0.45 ± 0.003 bA	0.45 ± 0.005 bA
	Mg EDTA	0.36 ± 0.005 bC	0.37 ± 0.005 bC	0.38 ± 0.005 bBC	0.38 ± 0.005 cBC	0.39 ± 0.008 bBC	0.40 ± 0.005 cAB	0.40 ± 0.003 cAB	0.42 ± 0.005 cA
	Mg-NPs	0.31 ± 0.008 cBC	0.30 ± 0.005 cC	0.33 ± 0.005 cABC	0.32 ± 0.005 dBC	0.34 ± 0.005 cAB	0.33 ± 0.005 dABC	0.35 ± 0.005 dA	0.34 ± 0.005 dAB

* The mean and standard error of the mean are used to represent the data. Tukey's HSD test at $p < 0.05$ for mean separation among columns (small letters) and rows (capital letters). Data were obtained at various stages of berry growth.

Table 8. The impact of various magnesium fertilization types ($MgSO_4$, Mg-EDTA, and Mg-NPs) on 'Superior seedless' vines on yield, berries proprieties, and fruit quality of 'Superior seedless' vine. Treatments were used four times on various phases during berry growth (flowering, fruit set, version, and at harvest time) throughout two summers (2020 and 2021).

Treatments	Cluster Weight (Kg)		Cluster Number Vine^{-1}		Yield Vine^{-1} (Kg)	Wood Pruned Weight (Kg)	Berry Weight (g)	Berry Size (Cm^3)	Total Soluble Solid (SSC %)	Total Acidity (TA %)	SSC:TA-Ratio
	Yield and Berry Properties								Berry Juice Proprieties		
Control	0.440 ± 0.002 ^{d}	13.56 ± 0.233 ^{c}	5.91 ± 0.073 ^{d}	14.19 ± 0.134 ^{d}	3.46 ± 0.029 ^{d}	3.31 ± 0.008 ^{d}	15.95 ± 0.014 ^{d}	0.713 ± 0.001 ^{a}	22.35 ± 0.062 ^{d}	0.713 ± 0.001 ^{a}	22.35 ± 0.062 ^{d}
$MgSO_4$	0.502 ± 0.003 ^{c}	14.71 ± 0.020 ^{b}	7.39 ± 0.055 ^{c}	14.67 ± 0.023 ^{c}	3.85 ± 0.020 ^{c}	3.66 ± 0.005 ^{c}	16.71 ± 0.028 ^{b}	0.684 ± 0.002 ^{c}	24.41 ± 0.029 ^{b}	0.684 ± 0.002 ^{c}	24.41 ± 0.029 ^{b}
Mg-EDTA	0.525 ± 0.002 ^{b}	15.51 ± 0.340 ^{a}	8.15 ± 0.210 ^{b}	15.94 ± 0.086 ^{b}	4.16 ± 0.029 ^{b}	4.17 ± 0.023 ^{b}	16.33 ± 0.014 ^{c}	0.699 ± 0.000 ^{b}	23.36 ± 0.020 ^{c}	0.699 ± 0.000 ^{b}	23.36 ± 0.020 ^{c}
Mg-Nano	0.582 ± 0.002 ^{a}	15.98 ± 0.015 ^{a}	9.13 ± 0.038 ^{a}	16.85 ± 0.272 ^{a}	4.67 ± 0.021 ^{a}	4.56 ± 0.008 ^{a}	17.38 ± 0.038 ^{a}	0.661 ± 0.000 ^{d}	26.30 ± 0.066 ^{a}	0.661 ± 0.000 ^{d}	26.30 ± 0.066 ^{a}

The main data of two seasons are analyzed using one-way (complete block randomized design) on 'Superior seedless' vines. Each value represents mean and ±SE (n = 4) replicates. The superscript letters differ ($p < 0.05$) and represent the significance between treatments using Tukey's HSD test at $p \leq 0.05$. Data were collected at different berry developmental stages.

3.7. *Multivariate Analysis of Leaf Parameters*

A PCA for physiological and biochemical variables data obtained from leaves was conducted from the tested different foliar magnesium fertilization forms (MgSO4, Mg-EDTA, and Mg-NPs) applied four times on different fruit developmental stages (flowering, fruit set, version, and at harvest time) throughout two growth seasons (2020 and 2021) of 'Superior Seedless' vines. The PCA separated the effect of magnesium forms under each seasonal stage. The PC1 explained 70.9% of the variability in the data, while PC2 explained 16.1% of the variability (Figure 2A). Figure 2B displays the negative correlation between MD-index with all the parameters except for EL%, MDA, Na+ %, and Cl− %. Chlorophyll a and b and total chlorophyll contents were negatively correlated with chlorophyll fluorescence variables (Fv/Fm; Fm, and F0). These four valuables (MD, MDA, Na+ %, and Cl− %) had a negative correlation with the other variables. Chl B showed negative correlation with Chl A:B. Chl A:B was positively correlated with Chls:Caro and Fv/Fm, whereas it had a negative correlation with the other valuables. Pearson's correlation matrix among the examined parameters shows the correlation and shows these results (Table 9).

(A) (B)

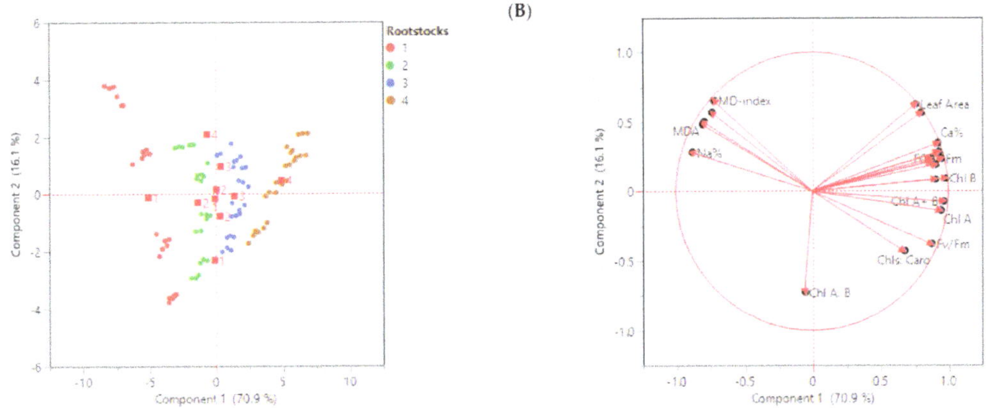

Figure 2. Principal Component Analysis (PCA) representing seasons and magnesium application forms to 'Superior seedless' vine grown in sandy soil and salt conditions, plotted with the contribution of each parameter on the two PCA axes (**A**) and all the physiological and biochemical parameters measured in leaf during the growing season (**B**). Principal Component Analysis (PCA)-Variable correlation of 7th leaf.

Table 9. Pearson's correlation pattern among the considered variables of 'Superior seedless' vines under four levels of magnesium foliar application.

Variables	MD−Index	Chl A	Chl B	Chl A+ B	Chl A:B	Caro	Chls: Caro	Fv/Fm	Fm	F0	Leaf Area	Shoot Car.	IL%	MDA	N%	P%	K%	Ca%	Mg%	Cl%	Na%
MD−index	* 1.0000																				
Chl A	−0.7543	1.0000																			
Chl B	−0.6424	0.9441	1.0000																		
Chl A+ B	−0.7307	0.9955	0.9712	1.0000																	
Chl A:B	−0.3765	0.2037	−0.1244	0.1114	1.0000																
Caro	−0.4960	0.8004	0.8970	0.8375	−0.2487	1.0000															
Chls:Caro	−0.7281	0.8552	0.7100	0.8232	0.4610	0.3803	1.0000														
Fv/Fm	−0.8758	0.8250	0.7977	0.8267	0.1369	0.7291	0.6523	1.0000													
Fm	−0.5258	0.8958	0.9557	0.9234	−0.1487	0.9469	0.5785	0.7285	1.0000												
F0	−0.4961	0.8148	0.8561	0.8361	−0.0965	0.9203	0.4576	0.6603	0.9097	1.0000											
Leaf Area	−0.1279	0.6528	0.8011	0.7032	−0.4336	0.8714	0.2790	0.4121	0.8757	0.7871	1.0000										
Shoot Car	−0.2059	0.6896	0.8255	0.7368	−0.3968	0.8856	0.3232	0.4662	0.8955	0.8277	0.9790	1.0000									
IL%	0.9348	−0.6787	−0.6213	−0.6700	−0.2364	−0.5626	−0.5611	−0.9008	−0.5262	−0.5085	−0.1789	−0.2593	1.0000								
MDA	0.9291	−0.7360	−0.7122	−0.7377	−0.1362	−0.6206	−0.6101	−0.9185	−0.5984	−0.5474	−0.2709	−0.3283	0.9527	1.0000							
N%	−0.4926	0.7260	0.8258	0.7632	−0.2568	0.8262	0.4332	0.6697	0.8156	0.7583	0.7545	0.7847	−0.5803	−0.6122	1.0000						
P%	−0.4905	0.7736	0.8962	0.8180	−0.3349	0.9253	0.4230	0.7248	0.9070	0.8319	0.8593	0.8810	−0.5735	−0.6271	0.9067	1.0000					
K%	−0.5934	0.7846	0.8409	0.8099	−0.1109	0.8390	0.5053	0.7460	0.8091	0.7626	0.7150	0.7489	−0.6685	−0.6911	0.9052	0.9047	1.0000				
Ca%	−0.4310	0.8136	0.9189	0.8534	−0.2780	0.9441	0.4630	0.6707	0.9491	0.8796	0.9144	0.9243	−0.4833	−0.5620	0.8649	0.9440	0.8655	1.0000			
Mg%	−0.5068	0.9090	0.9481	0.9308	−0.1059	0.8098	0.7315	0.6645	0.9205	0.7906	0.8300	0.8466	−0.4472	−0.5441	0.7670	0.8473	0.7656	0.8847	1.0000		
Cl%	0.8988	−0.7761	−0.7405	−0.7748	−0.1586	−0.5961	−0.7000	−0.8990	−0.6271	−0.4973	−0.3131	−0.3722	0.8944	0.9394	−0.5932	−0.6408	−0.6783	−0.5710	−0.6240	1.0000	
Na%	0.8310	−0.7779	−0.7937	−0.7915	−0.0073	−0.7709	−0.5462	−0.9271	−0.7360	−0.6594	−0.4830	−0.5469	0.9058	0.9311	−0.7271	−0.7797	−0.7809	−0.7134	−0.6535	0.9181	1.0000

* Values represent average values per season, berry developmental phases, and magnesium foliar application treatments. Chl A—Chlorophyll a content; Chl B—Chlorophyll b content; Chl A + B—Total chlorophyll content; Chl A:B—The ratio between chlorophyll A and B; Car—Carotene content; Chls:Car—The ration between total chlorophyll and Carotene; Fv/Fm—Chl fluorescence ratios; Fm—Maximum Chl fluorescence in the light-adapted state; F0—Ground fluorescence; IL%—Ion leakage percentage; MDA—Malondialdehyde accumulation; N%—Nitrogen content; P%—Phosphor content; K%—Potassium content; Ca%—Calcium content; Mg%—Magnesium content; Cl%—Chloride content; Na%—Sodium content.

4. Discussion

Magnesium is involved in a number of biochemical and physiological processes that influence plant growth and development [51]. As a result, the wounded bunches' early-stage leaves fall off throughout the growing season. However, under soil salinity conditions, a variety of mechanisms occur that result in Mg loss [52]. As a result, Mg insufficiency occurred on control vines earlier in the growth season than on vines treated with other Mg treatments. This can be seen in the slower transport of Mg through the soil profile, which results in more Mg adsorption [53]. In addition, changes in Ca and K content across Mg application rates suggest that Mg and two other cations interact throughout the season [54]. Foliar spraying is a common way for plants to adjust for nutritional deficiencies in the soil [55]. During the trial period, the efficiency of the nano-magnesium image revealed the fewest symptoms on the leaves. This result could be attributed to magnesium absorption being faster than the rest of the pictures, resulting in better photosynthetic efficiency [56]. These conclusions were reached because of the results shown in the graph. The presence of EDTA in chelated Mg form, on the other hand, has been shown to improve vine growth and biomass [57], and the sulfate part plays a critical role in the catalytic or electrochemical functions of the biomolecules in the cells [58].

Chlorophylls (Chls) are reputedly the most outstanding natural syntheses on the planet, as they are required for the photosynthesis process [59,60]. This method of vegetation occurs primarily based on gaining light rays by chlorophyll and especially chlorophyll A [61]. Photosynthesis is a very powerful method wherein it is supplied with 5 to 11 μmol CO_2 m^{-2} s^{-1}. This process is involved in the biosynthesis of essential organic molecules required for plant growth and development [62]. The photosynthesizing cells need a large amount of assimilatory pigment that reaches up to 5% of typical dry matter [63,64]. Most plant species have photosynthetic pigment content in their leaves (chlorophyll and carotene), which plays a fundamental function in the physiological overall performance of plants [65,66]. Mg participates in a variety of biochemical and physiological processes that contribute to vine growth. It is a critical component of the chlorophyll molecule, affecting both its structure and function [67]. Foliar magnesium fertilization compensates for deficits in the vines' growth stages. Additionally, it reflected the quantity and activity of photosynthetic pigments [54]. Mg is a mineral activator constituent of the chlorophyll molecule, which is responsible for photosynthetic regulation [68]. As a result, as compared to other Mg forms, the usage of Mg-NPs increased the chlorophyll components and carotene content [69]. Our findings corroborated those published in Tables 1 and 2. This comparison most likely reflects Mg-NPs' superior mobility and absorption capacity when compared to other forms [70].

Chlorophylls are critical functional and structural cofactors for all photosynthetic pigment proteins involved in oxygenic and anoxic photosynthesis, and so magnesium fertilization throughout the growing season contributes to photosynthesis's efficacy. The pigments' distinctiveness is owing to the porphyritic chromophore's extensive electron system, which chelates the Mg^{2+} ion in the center [71]. The results in Table 3 can be clarified by the variation in the Fv/Fm ratios of the various forms of foliar magnesium fertilization applied at various growth stages. In comparison to other forms, Mg-NPs dramatically boosted nucleic acid and carbohydrate enzymes [68]. However, the onset of magnesium deficit during the growing season may result in a reduction in chlorophyll and carotene levels [72]. Our findings established that Mg-NPs boosted photosynthetic pigment in comparison to other Mg forms, and our findings corroborated those of [56].

Since magnesium is required for carbohydrate accumulation in plants, its absence has an effect on the overall biomass production and distribution among plant sections [73,74]. This shows that three major factors could influence Mg effects. These are the magnesium forms, mobility, and absorption capacity of magnesium [75]. Our data indicated that the Mg-NPs increased the leaf area and carbohydrate content of the shoots during the growing season, owing to the higher photosynthesis performance. We observed reduced values for ion leakage and MDA quantity when vines were treated with Mg-NPs compared to other types. One may argue that increasing magnesium absorption in nano form [69] resulted in a reduction in the size of the cell wall, which was most likely because of its role in ion transport across the membrane and involvement in membrane-center ATPase activity [76,77]. This conclusion was consistent with previous research on citrus [78], banana [19], and coffee [79]. On the current experiment, we discovered a similar pattern of carbohydrate accumulation in vines stressed with evident leaf symptoms in the presence of a magnesium deficiency.

Normally, in plants, an element's uptake and distribution are controlled by both its supply conditions and interactions with other elements [80]. Mg, K, and Ca have been considered to exhibit opposing interactions as cation ions. Mg absorption was restricted when K or Ca concentrations increased and vice versa [81,82]. However, under salinity stress, the application of Mg-NPs increases the content of macro and micro-nutrients (Tables 5 and 6). This may be explained by the inaction between Ca^{++} and K^{+} and Mg^{++}, which increased the abortion of both cations by using Mg-NPs more than other forms [52]. The achieved outcomes regarding the effect of foliar Mg forms on leaf mineral content proved that the magnesium nano form has a pronounced effect on micronutrient status. The results agree with the findings of [56]. In addition, the foliar magnesium fertilizer improved the leaf mineral content of the mentioned fruit crop species.

This could be explained by the fact that the Mg-NPs enhanced photosynthesis during the growth stages [54]. As a result, the carbohydrate content of the product increased [7]. Our findings established that Mg-NPs raised carbohydrate content more than other forms (Table 4) and wood-trimmed weight more than other forms (Table 8). However, Mg-NPs had a considerably greater effect on berry quality features than other treatments, as measured by SSC percent (17.50%), TA percent (0.805%), and SSC:TA ratio (21.63%) (Table 7). The lowest SSC:TA ratio observed with Mg-NPs application might be read as indicating that bunches collected from vines treated with other Mg forms had a significantly longer shelf life. Additionally, magnesium has a role in protein synthesis as a bridge element that aids in ribosome assembly [83]. Additionally, it catalyzes about 300 enzymes, including phosphoenolpyruvate carboxylase, glutathione synthase, phosphatases, kinases, RNA polymerases, and ATPases [74].

A negative connection was detected between Chl B and Chl A:B. Chls:Caro and Fv/Fm were positively linked with Chl A:B but negatively with the other assets. Our observations were acknowledged by both parties [19].

5. Conclusions

The outcomes of this research recommend that the Fv/Fm ratio during the growth season of 'Superior Seedless' vines may be a good tool to assess magnesium fertilization effects before visible deficiency symptoms appear. Mg-NPs are more effective at improving 'Superior Seedless' vine growth than the other magnesium forms. Moreover, a comparison validated that the application of different forms of Mg foliar fertilization for 'Superior Seedless' vines does affect the yield and berry quality at harvest time as a final determination of the impact of Mg foliar fertilization. Overall, Mg-NPs are the most effective form for application to 'Superior Seedless' vines when compared to other Mg forms under saline soil. It enhanced biochemical and bunched quality variables.

Author Contributions: Conceptualization, S.M.A.-Q.; Data curation, S.F.A.E.-E; Formal analysis, H.M.A.; Funding acquisition, N.A.A.-H.; Investigation, Z.A.A.; Methodology, S.F.A.E.-E.; Project administration, S.M.A.-Q.; Resources, Z.A.A.; Software, L.A.A., N.A.A.-H. and H.M.A.; Supervision, L.A.A. and M.A.A.; Writing—original draft, M.A.A. All authors have read and agreed to the published version of the manuscript.

Funding: This research received no external funding.

Institutional Review Board Statement: Not applicable.

Informed Consent Statement: Not applicable.

Data Availability Statement: Relevant data applicable to this research are within the paper.

Acknowledgments: The author Lo'ay, A.A. extend their thanks, appreciation, and gratitude to Sally F. Abo EL-Ezz for their constructive cooperation during the research stages. This research is presented as a tribute to the soul of our dead colleague, Sally F. Abo EL-Ezz.

Conflicts of Interest: The authors declare no conflict of interest.

References

1. Wen, J.; Lu, L.-M.; Nie, Z.-L.; Liu, X.-Q.; Zhang, N.; Ickert-Bond, S.; Gerrath, J.; Manchester, S.R.; Boggan, J.; Chen, Z.-D. A new phylogenetic tribal classification of the grape family (Vitaceae). *J. Syst. Evol.* **2018**, *56*, 262–272. [CrossRef]
2. FAOSTAT. 2020. Available online: http://www.fao.org/faostat/en/#data/QC (accessed on 6 December 2021).
3. Asghari, M.; Rezaei-Rad, R. 24-Epibrassinolide enhanced the quality parameters and phytochemical contents of table grape. *J. Appl. Bot. Food Qual.* **2018**, *91*, 226–231. [CrossRef]
4. Wang, Z.; Hassan, M.U.; Nadeem, F.; Wu, L.; Zhang, F.; Li, X. Magnesium Fertilization Improves Crop Yield in Most Production Systems: A Meta-Analysis. *Front. Plant Sci.* **2020**, *10*, 1727. [CrossRef] [PubMed]
5. Farhat, N.; Elkhouni, A.; Zorrig, W.; Smaoui, A.; Abdelly, C.; Rabhi, M. Effects of magnesium deficiency on photosynthesis and carbohydrate partitioning. *Acta Physiol. Plant.* **2016**, *38*, 145. [CrossRef]
6. Tränkner, M.; Tavakol, E.; Jákli, B. Functioning of potassium and magnesium in photosynthesis, photosynthate translocation and photoprotection. *Physiol. Plant.* **2018**, *163*, 414–431. [CrossRef] [PubMed]
7. Delfani, M.; Baradarn Firouzabadi, M.; Farrokhi, N.; Makarian, H. Some physiological responses of black-eyed pea to iron and magnesium nanofertilizers. *Commun. Soil Sci. Plant Anal.* **2014**, *45*, 530–540. [CrossRef]
8. Monteiro, A.I.; Malheiro, A.C.; Bacelar, E.A. Morphology, Physiology and Analysis Techniques of Grapevine Bud Fruitfulness: A Review. *Agriculture* **2021**, *11*, 127. [CrossRef]
9. Gransee, A.; Führs, H. Magnesium mobility in soils as a challenge for soil and plant analysis, magnesium fertilization and root uptake under adverse growth conditions. *Plant Soil* **2013**, *368*, 5–21. [CrossRef]
10. Taiz, L.; Zeiger, E. *Plant Physiology*; Sinauer Associates: Sunderland, UK, 2010.
11. Grigore, M.N.; Boscaiu, M.; Llinares, J.; Vicente, O. Mitigation of salt stress-induced inhibition of Plantago crassifolia reproductive development by supplemental calcium or magnesium. *Not. Bot. Horti Agrobot. Cluj-Napoca* **2012**, *40*, 58–66. [CrossRef]
12. Li, D.; Ma, W.; Wei, J.; Mao, Y.; Peng, Z.; Zhang, J.; Kong, X.; Han, Q.; Fan, W.; Yang, Y.; et al. Magnesium promotes root growth and increases aluminum tolerance via modulation of nitric oxide production in Arabidopsis. *Plant Soil* **2020**, *457*, 83–95. [CrossRef]
13. Pang, J.-J.; Shin, J.-S.; Li, S.-Y. The Catalytic Role of RuBisCO for in situ CO2 Recycling in Escherichia coli. *Front. Bioeng. Biotechnol.* **2020**, *8*, 543807. [CrossRef]
14. Jamali Jaghdani, S.; Jahns, P.; Tränkner, M. The impact of magnesium deficiency on photosynthesis and photoprotection in Spinacia oleracea. *Plant Stress* **2021**, *2*, 100040. [CrossRef]
15. Jamali Jaghdani, S.; Jahns, P.; Tränkner, M. Mg deficiency induces photo-oxidative stress primarily by limiting CO2 assimilation and not by limiting photosynthetic light utilization. *Plant Sci.* **2021**, *302*, 110751. [CrossRef] [PubMed]
16. Pogłodziński, R.; Barłóg, P.; Grzebisz, W. Effect of nitrogen and magnesium sulfate application on sugar beet yield and quality. *Plant Soil Environ.* **2021**, *67*, 507–513. [CrossRef]
17. Laing, W.; Greer, D.; Sun, O.; Beets, P.; Lowe, A.; Payn, T. Physiological impacts of Mg deficiency in Pinus radiata: Growth and photosynthesis. *New Phytol.* **2000**, *146*, 47–57. [CrossRef]
18. Ling, L.L.; Peng, L.Z.; Cao, L.; Jiang, C.L.; Chun, C.P.; Zhang, G.Y.; Wang, Z.X. Effect of magnesium deficiency on photosynthesis characteristic of Beibei 447 Jinchen orange. *J. Fruit Sci.* **2009**, *26*, 275–280.
19. He, H.; Jin, X.; Ma, H.; Deng, Y.; Huang, J.; Yin, L. Changes of plant biomass partitioning, tissue nutrients and carbohydrates status in magnesium-deficient banana seedlings and remedy potential by foliar application of magnesium. *Sci. Hortic.* **2020**, *268*, 109377. [CrossRef]
20. Yang, G.-H.; Yang, L.-T.; Jiang, H.-X.; Li, Y.; Wang, P.; Chen, L.-S. Physiological impacts of magnesium-deficiency in Citrus seedlings: Photosynthesis, antioxidant system and carbohydrates. *Trees* **2012**, *26*, 1237–1250. [CrossRef]
21. Chaudhry, A.H.; Nayab, S.; Hussain, S.B.; Ali, M.; Pan, Z. Current Understandings on Magnesium Deficiency and Future Outlooks for Sustainable Agriculture. *Int. J. Mol. Sci.* **2021**, *22*, 1819. [CrossRef]
22. Sattari Vayghan, H.; Nawrocki, W.J.; Schiphorst, C.; Tolleter, D.; Hu, C.; Douet, V.; Glauser, G.; Giovanni, F.; Croce, R.; Wientjes, E.; et al. Photosynthetic light harvesting and thylakoid organization in a CRISPR/Cas9 *Arabidopsis thaliana* LHCB1 knockout mutant. *bioRxiv* **2021**, *12*, 22–473855. [CrossRef]
23. El-Fouly, M.; Rezk, A.I.; Nofal, O.; Abou El-Nour, E.-Z. Depletion of magnesium in Egyptian soils, its content in crops and estimated needs. *Afr. J. Agric. Res.* **2010**, *5*, 1060–1067.
24. Esteves, E.; Maltais-Landry, G.; Zambon, F.; Ferrarezi, R.S.; Kadyampakeni, D.M. Nitrogen, Calcium, and Magnesium Inconsistently Affect Tree Growth, Fruit Yield, and Juice Quality of Huanglongbing-affected Orange Trees. *HortScience* **2021**, *56*, 1269–1277. [CrossRef]

25. Hu, W.; Yang, B.; He, Z.; Li, G. Magnesium may be a key nutrient mechanism related to Fusarium wilt resistance: A new banana cultivar (Zhongjiao No. 9). *PeerJ* **2021**, *9*, e11141. [CrossRef] [PubMed]
26. Rustioni, L.; Grossi, D.; Brancadoro, L.; Failla, O. Iron, magnesium, nitrogen and potassium deficiency symptom discrimination by reflectance spectroscopy in grapevine leaves. *Sci. Hortic.* **2018**, *241*, 152–159. [CrossRef]
27. Mostafa, E.A.M.; Sakeg, M.M.S.; El-Migeed Abd, M.M.M. Response of banana plants to soil and foliar application of magnesium. *Am. Eurasian J. Agric. Environ. Sci.* **2007**, *2*, 141–146.
28. Fawzi, M.; Shahin, F.; Elham, A.; Kandil, E. Effect of organic, biofertilizers and magnesium sulfate on growth, yield, chemical composition and fruit quality of "Le Conte" pear trees. *Nat. Sci.* **2010**, *8*, 273–280.
29. Hanafy, A.H.; Khalil, M.K.; El-Rahman, A.A.; Nadia, A.M. Effect of magnesium, copper and growth regulators on growth, yield and chemical composition of Washington Navel orange trees. *J. Appl. Sci. Res.* **2012**, *8*, 1271–1288.
30. El-Badawy, H.E.M. Implication of Using Potassium and Magnesium Fertilization to Improve Growth, Yield and Quality of Crimson Seedless Grapes (Vitis vinifera L). *J. Plant Prod. Mansoura Univ.* **2019**, *10*, 133–141. [CrossRef]
31. Artés-Hernández, F.; Tomás-Barberán, F.A.; Artés, F. Modified atmosphere packaging preserves quality of SO2-free 'Superior seedless' table grapes. *Postharvest Biol. Technol.* **2006**, *39*, 146–154. [CrossRef]
32. Menora, N.; Joshi, V.; Kumar, V.; Vijaya, D.; Debnath, M.; Pattanashe, S.; Padmavatha, A.S.; Variath, M.; Biradar, S.; Khadakabhavi, S. Influence of Rootstock on Bud Break, Period of Anthesis, Fruit Set, Fruit Ripening, Heat Unit Requirement and Berry Yield of Commercial Grape Varieties. *Int. J. Plant Breed. Genet.* **2015**, *9*, 126–135. [CrossRef]
33. Gowda, V.N.; Keshava, S.A.; Shyamalamma, S. Growth, yield and quality of Bangalore Blue grapes as influenced by foliar applied Polyfeed and Multi-K. *Acta Hortic.* **2008**, *785*, 207–212. [CrossRef]
34. Lo'ay, A.A.; Taha, N.A. Evaluation rachis browning phenomena of 'Superior Seedless' vines grafted on different rootstocks during shelf life. *Sci. Hortic.* **2020**, *261*, 109040. [CrossRef]
35. Jackson, M.L. *Soil Chemical Analysis*; Prentice Hall of India (Pvt.) Ltd.: New Delhi, India, 1973.
36. Black, C.A. Method of Soil Analysis Part 2. *Chem. Microbiol. Prop.* **1965**, *9*, 1387–1388.
37. Christensen, L.P.; Peacock, W.L. Mineral nutrition and fertilization. In *Raisin. Production Manual*; Christensen, L.P., Ed.; University of California: Oakland, CA, USA, 2000; pp. 102–114.
38. Porra, R.J.; Thompson, W.A.; Kriedemann, P.E. Determination of accurate extinction coefficients and simultaneous equations for assaying chlorophylls a and b extracted with four different solvents: Verification of the concentration of chlorophyll standards by atomic absorption spectroscopy. *Biochim. Biophys. Acta* **1989**, *975*, 384–394. [CrossRef]
39. DeEll, J.R.; Toivonen, P.M.A. Chlorophyll fluorescence as a nondestructive indicator of broccoli quality during storage in modified atmosphere packaging. *HortScience* **2000**, *35*, 256–259. [CrossRef]
40. Lo'ay, A.A.; El-Ezz, S.F.A.; Awadeen, A.A. Effect of different foliar potassium fertilization forms on vegetative growth, yield, and fruit quality of kaki trees grown in sandy soil. *Sci. Hortic.* **2021**, *288*, 110420. [CrossRef]
41. DuBois, M.; Gilles, K.A.; Hamilton, J.K.; Rebers, P.A.; Smith, F. Colorimetric method for determination of sugars and related substances. *Anal. Chem* **1956**, *28*, 350–356. [CrossRef]
42. Krauss, S.; Schnitzler, W.H.; Grassmann, J.; Woitke, M. The Influence of Different Electrical Conductivity Values in a Simplified Recirculating Soilless System on Inner and Outer Fruit Quality Characteristics of Tomato. *J. Agric. Food Chem.* **2006**, *54*, 441–448. [CrossRef] [PubMed]
43. Lo'ay, A.A.; Ameer, N.M. Performance of calcium nanoparticles blending with ascorbic acid and alleviation internal browning of 'Hindi Be-Sennara' mango fruit at a low temperature. *Sci. Hortic.* **2019**, *254*, 199–207. [CrossRef]
44. Iturbe-Ormaetxe, I.; Escuredo, P.R.; Arrese-Igor, C.; Becana, M. Oxidative damage in pea plants exposed to water deficit or paraquat. *Plant Physiol.* **1998**, *116*, 173–181. [CrossRef]
45. Lo'ay, A.A.; Doaa, M.H. The potential of vine rootstocks impacts on 'Flame Seedless' bunches behavior under cold storage and antioxidant enzyme activity performance. *Sci. Hortic.* **2020**, *260*, 108844. [CrossRef]
46. Pregl, F. *Quantitative Organic Micro Analysis*, 4th ed.; J. and A. Churchill Ltd.: London, UK, 1945; p. 203.
47. Snell, F.D.; Snell, C.T. *Colorimetric Method of Analysis*; D. Van Nostrand Company: Princeton, NJ, USA, 1967; pp. 551–552.
48. Wilde, A.A.; Corey, R.B.; Lyer, J.G.; Voigt, G.K. *Soil and Plant Analysis for Tree Culture*, 3rd ed.; Oxford IBH Publishing Co.: New Delhi, India, 1985; pp. 64–115.
49. AOAC. *Official Methods of Analysis of AOAC International*, 19th ed.; AOAC International: Gaithersburg, MD, USA, 2012.
50. El-Banna, M.F.; Lo'ay, A.A. Evaluation berries shattering phenomena of 'Flame seedless' vines grafted on different rootstocks during shelf life. *Sci. Hortic.* **2019**, *246*, 51–56. [CrossRef]
51. Kirkby, E. Chapter 1—Introduction, Definition and Classification of Nutrients. In *Marschner's Mineral Nutrition of Higher Plants*, 3rd ed.; Marschner, P., Ed.; Academic Press: San Diego, CA, USA, 2012; pp. 3–5.
52. Härdter, R.; Rex, M.; Orlovius, K. Effects of different Mg fertilizer sources on the magnesium availability in soils. *Nutr. Cycl. Agroecosys.* **2004**, *70*, 249–259. [CrossRef]
53. Tang, M.K.; Mohd, N.; Loong, S.G. Oil palm responses to different sources of magnesium on an inland reworked soil in Peninsular Malaysia. In Proceedings of the 2001 PIPOC International Palm Oil Congress (Agriculture), Kuala Lumpur, Malaysia, 20–23 August 2001; pp. 261–271.
54. Jayaganesh, S.; Venkatesan, S.; Senthurpandian, V.K. Impact of different sources and doses of magnesium fertilizer on biochemical constituents and quality parameters of black tea. *Asian J. Biochem.* **2011**, *6*, 273–281. [CrossRef]

55. Jezek, M.; Geilfus, C.-M.; Bayer, A.; Mühling, K.-H. Photosynthetic capacity, nutrient status, and growth of maize (*Zea mays* L.) upon $MgSO_4$ leaf-application. *Front. Plant Sci.* **2015**, *5*, 781. [CrossRef]
56. Ali, E.A.M.; Aml, R.M.; Yousef, D.; Ahmed, M.M.; Abd El-Hady, M. Influence of foliar applications of magnesium sources on improving nutrients status, yield and fruit quality of murcott mandarins. *Middle East J. Appl. Sci* **2017**, *7*, 361–372.
57. Székely, A.; Balota, D.A.; Duchek, J.M.; Nemoda, Z.; Vereczkei, A.; Sasvari Szekely, M. Genetic factors of reaction time performance: DRD47 repeat allele associated with slower responses. *Genes Brain Behav.* **2011**, *10*, 129–136. [CrossRef]
58. Saito, K. Sulfur assimilatory metabolism. The long and smelling road. *Plant Physiol.* **2004**, *136*, 2443–2450. [CrossRef]
59. Dalal, V.K.; Tripathy, B.C. Modulation of chlorophyll biosynthesis by water stress in rice seedlings during chloroplast biogenesis. *Plant Cell Environ.* **2012**, *35*, 1685–1703. [CrossRef]
60. Mochizuki, N.; Tanaka, R.; Grimm, B.; Masuda, T.; Moulin, M.; Smith, A.G.; Tanaka, A.; Terry, M.J. The cell biology of tetrapyrroles: A life and death struggle. *Trends Plant Sci.* **2010**, *15*, 488–498. [CrossRef]
61. Harmatys, K.M.; Overchuk, M.; Zheng, G. Rational Design of Photosynthesis-Inspired Nanomedicines. *Acc. Chem. Res.* **2019**, *52*, 1265–1274. [CrossRef]
62. Popescu, M.; Popescu, G.C. Diurnal changes in leaf photosynthesis and relative water content of grapevine. *Curr. Trends Nat. Sci.* **2014**, *3*, 74–81.
63. Wang, J.; Lu, W.; Tong, X.Y.; Yang, Q.C. Leaf morphology, photosynthetic performance, chlorophyll fluorescence, stomatal development of lettuce (*Lactuca sativa* L.) exposed to different ratios of red light to blue light. *Front. Plant Sci.* **2016**, *7*, 250. [CrossRef] [PubMed]
64. Yang, Z.; He, W.; Mou, S.; Wang, X.; Chen, D.; Hu, X.; Chen, L.; Bai, J. Plant growth and development of pepper seedlings under different photoperiods and photon flux ratios of red and blue LEDs. *Trans. Chin. Soc. Agric. Eng.* **2017**, *33*, 173–180. [CrossRef]
65. Peter, E.; Rothbart, M.; Oelze, M.L.; Shalygo, N.; Dietz, K.J.; Grimm, B. Mg protoporphyrin monomethylester cyclase deficiency and effects on tetrapyrrole metabolism in different light conditions. *Plant Cell Physiol.* **2010**, *51*, 1229–1241. [CrossRef] [PubMed]
66. Croft, H.; Chen, J.M.; Luo, X.; Bartlett, P.; Chen, B.; Staebler, R.M. Leaf chlorophyll content as a proxy for leaf photosynthetic capacity. *Glob. Change Biol.* **2017**, *23*, 3513–3524. [CrossRef] [PubMed]
67. Fiedor, L.; Kania, A.; Myśliwa-Kurdziel, B.; Orzeł, Ł.; Stochel, G. Understanding chlorophylls: Central magnesium ion and phytyl as structural determinants. *Biochim. Biophys. Acta (BBA)-Bioenerg.* **2008**, *1777*, 1491–1500. [CrossRef] [PubMed]
68. Venkatesan, S. Notes and amendments to the recommendations on manuring of tea in South India. In *Handbook of Tea Culture*; Section-11; UPASI Tea Research Institute: Valparai, India, 2006.
69. Stagnari, F.; Onofri, A.; Pisante, M. Response of French bean (*Phaseolus vulgaris* L.) cultivars to foliar applications of magnesium. *Ital. J. Agron.* **2009**, *3*, 101–110. [CrossRef]
70. Takacs-Hajus, M.; Kiss, A.S. The effect of Mg-sulphate foliar fertilization on economic qualities of different garden pea varieties. *Acta Agron.* **2004**, *363*, 44–50.
71. Roca, M.; Chen, K.; Pérez-Gálvez, A. 6-Chlorophylls. In *Handbook on Natural Pigments in Food and Beverages*; Carle, R., Schweiggert, R.M., Eds.; Woodhead Publishing: Cambridge, UK, 2016; pp. 125–158.
72. Broadley, M.R.; Hammond, J.P.; King, G.J.; Astley, D.; Bowen, H.C.; Meacham, M.C.; Mead, A.; Pink, D.A.C.; Teakle, G.R.; Hayden, R.M. Shoot calcium and magnesium concentrations differ between subtaxa, are highly heritable, and associate with potentially pleiotropic loci in Brassica oleracea. *Plant Physiol.* **2008**, *146*, 1707–1720. [CrossRef] [PubMed]
73. Cakmak, I.; Yazici, A.M. Magnesium: A forgotten element in crop production. *Better Crops* **2010**, *94*, 23–25.
74. Verbruggen, N.; Hermans, C. Physiological and molecular responses to magnesium nutritional imbalance in plants. *Plant Soil* **2013**, *368*, 87–99. [CrossRef]
75. Cakmak, I.; Kirkby, E.A. Role of magnesium in carbon partitioning and alleviating photooxidative damage. *Physiol. Plant.* **2008**, *133*, 692–704. [CrossRef] [PubMed]
76. Shabala, S.; Haiadi, Y. Effects of magnesium availability on the activity of plasma membrane ion transporters and light-induced responses from broad bean leaf mesophyll. *Planta* **2005**, *221*, 56–65. [CrossRef] [PubMed]
77. Lo'ay, A.A.; El-Ezz, S.F.A. Performance of 'Flame seedless' grapevines grown on different rootstocks in response to soil salinity stress. *Sci. Hortic.* **2021**, *275*, 109704. [CrossRef]
78. Meireles da Silva, D.; Brandão, I.R.; Alves, J.D.; de Santos, M.O.; de Souza, K.R.D.; de Silveira, H.R.O. Physiological and biochemical impacts of magnesium-deficiency in two cultivars of coffee. *Plant Soil* **2014**, *382*, 133–150. [CrossRef]
79. White, P.J. Chapter 2-Ion Uptake Mechanisms of Individual Cells and Roots: Short-distance Transport. In *Marschner's Mineral Nutrition of Higher Plants*, 3rd ed.; Marschner, P., Ed.; Academic Press: San Diego, CA, USA, 2012; pp. 7–47.
80. Jakobsen, S.T. Interaction between Plant Nutrients: III. Antagonism between Potassium, Magnesium and Calcium. *Acta Agric. Scand. Sect. B—Soil Plant Sci.* **1993**, *43*, 1–5. [CrossRef]
81. Karley, A.J.; White, P.J. Moving cationic minerals to edible tissues: Potassium, magnesium, calcium. *Curr. Opin. Plant Biol.* **2009**, *12*, 291–298. [CrossRef]
82. Hawkesford, M.; Horst, W.; Kichey, T.; Lambers, H.; Schjoerring, J.; Møller, I.S.; White, P. Chapter 6-Functions of Macronutrients. In *Marschner's Mineral Nutrition of Higher Plants*, 3rd ed.; Marschner, P., Ed.; Academic Press: San Diego, CA, USA, 2012; pp. 135–189.
83. Bianchi, D.; Grossi, D.; Simone Di Lorenzo, G.; Zi Ying, Y.; Rustioni, L.; Brancadoro, L. Phenotyping of the "G series" Vitis hybrids: First screening of the mineral composition. *Sci. Hortic.* **2020**, *264*, 109155. [CrossRef]

coatings

Article

Effect of Copper Nanoparticles and Ions on Epididymis and Spermatozoa Viability of Chinese Soft-Shelled Turtles *Pelodiscus sinensis*

Li Yang [1,2,3], Yating Wei [1,3], Shuai Gao [4], Qifei Wang [1,3], Jiaqi Chen [4], Boping Tang [1,2,3,*] and Xunguang Bian [1,4,*]

1 Jiangsu Provincial Key Laboratory of Coastal Wetland Bioresources and Environmental Protection, Yancheng 224007, China; yangli86716@163.com (L.Y.); ttzdsys@yctu.edu.cn (Y.W.); huabzhang@163.com (Q.W.)
2 Jiangsu Synthetic Innovation Center for Coastal Bioagriculture, Yancheng 224007, China
3 School of Wetlands, Yancheng Teachers University, Yancheng 224007, China
4 College of Ocean and Biology Engineering, Yancheng Teachers University, Yancheng 224007, China; ahgebaoming@163.com (S.G.); jlcbe@126.com (J.C.)
* Correspondence: tangbp@yctu.edu.cn (B.T.); bxguang8311@163.com (X.B.)

Abstract: Copper nanoparticles (CuNPs) have been widely used in various industrial and commercial applications, which become a potential threat to aquatic organisms. Nevertheless, their potential toxicity to the epididymis and sperm remains little known. In this study, we evaluated the effect of CuNPs and copper ions ($CuSO_4$) on the spermatozoa viability, epididymal structure, antioxidant enzyme activity, and inflammatory cytokines in cauda epididymis of the Chinese soft-shelled turtle. Results showed that the spermatozoa viability of Chinese soft-shelled turtles decreased significantly with an increase in CuNPs or Cu ions concentrations. The epithelial cells of the epididymal duct of the Chinese soft-shelled turtles with the treatment of 5 mg kg^{-1} CuNPs were slightly swollen, and the connective tissue between the epididymal ducts was loose. The epithelial structure of the epididymal tube was severely damaged with an increase in Cu ion concentrations. Compared to the control, the antioxidative enzymes activities and the expression of IL-1β, TNF-α, and IL-6 mRNA in the epididymis significantly increased with the treatment of CuNPs or $CuSO_4$. The present study revealed that Cu ions exert more harmful effect on the epididymis and spermatozoa viability of Chinese soft-shelled turtles than copper nanoparticles.

Keywords: CuNPs; epididymis; sperm; Chinese soft-shelled turtles; toxicity

Citation: Yang, L.; Wei, Y.; Gao, S.; Wang, Q.; Chen, J.; Tang, B.; Bian, X. Effect of Copper Nanoparticles and Ions on Epididymis and Spermatozoa Viability of Chinese Soft-Shelled Turtles *Pelodiscus sinensis*. *Coatings* **2022**, *12*, 110. https://doi.org/10.3390/coatings12020110

Academic Editor: Ajay Vikram Singh

Received: 17 December 2021
Accepted: 15 January 2022
Published: 19 January 2022

1. Introduction

Copper nanoparticles (CuNPs) are one of the most commonly used nanomaterials, which have received much attention in these years and have been widely used in fungicides, cosmetics, printers, and electronics, because of their antifungal, optical, and electrical properties [1,2]. The extensive use of CuNPs has led to direct contamination in the aquatic environment, as they can be released into aquatic ecosystems through indirect pathways, such as runoffs and leaching from industrial and agricultural sites, and also through direct pathways, with the use of antifouling paints [3,4], raising a global concern for ecological impacts of such contamination [5]. The aquatic environment is considered to be a sink for pollutants [6]. Although copper is an essential trace metal required by all living organisms, it can be toxic to organisms when it excesses the physiological requirement of organisms. The application of CuNPs has caused adverse effects on aquatic organisms, such as fish, daphnias, and algaes, due to their dynamic physicochemical processes in an aquatic environment [7,8]. The LC_{50} value represents a common point in the lethal physiological response to toxicity, which has been abundant in data in many crustaceans. For example, the 96 h-LC_{50} value for shrimps of *Exopalaemon carinicauda*, *Echinogammars olivii*, *Sphaeroma serratum*, and *Palaemon elegans* was 0.712 mg Cu/L, 0.25 mg Cu/L, 1.98 mg Cu/L, and 2.52 mg Cu/L, respectively [9,10]. In addition, for paddy field crab *Paratelphusa hydrodromus* and freshwater crab, *Barytelphusa cunicularis*, the 96 h-LC_{50} values recorded were 15.70 mg Cu/L and 215 mg Cu/L, respectively [11,12]. Likewise, in freshwater crayfish, *Procambarus clarkia*, the 96 h-LC_{50} value reached 162 mg Cu/L [13]. These large variations in sub-lethal effects to Cu toxicity in crustaceans appear to be species specific. There are a large number of studies on the effects of copper on fish toxicity. According to a review by Malhotra et al. [14], it can be seen that the acute and sub-chronic exposure concentrations of different fish to copper vary significantly, ranging from micrograms to approximately 250 g per liter.

The negative impacts of CuNPs on different aquatic organisms have been reported by many researchers; however, it remains difficult to conclude what is the major cause of NPs toxicity [15]. Most studies showed that the toxicity of metal nanoparticles (MNPs) to organisms was attributed to their release of soluble ions [7,16,17], while some other studies demonstrated that CuNPs could penetrate into cells or move to different organs and tissues, thus damaging the organisms [18,19]. There are also some reports indicating that the toxicity of CuNPs was caused by the combination of nanoparticles and the released copper ions [2,20]. Zhang et al. [21] illustrated that the toxicity of CuNPs to zebrafish embryonic cells was not only caused by the released Cu ions, and CuNPs and Cu ions accounted for 48.4% and 51.6% of the total Cu content, respectively.

Sperm quality analysis is considered to be a potential tool for use in the assessment of environmental quality and has been successfully applied as a biomarker to evaluate the effects of contaminants in the aquatic environment on fish reproduction [22,23]. Contaminants in aquatic environments could affect the fertilization efficiency of aquatic organisms [24], which can decrease spermatozoa motility and fertilization efficiency. Histopathological studies can be used as a dependable tool to investigate the toxic effects of environmental pollutants and to assess symptoms of damages [25]. Gürkan et al. [26] reported that 10 days of Fe_2O_3 NPs exposure caused significant histopathological damages to rainbow trout (*Oncorhynchus mykiss*). Antioxidant enzymes are known to be the first line of response upon stress, playing an important role in maintaining normal metabolism and function of the organisms [27]; they play an important role in maintaining the ROS homeostasis of organisms [28]. The classic antioxidant enzymes include superoxide dismutase (SOD), catalase (CAT), glutathione peroxidase (GPx), etc. They are very sensitive to environmental pollutants, and thus, they are usually employed as important biomarkers in numerous environmental toxicology experiments [29,30].

Previous studies suggested that contaminants could cause a potential risk to the fish species during their reproduction season when fertilization occurs [31]. Research on the effect of contamination on fish spermatozoa has mainly focused on the influence of the ionic form of metals. Few studies have performed risk assessment of toxicity of nanomaterials to male gamete motility. $CuSO_4$, CuNPs, and CuO NPs in the aqueous environment primarily cause a reduction of spermatozoa velocity, and the effect of Cu ions on sea trout spermatozoa motility was more harmful than that of copper nanoparticles [24]. Our current understanding of the toxic of CuNPs to the sperm in reptiles is severely limited as compared to the fish. Chinese soft-shelled turtles (*Pelodiscus sinensis*) are one of the common species of reptiles and are widely distributed in China, which is famous for their economic and pharmacologic value. Freshwater turtles play a major role in maintaining balance in the ecosystem of any wetland. They act as scavengers by decaying dead organic matter and indicators of a healthy aquatic ecosystem. Some of them are carrion eating species that feed on aquatic weeds and reduce eutrophication. As a representative of freshwater turtles, Chinese soft-shelled turtles first appeared 120 million years ago and have an important position in the study of species evolution. Our previous results have shown that the epididymis of the Chinese soft-shelled turtle is similar to that of mammals and is divided into three parts: caput, corpus, and cauda. The cauda epididymis is the main site of sperm storage [32]. The Chinese turtles' epididymis is unique in that sperm can be stored in its epididymis long-term with a low apoptosis rate. It can be seen that the epididymis plays an important role in the storage of sperm and the reproduction of species.

In this study, we aimed to assess the in vivo effects of CuNPs and $CuSO_4$ on spermatozoa viability, epididymal structure, antioxidant enzyme activity, and inflammatory cytokines in cauda epididymis of the Chinese soft-shelled turtle. This study may provide preliminary information regarding the ecological impacts of CuNPs on the fertilization efficiency of aquatic organisms.

2. Materials and Methods

2.1. Chemicals and the Test Organism

CuNPs with a primary size of 40 nm were purchased from Sigma (St. Louis, MO, USA), the particle size was determined using Transmission Electron Microscopy (TEM, JEOL Ltd., Tokyo, Japan), and the pristine shape of CuNPs used in the experiment was spherical (Supplementary Materials). The stock suspension of CuNPs (100 mg L^{-1}) was prepared by suspending the powder in MilliQ water. The suspension was sonicated (20 kHz, 100 w, 20% cycle off at 25 °C) in an ice-cold bath for at least 40 min and stirred for 30 min. Cu ions stock solution (100 mg L^{-1}) was prepared by dissolving $CuSO_4$ (99% purity, Sigma-Aldrich) in deionized water. The concentration of the products tested (CuNPs and $CuSO_4$) was expressed as mg Cu L^{-1} in the media regardless of the dissolution and aggregation of the products.

A total of 45 adult male Chinese soft-shelled turtles (body weight 700 ± 50 g) were chosen as model species of aquatic organism, which were purchased from Yancheng, Jiangsu Province, China. The healthy individuals with basically the same size, having no damage on the body surface and showing normal activities, were selected in the experiment. Animals were treated in accordance with the Care and Use of Wild Animals in The People's Republic of China. In the laboratory, turtles were reared in big water tanks (30 × 40 × 40-cm) containing sand and filled with freshwater during the 2021 breeding season for 7 days prior to the beginning of experiments.

Turtles were divided into five groups (Control, 5 mg kg^{-1}, and 50 mg kg^{-1} CuNPs or $CuSO_4$), and each treatment had three replicates. We calculated the volume of the required stock solution based on the body weight of the turtle; then, we set the volume to 20 mL and injected it into the abdominal cavity of the turtle. The control group was replaced with PBS (sterilized distilled water) solution. The animals were sacrificed by cervical dislocation at 24, 48, and 72 h after intraperitoneal injection. The cauda epididymis was separated according to the anatomical structure. One of each pair of cauda epididymis was used for light microscopy; the other was used to collect sperm to test its viability. Subsequently, the cauda epididymis with sperm removed was cryopreserved in liquid nitrogen for other experiments.

2.2. Light Microscopy

The samples were fixed in neutral buffered formalin, embedded in paraffin, and serial sectioned (at 5 μm). These sections were stained with Harris' hematoxylin and eosin (HE) for light microscopic observation. Slides were assessed using an Olympus microscope (BX53, Olympus Corporation, Tokyo, Japan), camera (Olympus DP73, Olympus Corporation, Tokyo, Japan), and the AnalySIS image-analyzing system (AnalySIS).

To observe the integrity of the epididymal epithelial structure, 20 pictures of straight perpendicularly cut epithelium sections were taken at 400× magnification. In each picture, the epithelium was measured 3 times.

2.3. Epididymal Sperm Preparation and Viability Test

The epididymal sample was placed in phosphate-buffered saline and exsanguinated for 5–10 min. After transferring to fresh PBS, the connective tissue around the epididymis was removed, and the sperm in the epididymal tubules flew into PBS for 10-15 min to ensure that all sperm are obtained as much as possible. Then, we transferred the PBS buffer containing sperm to a centrifuge tube and concentrated it by centrifugation at 700× *g* for

10 min at 4 °C. Sperm were suspended in sterile PBS. We diluted the sperm concentration to 10^6 mL^{-1}.

Sperm viability was quantified with double staining of fluorescein isothiocyanate (FITC) conjugated annexin-V and propidium iodide (PI) (Biouniquer, BU-AP0103, Biouniquer Technology, Nanjing, China). Ten thousand cells per sample were acquired with an FACS scan flow cytometer (FACScan, Becton Dickinson, Franklin Lakes, NJ, USA). Take 1 mL of the prepared sperm suspension, which was washed twice with binding buffer and processed following the manufacturer's instructions. Cells fluorescence was analyzed with flow cytometry using the Cell Quest Pro software (Beckman Coulter, Becton Dickinson, Franklin Lakes, NJ, USA).

2.4. Biochemical Assay

The epididymal tube sample from which the sperm has been removed was homogenized with a glass homogenizer in ice-cold conditions by adding Tris-HCl buffer and a small amount of protease inhibitor. Then, homogenates were centrifuged at 9000× *g* for 30 min at 4 °C. The supernatant was collected for measuring oxidative stress markers. SOD, CAT, and GSH activities were determined using the kits (Nanjing Jiancheng Bioengineering Institute, Nanjing, China) and according to the manufacturer's instructions.

2.5. Analysis of mRNA Expression Levels

Primers were designed using Primer 3 software (http://bioinfo.ut.ee/primer3/, accessed on 19 February 2021). The expression of IL-1β, TNF-α, and IL-6 mRNA in the cauda epididymis of Cu-treated soft-shelled turtles was examined using real-time PCR. Total RNA was isolated from tissue samples using the RNeasy Mini Kit Qiagen (Catalog no. 74104).

Real-time reverse transcription polymerase chain reaction (RT-PCR) was used to detect the mRNA expression levels of IL-1β, TNF-α, and IL-6. Total RNAs were extracted from epididymal epithelial cells by adding 1 mL of TRIzol Reagent (Invitrogen, Waltham, CA, USA) and reverse-transcribed to form cDNAs using the ABI7500 Real-Time PCR system (Applied Biosystems, Foster City, CA, USA) according to the manufacturer's instructions. β-actin was used as an endogenous quantity control. The primer sequences used in the quantitative real-time PCR reactions were as shown in Table 1. The total volume of 20 μL PCR reactions was prepared by mixing 1.0 μL of cDNA sample, 10.0 μL of 2× SYBR Premix Ex Taq II (Tli RNaseH Plus, Takara Bio Inc., Shiga, Japan), 0.4 L of forward primer (10 M), 0.4 L of reverse primer (10 M), and 8.2 μL of ddH_2O. The final results were presented as the ratios of the relative amount of the target gene to the control gene using the $2^{-\Delta\Delta CT}$ equation, where $\Delta\Delta CT$ = CT (target gene) − CT (internal control gene). The results were analyzed using the Rotor-Gene 6.0 software (Corbett Research, Mortlake, Australia). The results were comparable to those normalized to β-actin, and the statistical analysis was performed using the average relative mRNA levels from three independent samples.

Table 1. Primers and annealing temperature for real-time PCR.

Genes Primer Sequence (5′-3′)	Annealing (°C)	Length (bp)	GenBank Accession Number
psIL-1β Fwd GACAATGACTTGAGCAGCA psIL-1β Rev AGCCATGTTCAGCCTCACTT	60	226	JX846915
psTNF-α Fwd CAGAGAGCAGGCCCTATGAC psTNF-α Rev TTGTTCCCCTTGAAGACCAC	60	218	XM_006127997
psIL-6 Fwd GGAGATGCTTGCCCAAAAT psIL-6 Rev TGCACCAGGATTTTTCACC	60	244	XM_006138351
β-actin-Fwd TGTTACCCATACTGTGCCCATC β-actin-Rev TAGCCATCTCCTGTTCAAAATCC	60	212	EU727174

2.6. Statistical Analysis

All data were expressed as means ± SE. Significant differences were analyzed by one-way ANOVA followed by Tukey's test using SPSS (version 19.0). $p < 0.05$ was considered significant.

3. Results and Discussion

The widespread use of CuNPs inevitably results in their release into the aquatic environments; therefore, it is essential to evaluate their potential toxic effects on aquatic organisms. In our study, the response of the Chinese turtle to the external stimulus has gradually deteriorated, and the degree of injury was positively correlated with the increase in Cu concentration and time. At 50 mg kg^{-1}, although the animals still survive, their physiological state has been very poor; there was almost no response to external stimuli, as the head and neck remained outside with the treatment of 50 mg kg^{-1} $CuSO_4$ for 72 h. In addition, there were no changes in the weight and color of the turtles with the treatment of $CuSO_4$ or CuNPs for 72 h.

3.1. Effect of CuNPs and Cu Ions on Sperm Viability

Sperm analysis is considered to be useful for determination of the male infertility cause [33]. Some reports showed that Cu can negatively affect sperm quality parameters in rats [34] and rabbits [35] under acute exposure. In the present study, sperm viability was detected by the flow cytometry. Results showed that the treatment of CuNPs and $CuSO_4$ resulted in a dramatic decrease in sperm viability after 24 h of injection compared with the control (Figure 1B–E). The sperm viability of Chinese soft-shelled turtles decreased significantly with a further increase in CuNPs or Cu ion concentrations (Figure 1F). Sperm viability was nearly 18.2% and 40.4% for 50 and 5 mg kg^{-1} $CuSO_4$ after 72 h, respectively. Meanwhile, the viability was much higher with the treatment of 50 and 5 mg kg^{-1} CuNPs (Figure 1F), suggesting that Cu ions were more toxic to the sperm. The amount of soluble ions released from CuNPs is much lower than the concentration of corresponding $CuSO_4$. The toxicity of copper ions on the organisms was much higher than that of the nanoparticles, although nanoparticles could cause toxicity to the organisms. Kowalska-Góralska et al. [24] reported that CuNPs and CuO NPs exerted less harmful effects than Cu ions on spermatozoa motility of sea trout. In line with this, Adam et al. [16] illustrated that the toxicity of CuO NPs to aquatic organisms was mainly caused by the released Cu ion.

3.2. Effect of CuNPs and Cu Ions on the Histological Analysis

Compared with the control, the epithelial cells of the epididymal duct of the Chinese soft-shelled turtle with the treatment of 5 mg kg^{-1} CuNPs were slightly swollen, and the connective tissue between the epididymal ducts was slightly loose (Figure 2A); meanwhile, we found intravascular congestion, and a small amount of blood appeared in the lumen blood cells in the epididymis of Chinese soft-shelled turtle with the treatment of 50 mg kg^{-1} CuNPs; however, the overall structure of the epididymal duct is basically intact (Figure 2B). Part of the epithelium of the epididymal tube was peeled off, and the connective tissue between the epididymal tubes was very loose in the epididymis with the treatment of 5 mg kg^{-1} $CuSO_4$ (Figure 2C,D). The epithelial structure of the epididymal tube was severely damaged with an increase in Cu ion concentrations (from 5 to 50 mg kg^{-1}), cell connections were broken, and sperm in the cavity may flow to the knot tissue through the rupture (Figure 2E).

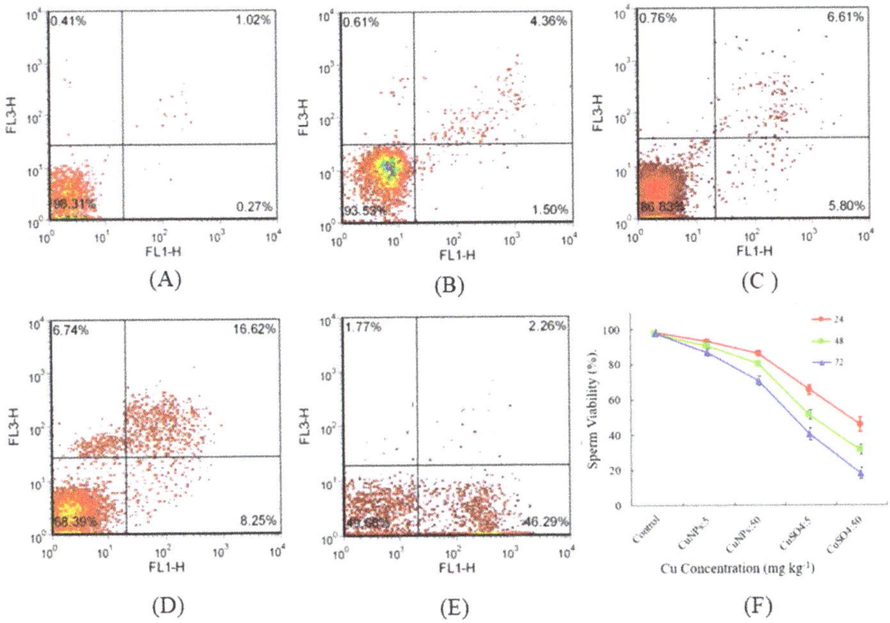

Figure 1. Flow cytometry analysis using Annexin V and propidium iodide (PI) staining illustrating sperm viability at 24 h. Note: (**A**) Control; (**B**) CuNPs 5 mg kg^{-1}; (**C**) CuNPs 50 mg kg^{-1}; (**D**) $CuSO_4$ 5 mg kg^{-1}; (**E**) $CuSO_4$ 50 mg kg^{-1}; (**F**) sperm viability.

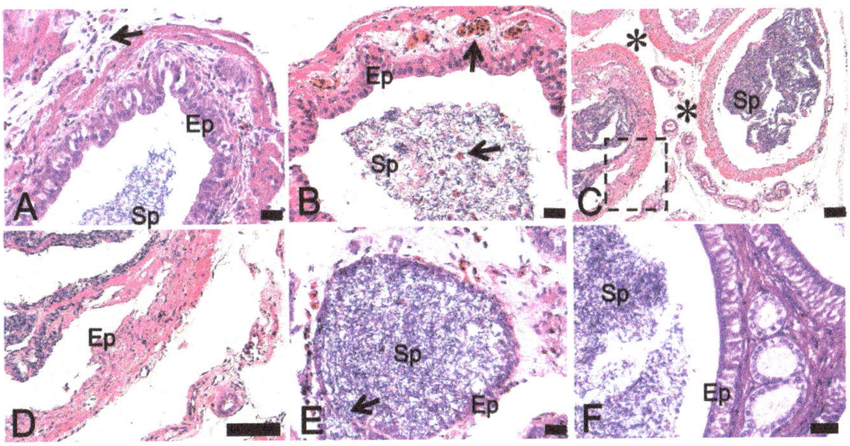

Figure 2. Photomicrograph of cauda epididymis in Chinese soft-shelled turtle with different treatment. Note: (**A**) Control; (**A**) CuNPs 5 mg kg^{-1}; (**B**) CuNPs 50 mg kg^{-1}; (**C**) $CuSO_4$ 5 mg kg^{-1}; (**D**) Higher magnification of the boxed area in (**C**); (**E**) $CuSO_4$ 50 mg kg^{-1}; (**F**) Control. The arrow in (**A**) shows connective tissue, the arrow in (**B**) shows blood cells, the asterisk in C also shows connective tissue, and the arrow in (**E**) shows sperm outflow. Ep, epithelium; Sp, Sperm. Scale bar: (**A**,**B**,**E**) = 20 μm; (**C**,**D**) = 100 μm; (**F**) = 50 μm.

The plasma membrane is an important structure for maintaining a highly controlled intracellular environment that enables biochemical reactions and enzymes to function at optimum rates; it also plays an important role in sending and receiving cellular signals, and it takes part in the process of nutrient absorption [36]. With this in mind, the reduction in spermatic plasma membrane integrity may be directly associated with cellular damage. It is not difficult to conclude that exposure to CuNPs and Cu ions have different effects on the structure of epididymis, although so far, there are few reports on the damage of copper nanoparticles on the epididymis structure of Chinese soft-shelled turtles.

3.3. Effect of CuNPs and Cu Ions on the Antioxidant Enzyme Activity

NPs can cause oxidative damage to the aquatic tissue; they would induce the generation of reactive oxygen species (ROS) and oxidative damage. To lessen the damage, organisms have developed a detoxification system with antioxidant enzyme components, such as SOD, CAT, and GSH, which protect cells against oxidative damage and remove superoxide radicals and hydrogen peroxide [37]. In this study, the antioxidant enzymes activities were determined in the epididymis of Chinese soft-shelled turtles with the treatment of 5 and 50 mg kg^{-1} CuNPs and $CuSO_4$. The results showed that the injection of CuNPs and $CuSO_4$ resulted in a massive ROS production, and it increased with the increasing concentration of both CuNPs and $CuSO_4$ (Figure 3A). The ROS level was not significantly affected by time, except for 50 mg kg^{-1} $CuSO_4$. ROS decreased at 48 h and 72 h compared to 24 h (Figure 3A). This may be due to the epididymal epithelial necrosis rate, which increased with the treatment of $CuSO_4$ for time. SOD, CAT, and GSH activities significantly increased with the treatment of both CuNPs and $CuSO_4$ (Figure 3B,C), as the SOD-CAT system is the first line to defend against oxidative stress [38]. Compared to the control, SOD activity even with a low dose (5 mg kg^{-1}) of CuNPs and $CuSO_4$ significantly increased ($p < 0.05$). It increased by 37.6% and 197.0% after 72 h injection with 5 mg kg^{-1} CuNPs and $CuSO_4$, respectively. There was no significant difference of antioxidant enzymes activities at different times, except for the treatment of 50 mg kg^{-1} $CuSO_4$ (Figure 3). SOD, CAT, and GSH activities in the epididymis of the Chinese soft-shelled turtle with the treatment of 50 mg kg^{-1} $CuSO_4$ significantly increased; however, they significantly decreased with time. This may be explained by the broken epithelial cells with the treatment of 50 mg kg^{-1} $CuSO_4$ (Figure 3). The ROS levels and CAT activity decreased significantly at 48 h and 72 h. Combined with histological results, it can be seen that with time, the epididymal epithelial necrosis rate increased, resulting in a decrease in CAT activity. According to the related literature [39], ROS plays an important role in the process of removing necrotic cells. We speculate that this may be one of the important reasons for the decline in ROS levels. GSH is involved in cell protection by providing reducing capacity through the transformation of reduced to oxidized GSH and binding to metals during the detoxification process [30]. These results showed that the antioxidant capacity was significantly impaired in the epididymis due to massive ROS production, which resulted in oxidative stress. The accumulation of ROS in sperm leads to a reduction in sperm motility, viability, and enzymatic activity [40]. The accumulation of ROS to high concentrations leads to oxidative stress and subsequent lipid peroxidation (LPO), which provokes an irreversible loss of motility and inhibition of fructolysis and respiration in spermatozoa [41]. This could be an explanation of the decrease in sperm motility, viability, and enzymatic activity. In addition, sperm motility is particularly associated with mitochondria activity, as mitochondria represent the cell energy generator and concomitantly the major site of intracellular ROS formation, resulting in a disruption of electron transport. This disruption also participates in the decline of motility and amplifies LPO in the spermatozoa membrane.

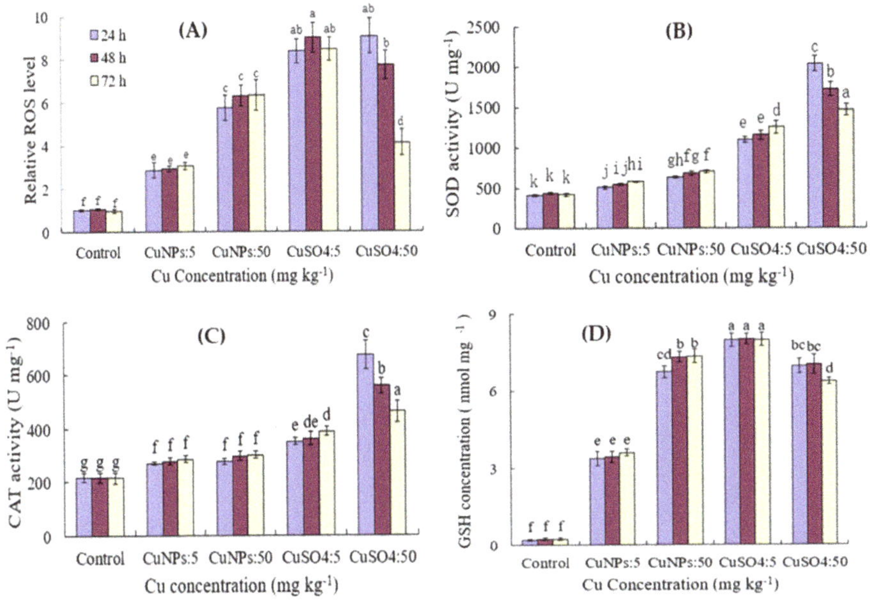

Figure 3. ROS and antioxidant enzyme activities with the treatment of CuNPs and $CuSO_4$. (**A–D**) represent the level of ROS and the activity of antioxidant enzymes SOD, CAT, and GSH, respectively. Different letters at the top of the histogram indicate significant differences between groups ($p < 0.05$).

3.4. Effect of CuNPs and Cu Ions on the Relative mRNA Expression

The expression of IL-1β, TNF-α, and IL-6 mRNA in the cauda epididymis of the Chinese soft-shelled turtle treated with CuNPs and $CuSO_4$ was detected by qPCR. The mRNA expression of the three inflammatory factors significantly increased with the treatment of CuNPs and $CuSO_4$ (Figure 4). They increased with the increase in the CuNPs concentrations and time (Figure 4). Different from CuNPs treatment, the relative expression of IL-1β and IL-6 mRNA caused by 5 mg kg^{-1} $CuSO_4$ treatment increased progressively with time, peaking at 48 h after injection. However, the injection of 50 mg kg^{-1} $CuSO_4$ resulted in a significant decrease in the relative expression of IL-1β and IL-6 mRNA with time (Figure 4). IL-1β and TNFα are well-known pro-inflammatory cytokines and are generally considered to be markers to assess the inflammatory response when there is an immune stimulus in vertebrates [41]. Compelling evidence suggests that the two cytokines have similar functions to their mammalian homologues [42,43]. Our results showed that IL-1β and TNFα expression were greatly enhanced after CuNPs and $CuSO_4$ treatment, which suggests that the inflammatory response to the epididymis of the Chinese soft-shelled turtle significantly increased after the stress of CuNPs and $CuSO_4$. The difference of CuNPs and $CuSO_4$ in mRNA expression of inflammatory factors may be caused by the amount of soluble ions released from CuNPs being much lower than the concentration of corresponding $CuSO_4$. The cytokine IL-6 possesses pro-inflammatory activity, and it is mainly produced by T cells and macrophages. In reptiles, there is no report on IL-6 expression after CuNPs treatment. Our results show that CuNPs and $CuSO_4$ treatment can also cause significant expression of the inflammatory factor IL-6 (Figure 4), but the underlying mechanism needs to be further studied.

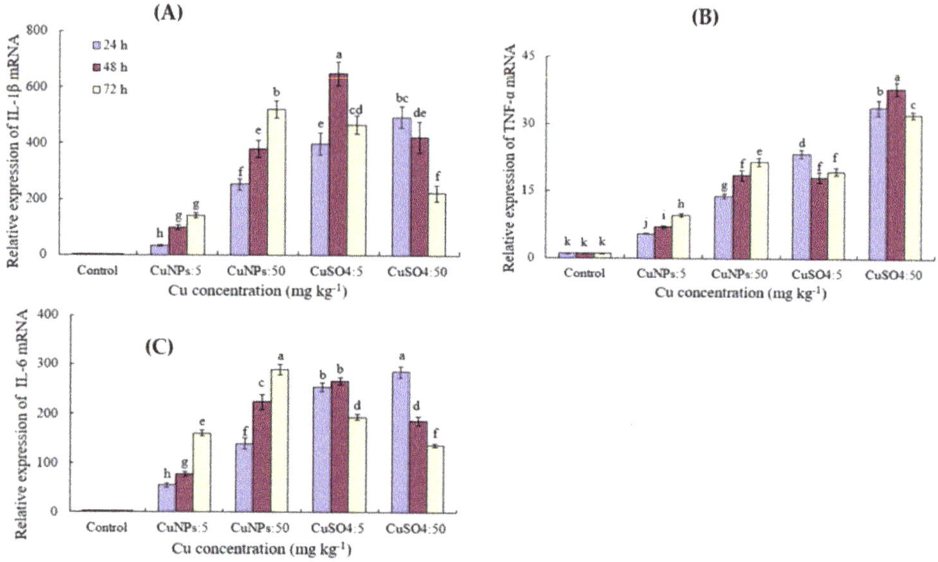

Figure 4. The relative inflammatory factors mRNA expression with the treatment of CuNPs and $CuSO_4$. (**A–C**) represent the relative expression levels of IL-1β, TNFα, and IL-6 mRNA, respectively. Different letters at the top of the histogram indicate significant differences between groups ($p < 0.05$).

4. Conclusions

This study showed that Cu ions exert a more harmful effect on the epididymis and spermatozoa viability of Chinese soft-shelled turtles than copper nanoparticles. Contamination of the aqueous environment with copper primarily caused a reduction in sperm viability, which may impair reproduction. The antioxidative enzymes and the relative inflammatory factors (IL-1β, TNFα, and IL-6 mRNA) expression in the epididymis of Chinese soft-shelled turtles significantly increased with the treatment of CuNPs and $CuSO_4$. The exposure to $CuSO_4$ caused obvious morphological damage to the epididymis of Chinese soft-shelled turtles, such as, the epithelium of the epididymal tube peeled off. These findings in this work may provide preliminary information regarding the ecological impacts of CuNPs on the fertilization efficiency of aquatic organisms.

Supplementary Materials: The following supporting information can be downloaded at: https://www.mdpi.com/article/10.3390/coatings12020110/s1, Figure S1: Representative TEM image (A) and particle size distribution (B) of CuNPs in Milli-Q water.

Author Contributions: L.Y. Conceptualization, Methodology, Writing—Original draft preparation; Y.W. Software; S.G. Data curation; Q.W. Investigation; J.C. Validation; B.T. Resources; X.B. Editing, Visualization. All authors have read and agreed to the published version of the manuscript.

Funding: This research was funded by the sponsorship of Jiangsu overseas visiting scholar program for university prominent young and middle-aged teachers and presidents and the natural science foundation of the Jiangsu higher education institutions of China (20KJA180006), innovation and entrepreneurship training program for college students in Jiangsu Province (202110324018Z), (202110324022Y), and foundation of Jiangsu Key Laboratory for Bioresources of Saline Soils (JKLBS2021007).

Institutional Review Board Statement: Not applicable.

Informed Consent Statement: Not applicable.

Data Availability Statement: Data are contained within the article or Supplementary Materials.

Conflicts of Interest: The authors declare no conflict of interest.

References

1. Hou, J.; Wang, X.X.; Hayat, T.; Wang, X.K. Ecotoxicological effects and mechanism of CuO nanoparticles to individual organisms. *Environ. Pollut.* **2017**, *221*, 209–217. [CrossRef] [PubMed]
2. Yang, L.; Wang, W.X. Comparative contributions of copper nanoparticles and ions to copper bioaccumulation and toxicity in barnacle larvae. *Environ. Pollut.* **2019**, *249*, 116–124. [CrossRef]
3. Adeleye, A.S.; Oranu, E.A.; Tao, M.; Keller, A.A. Release and detection of nanosized copper from a commercial antifouling paint. *Water Res.* **2016**, *102*, 374–382. [CrossRef] [PubMed]
4. Muller-Karanassos, C.; Arundel, W.; Lindeque, P.K.; Vance, T.; Turner, A.; Cole, M. Environmental concentrations of antifouling paint particles are toxic to sediment dwelling invertebrates. *Environ. Pollut.* **2020**, *268*, 115754. [CrossRef]
5. Wu, F.; Harper, B.J.; Crandon, L.E.; Harper, S.L. Assessment of cu and CuO nanoparticle ecological responses using laboratory small-scale microcosms. *Environ. Sci. Nano* **2020**, *7*, 105–115. [CrossRef]
6. Bashir, I.; Lone, F.A.; Bhat, R.A.; Mir, S.A.; Dar, Z.A.; Dar, S.A. Concerns and threats of contamination on aquatic ecosystems. In *Bioremediation and Biotechnology*; Springer: Cham, Switzerland, 2020; pp. 1–26.
7. Zhu, Y.C.; Xu, J.H.; Lu, T.; Zhang, M.; Ke, M.J.; Fu, Z.W.; Pan, X.L.; Qian, H.F. A comparison of the effects of Cu nanoparticles and Cu sulfate on *Phaeodactylum tricornutum* physiology and transcription. *Environ. Toxicol. Pharmacol.* **2017**, *56*, 43–49. [CrossRef] [PubMed]
8. Fang, R.; Gong, J.L.; Cao, W.C.; Chen, Z.P.; Huang, D.L.; Ye, J.; Cai, Z. The combined toxicity and mechanism of multi-walled carbon nanotubes and nano copper oxide toward freshwater algae: *Tetradesmus obliquus*. *J. Environ. Sci.* **2022**, *112*, 376–387. [CrossRef] [PubMed]
9. Zhang, C.; Li, F.; Xiang, J. Acute effects of cadmium and copper on survival, oxygen consumption, ammonia-N excretion, and metal accumulation in juvenile *Exopalaemon carinicauda*. *Ecotoxicol. Environ. Saf.* **2014**, *104*, 209–214. [CrossRef]
10. Bat, L.; Gündoğdu, A.; Sezgin, M.; Çulha, M.; Gönlügür, G.; Akbulut, M. Acute toxicity of zinc, copper and lead to three species of marine organisms from the Sinop Peninsula, Black Sea. *Turk. J. Biol.* **1999**, *23*, 537–544.
11. Vardhanan, Y.S.; Radhakrishnan, T. Acute toxicity evaluation of copper, arsenic and HCH to paddy field crab, *Paratelphusa hydrodromus* (Herb). *J. Environ. Biol.* **2002**, *23*, 387–392. [PubMed]
12. Chourpagar, A.R.; Kulkarni, G.K. Heavy metal toxicity to a freshwater crab, *Barytelphusa cunicularis* (westwood) from aurangabad region. *Recent Res. Sci. Technol.* **2011**, *3*, 1–5.
13. Bini, G.; Chelazzi, G. Acclimatable cardiac and ventilatory responses to copper in the freshwater crayfish *Procambarus clarkii*. *Comp. Biochem. Physiol. C Toxicol. Pharmacol.* **2006**, *144*, 235–241. [CrossRef]
14. Malhotra, N.; Ger, T.R.; Uapipatanakul, B.; Huang, J.C.; Chen, K.H.; Hsiao, C.D. Review of Copper and Copper Nanoparticle Toxicity in Fish. *Nanomaterials* **2020**, *10*, 1126. [CrossRef] [PubMed]
15. Boyle, D.; Clark, N.J.; Handy, R.D. Toxicities of copper oxide nanomaterial and copper sulphate in early life stage zebrafish: Effects of pH and intermittent pulse exposure. *Ecotoxicol. Environ. Saf.* **2020**, *190*, 109985. [CrossRef]
16. Adam, N.; Schmitt, C.; De Bruyn, L.; Knapen, D.; Blust, R. Aquatic acute species sensitivity distributions of ZnO and CuO nanoparticles. *Sci. Total Environ.* **2015**, *526*, 233–242. [CrossRef]
17. Jiang, C.J.; Castellon, B.T.; Matson, C.W.; Aiken, G.R.; Hsu-Kim, H. Relative contributions of Cu oxide nanoparticles and dissolved Cu to Cu uptake kinetics of gulf killifish (*Fundulus grandis*) embryos. *Environ. Sci. Technol.* **2017**, *51*, 1395–1404. [CrossRef] [PubMed]
18. Hua, J.; Vijver, M.G.; Richardson, M.K.; Ahmad, F.; Peijnenburg, W.J. Particlespecific toxic effects of differently shaped zinc oxide nanoparticles to zebrafish embryos (*Danio rerio*). *Environ. Toxicol. Chem.* **2014**, *33*, 2859–2868. [CrossRef] [PubMed]
19. Wang, N.; Liu, D.; Xie, M.W.; Li, Q.B.; Liu, Q.M. Behavior and toxicity of zinc oxide nanoparticles in aquatic environment. *Environ. Chem.* **2016**, *35*, 2528–2534.
20. Volland, M.; Hampel, M.; Katsumiti, A.; Yeste, M.P.; Gatica, J.M.; Cajaraville, M.; Blasco, J. Synthesis methods influence characteristics, behaviour and toxicity of bare CuO NPs compared to bulk CuO and ionic Cu after in vitro exposure of *Ruditapes philippinarum* hemocytes. *Aquat. Toxicol.* **2018**, *199*, 285–295. [CrossRef] [PubMed]
21. Zhang, Y.J.; Ding, Z.C.; Zhao, G.; Zhang, T.; Xu, Q.H.; Cui, B.; Liu, J.X. Transcriptional responses and mechanisms of Cu nanoparticle toxicology on zebrafish embryos. *J. Hazard Mater.* **2018**, *344*, 1057–1068. [CrossRef]
22. Lopes, F.M.; Varela Junior, A.S.; Corcini, C.D.; Da Silva, A.C.; Guazzelli, V.G.; Tavares, G.; Da Rosa, C.E. Effect of glyphosate on the sperm quality of zebrafish *Danio rerio*. *Aquat. Toxicol.* **2014**, *155*, 322–326. [CrossRef] [PubMed]
23. Devaux, A.; Bony, S.; Plenet, S.; Sagnes, P.; Segura, S.; Suaire, R.; Novack, M.; Gilles, A.; Olivier, J.M. Field evidence of reproduction impairment through sperm DNA damage in the fish nase (*Chondrostoma nasus*) in anthropized hydrosystems. *Aquat. Toxicol.* **2015**, *169*, 113–122. [CrossRef] [PubMed]
24. Kowalska-Góralska, M.; Dziewulsk, K.; Kulasz, M. Effect of copper nanoparticles and ions on spermatozoa motility of sea trout (*Salmo trutta m. Trutta* L.). *Aquat. Toxicol.* **2019**, *211*, 11–17. [CrossRef] [PubMed]
25. Mansouri, B.; Maleki, A.; Johari, S.A.; Shahmoradi, B.; Mohammadi, E.; Davari, B. Histopathological effects of copper oxide nanoparticles on the gill and intestine of common carp (*Cyprinus carpio*) in the presence of titanium dioxide nanoparticles. *Chem. Ecol.* **2017**, *33*, 295–308. [CrossRef]
26. Gürkan, M.; Gürkan, S.E.; Yılmaz, S.; Ate, M. Comparative toxicity of Alpha and Gamma iron oxide nanoparticles in Rainbow Trout: Histopathology, hematology, accumulation, and oxidative stress. *Water Air Soil Pollut.* **2021**, *232*, 37–52. [CrossRef]

27. Kim, E.H.; Jeong, J.A.; Choi, E.K.; Jeong, T.Y. Antioxidant enzyme activity in *Daphnia magna* under microscopic observation and shed carapace length as an alternative growth endpoint. *Sci. Total Environ.* **2021**, *794*, 148771. [CrossRef]
28. Chen, B.J.; Zhang, W.Y.; Niu, C.J.; Li, W.J.; Jia, H.; Storey, K.B. Antioxidant response to acute cold exposure and during recovery in juvenile Chinese soft-shelled turtles (*Pelodiscus sinensis*). *J. Exp. Biol.* **2019**, *222*, 197863. [CrossRef]
29. Sanpradit, P.; Buapet, P.; Kongseng, S.; Peerakietkhajorn, S. Temperature and concentration of ZnO particles affect life history traits and oxidative stress in *Daphnia magna*. *Aquat. Toxicol.* **2020**, *224*, 105517. [CrossRef]
30. Zhang, W.J.; Gao, J.L.; Lu, L.; Bold, T.; Li, X.; Wang, S.; Chang, Z.S.; Chen, J.; Kong, X.; Zheng, Y.X.; et al. Intracellular GSH/GST antioxidants system change as an earlier biomarker for toxicity evaluation of iron oxide nanoparticles. *NanoImpact* **2021**, *23*, 100338. [CrossRef]
31. Zarski, D.; Cejko, B.I.; Krejszeff, S.; Palinska-Zarska, K.; Horvath, A.; Sarosiek, B.; Judycka, S.; Kowalski, R.K.; Laczynska, B.; Kucharczyk, D. The effect of osmolality on egg fertilization in common carp, *Cyprinus carpio* Linnaeus, 1758. *J. Appl. Ichthyol.* **2015**, *31*, 159–163. [CrossRef]
32. Bian, X.G.; Gandahi, J.A.; Liu, Y.; Yang, P.; Liu, Y.; Zhang, L.L.; Zhang, Q.; Chen, Q.S. The ultrastructural characteristics of the spermatozoa stored in the cauda epididymidis in Chinese soft-shelled turtle *Pelodiscus sinensis* during the breeding season. *Micron* **2013**, *44*, 202–209. [CrossRef]
33. Zebral, Y.D.; Anni, I.S.A.; Junior, A.S.V.; Corcini, C.D.; Silva, J.C.; Caldas, J.S.; Acosta, I.B.; Afonso, S.B.; Bianchini, A. Life-time exposure to waterborne copper IV: Sperm quality parameters are negatively affected in the killifish *Poecilia vivipara*. *Chemosphere* **2019**, *236*, 124332. [CrossRef] [PubMed]
34. Sakhaee, E.; Emadi, L.; Abshenas, J.; Kheirandish, R.; Azari, O.; Amiri, E. Evaluation of epididymal sperm quality following experimentally induced copper poisoning in male rats. *Andrologia* **2012**, *44*, 110–116. [CrossRef] [PubMed]
35. Roychoudhury, S.; Massanyi, P. In vitro copper inhibition of the rabbit spermatozoa motility. *J. Environ. Sci. Health Part A* **2008**, *43*, 651–656. [CrossRef]
36. Zhang, Y.; Chen, X.; Gueydan, C.; Han, J. Plasma membrane changes during programmed cell deaths. *Cell Res.* **2018**, *28*, 9. [CrossRef]
37. Yuan, J.L.; Gu, Z.M.; Zheng, Y.; Zhang, Y.Y.; Gao, J.G.; Chen, S.; Wang, Z.Z. Accumulation and detoxification dynamics of microcystin-LR and antioxidant responses in male red swamp crayfish *Procambarus clarkii*. *Aquat. Toxicol.* **2016**, *177*, 8–18. [CrossRef]
38. Mahmoud, A.E.; Mahmoud, M.A.D.; Amaal, M.; Radwa, M.S. Effects of titanium dioxide nanoparticles on red swamp crayfish, *Procambarus clarkii*: Bioaccumulation, oxidative stress and histopathological biomarkers. *Egypt. J. Aquat. Res.* **2019**, *45*, 11–18.
39. Bagaitkar, J.; Huang, J.; Zeng, M.Y.; Pech, N.K.; Monlish, D.A.; Perez-Zapata, L.J.; Miralda, I.; Schuettpelz, L.G.; Dinauer, M.C. NADPH oxidase activation regulates apoptotic neutrophil clearance by murine macrophages. *Blood* **2018**, *131*, 2367–2378. [CrossRef]
40. Mostek, A.; Słowińska, M.; Judycka, S.; Karol, H.; Ciereszko, A.; Dietrich, M.A. Identification of oxidatively modified proteins due to cryopreservation of carp semen. *J. Anim. Sci.* **2018**, *96*, 1453–1465. [CrossRef]
41. Varfolomeev, E.E.; Ashkenazi, A. Tumor Necrosis Factor. *Cell* **2004**, *116*, 491–497. [CrossRef]
42. Reis, M.I.R.; Vale, A.D.; Pereira, P.J.B.; Azevedo, J.E.; dos Santo, N.M.S. Caspase-1 and IL-1β Processing in a Teleost Fish. *PLoS ONE* **2012**, *7*, 50450.
43. Zhang, A.; Chen, D.; Wei, H.; Du, L.; Zhao, T.; Wang, X.; Hong, Z. Functional characterization of TNF-α in grass carp head kidney leukocytes: Induction and involvement in the regulation of NF-κB signaling. *Fish Shellfish Immun.* **2012**, *33*, 1123–1132. [CrossRef]

coatings

MDPI

Article

The Application of Carotenoid-Coated Chitosan Nanoparticles to Reduce the PAHs Stress on Spinach Growth

Jin Zhang [1,2,†], Menghan Cui [1,2,†], Ran Tao [1,2], Yifan Yao [1], Jiangang Han [1,2,*] and Yu Shen [1,2]

1 Co-Innovation Center for Sustainable Forestry in Southern China, College of Biology and the Environment, Nanjing Forestry University, Nanjing 210037, China; 18262397885@163.com (J.Z.); mhancui@163.com (M.C.); nature19980626@gmail.com (R.T.); ivanyyfan@163.com (Y.Y.); sheyttmax@hotmail.com (Y.S.)
2 National Positioning Observation Station of Hung-tse Lake Wetland Ecosystem in Jiangsu Province, Nanjing 210037, China
* Correspondence: jianganghan310@outlook.com
† These authors contributed equally to this work.

Abstract: Polycyclic aromatic hydrocarbons (PAHs) pose risks to human and animal health, and their accumulation in crops is a concern for the food chain in the environment. Nanoparticles (NPs) have shown potential for chemical delivery and can be used to enhance plant resistance to PAHs. In this study, carotenoid-coated chitosan nanoparticles (CCNPs) loaded with β-carotene were prepared and applied to spinach grown in PAH-contaminated soil. The size of the CCNPs varied based on reaction conditions with temperature, TPP, and pH, with sizes ranging from 260 to 682 nm. After four weeks of treatment, the spinach showed varying growth responses depending on the specific CCNP treatment. The treatment with CCNPs prepared at 20 °C, pH 6, and 10 mg/mL TPP resulted in the best spinach growth, while the treatment at 40 °C, pH 6, and a TPP concentration of 20 mg/mL hindered growth; and the growth ration increased by over 47.4% compared to the normal growing spinach, the final biomass reached 2.53 g per plant. In addition, phenanthrene (PHE) and pyrene (PYR) predominantly accumulated more in the spinach roots, with variations depending on the specific CCNP treatment. The exogenous application of CCNPs can reduce the PAH transfer to the shoots. The bioconcentration factors and transfer factors of PYR and PHE reduced differential movement within the spinach plants, and the spinach prefers PYR to PHE in biological accumulation. This study offers a new understanding of the mechanisms underlying NPs and PAHs interactions and NP's implications for crop protection and food safety.

Keywords: chitosan nanoparticles; coatings; carotenoid; PAHs; spinach

Citation: Zhang, J.; Cui, M.; Tao, R.; Yao, Y.; Han, J.; Shen, Y. The Application of Carotenoid-Coated Chitosan Nanoparticles to Reduce the PAHs Stress on Spinach Growth. *Coatings* **2023**, *13*, 1404. https://doi.org/10.3390/coatings13081404

Academic Editor: M. Shaheer Akhtar

Received: 18 July 2023
Revised: 7 August 2023
Accepted: 8 August 2023
Published: 10 August 2023

1. Introduction

Polycyclic aromatic hydrocarbons (PAHs) are a group of organic compounds composed of multiple fused aromatic rings [1]. They are formed during the incomplete combustion or pyrolysis of organic materials naturally in coal, crude oil, gasoline, etc. [2]. PAHs are ubiquitous in the environment and can be found in the air, water, soil, and various food sources, including agricultural products [3,4]. They are known to be persistent in the environment, bioaccumulate in organisms, and pose potential health risks to both humans and animals [5]. Thus, it is important to reduce the bioaccumulation of PAHs in crops.

PAHs can have detrimental effects on agricultural ecosystems [6]; PAHs can be absorbed by plant roots from contaminated soil or water and accumulate in plant tissues [7,8], leading to a decrease in the biomasses of both roots (21.0%–42.7%) and leaves (6.4%–22.1%) grown in PAH-contaminated soil [9]. If PAH-contaminated soil is present, plants growing in the soil can uptake PAHs through their root system [10]; it was found that the root can transfer PAHs through root pressure and transpiration from the root to the shoot [11], and the root xylem plays an important role in the PAHs transfer [12]. Moreover, it was reported that the PAHs accumulation in roots can reach 203 ng g^{-1}, which is two times higher than that in the shoot tissue in crops [13]; and the cell wall fraction of the root is the major part of over 45% of PAHs adsorption at the subcellular level [14]. How to control and reduce the adsorption of PAHs from the soil is a breakthrough for crop protection in PAH-contaminated environments.

Nanoparticles (NPs) have emerged as valuable tools for chemical delivery due to their unique properties and potential applications [15,16]. The NPs size, from 1 to 100 nanometers, offers several advantages for delivering chemicals efficiently and effectively, and this technology holds the promise of the controlled release of agrochemicals and site-targeted delivery of various macromolecules needed for improved plant disease resistance, efficient nutrient utilization, and enhanced plant growth [17]. For example, a kind of poly(lactic-co-glycolic) acid NP was designed for the membrane of neural stem cell delivery in the ischemic brain and established a novel formulation of glyburide [18]; the 400.57 nm chitosan nanoparticles (CNPs) coated with hyaluronic-acid can be used for the delivery of dexamethasone for patients, and the final loaded ration can reach around 72.95% and 14.51% those are very useful for medicinal use [19]. NPs dual-coated with chitosan and albumin allowed sustained insulin release following their passage to simulated intestinal conditions [20]. As described, chitosan NPs present the potential to overcome the barriers to the oral delivery of protein drugs, leading to the development of platforms capable of improving their bioavailability [21].

Chitosan (chitin) is an environmentally friendly material that comes from the outer skeleton of *Crustacea* [22]. In addition, chitosan has been established as a non-toxic, biodegradable, and biocompatible compound, as recognized by the United States Food and Drug Association (US FDA) [23]. Furthermore, chitosan production offers an environmentally sustainable solution by utilizing bio-waste generated from the crustacean production industries. Globally, chitosan production amounts to approximately 6–8 million tons per year, with 1.5 million tons produced by Southeast Asian countries [24]. This approach contributes to a "zero-waste" food industry, benefiting both the economy and the environment [25]. By repurposing these by-products, chitosan serves as a valuable resource in various applications, including the synthesis of CCNPs, and underscores the potential of eco-friendly practices in fostering a more sustainable future. Ionic gelation is the most commonly used method for synthesizing CNPs, and this kind of polymeric nanoparticles has gained significant importance as they are biodegradable, biocompatible, and because formulation methods are more widely available with a large surface area-to-volume ratio [26]. It was found that chitosan-coated mesoporous silica nanoparticles at the seedling stage led to a 70% increase in the fruit yield of uninfected watermelon because of their high surface area [27]. A carrier system for paraquat using polymeric nanoparticles composed of chitosan/TPP can make paraquat less toxic and, therefore, allows safer control of weeds in agriculture due to its controlled release [28]. The increased surface area allows for greater interaction between the nanoparticles and the target chemicals, enabling efficient loading, encapsulation, and controlled release. Thus, it is possible to figure out a kind of chemical to increase PAHs resistance in plants from root absorption.

In a previous study, it was found that β-carotene is a kind of antioxidant performing a positive role in ROS scavenging when the plants were treated with PAHs [29], and the root application of β-carotene could significantly increase wheat's resistance to the PAHs [30]. Indeed, the direct application of β-carotene is challenging due to its sensitivity to heat, light, and oxidation [31]. In this study, we addressed this issue by designing CNPs capable of encapsulating β-carotene for use in PAH-contaminated soil. The small size of these nanoparticles not only minimizes systemic toxicity but also enhances their bioavailability, making them more effective in improving the PAH resistance of crops. By entrapping β-carotene within CNPs, we provide a protective shield, safeguarding it from degradation caused by environmental factors. This innovative approach offers a promising solution for utilizing β-carotene's benefits in enhancing crop resilience under PAH contamination, contributing to sustainable agricultural practices. We selected the most common vegetable, spinach (*Spinacia oleracea*), for the study, which is a leafy, green vegetable that originated in Persia; it is considered healthy, as it is loaded with nutrients and antioxidants [32]. In addition, PAHs have stronger accumulation in leafy plants [33]. Therefore, with the best carrier and release performance of CNPs, β-carotene is expected to show its role in improving PAH resistance and protecting crop growth under a PAH-contaminated environment. Our study is important for PAH-contamination management in agricultural systems and minimizing the risk of plant uptake. Additionally, we study implementing good agricultural practices and avoiding the mitigation of PAH absorption by plants.

2. Material and Methods

2.1. Carotenoid-Coated Chitosan Nanoparticles (CCNPs) Preparation

The CNPs were prepared by ionotropic gelation with some modifications. First, 50 mL of a solution of chitosan (10 mg/mL; pH 4.0; 27 kDa; 75%–85% deacetylation), prepared in an aqueous solution of 1% acetic acid, was kept under vigorous stirring on a magnetic intelligent color display heating stirrer (TP-350$^+$; MIULAB Co., Ltd., Hangzhou, China). The reaction formula is presented below,

$$\text{C-NH}_3{}^+ - \text{Ac}^- + \text{Na}^+(\text{TPP} - \text{PO}^-) \rightarrow \text{C-NH}_3{}^+(\text{TPP} - \text{PO}^-) + \text{NaAC}$$

After the chitosan was dissolved totally, the tripolyphosphate (TPP) solution was added to the chitosan solution at the ratio of 1:5 (w/w), and the mixed solution was stirred for one hour at 1000 r/min to obtain chitosan nanoparticles (Figure 1a). In the preparation, two temperature levels (20 and 40 °C), two TPP concentrations (10 and 20 mg/mL), and two pH levels (5 and 6) were selected for the CCNP preparation (Table 1). Then the following, CCNPs were prepared with the CNPs and β-carotene, which were added to the CNP solution at a 1:2 (w/w) ratio with the ultrasonic of 40 KHz (KH-250E; Kunshan Hechuang Ultrasonic Instruments Co., Ltd., Shanghai, China) for 30 min. Then, the CCNP solution was filtered by the dialyzer (HCA000808; Union Carbide Co., Houston, TX, USA) at 10 kDa. Then, the CCNP solution was stored in the 4 °C fridge for the root ball application. The prepared CNPs and CCNPs were also characterized using transmission electron microscopy (TEM, JEM-2100 UHR, JEOL, Akishima, Japan). The NP solutions were dispersed properly with water, and a mixed nanoparticle solution was prepared by sonicating the solution for 10 min. We took a carbon-coated grid on a Whatman paper and added a 10 µL drop onto a grid using a micropipette for natural drying. The NP images are shown in Figure 1b,c.

2.2. Greenhouse Experiment

The spinach seeds used were commercial Changfeng seeds (Aishen Vegetable Seed Breeding Center, Qingxiang, China). The seeds were germinated in a nursery box at 25 °C and 70% humidity (RDN-1000D; Yanghui Equipment Co., Ltd., Ningbo, China). Every two seeds were placed in a plastic pot (9 cm × 7 cm × 6.5 cm) with four drainage holes of 1 cm diameter at the bottom, and the pots were placed in trays (51 cm × 28 cm × 6.5 cm) to prevent soil leakage. Two weeks later, when the seedlings had three true leaves, uniform seedlings were selected for the root ball application at Baima Greenhouse, Nanjing Forestry University, Nanjing, China.

PAH soil was prepared with phenanthrene (PHE) and pyrene (PYR) (Merck KGaA, Darmstadt, German); 500 ppm concentration of PHE and PYR were added into the peat soil (Pindstrup Mosebrug, Ryomgaard, Demark), and the final PAH concentration was 5 ppm. In the seedling transfer, the root ball application was treated with the CCNPs; in the process, the configured nano-colloid solution was ultrasonically shaken for 30 min, and 5 mL of 8 kinds of CCNPs solution was injected into the root ball; the root ball was completely infiltrated with CCNPs. The seedlings were subsequently placed on the skeletonized shelf until there was no dripping liquid and then transplanted to the pots (11 cm × 10 cm × 13 cm); during the transplanting, the bulb was completely embedded in the soil and mulched to complete the inter-root exposure of CCNPs. Each treatment has three biological replicates.

After four weeks, the treated spinach was harvested, and the phenotype was recorded at the same time; afterward, the spinach was carefully separated from the soil and gently rinsed four times with tap water, followed by deionized water to remove soil adhering to the plant surface. After drying, the shoots and roots of spinach were separated with scissors and weighed. Three replicates of each treatment were applied, and a nutrient solution was applied once in the second week after treatment.

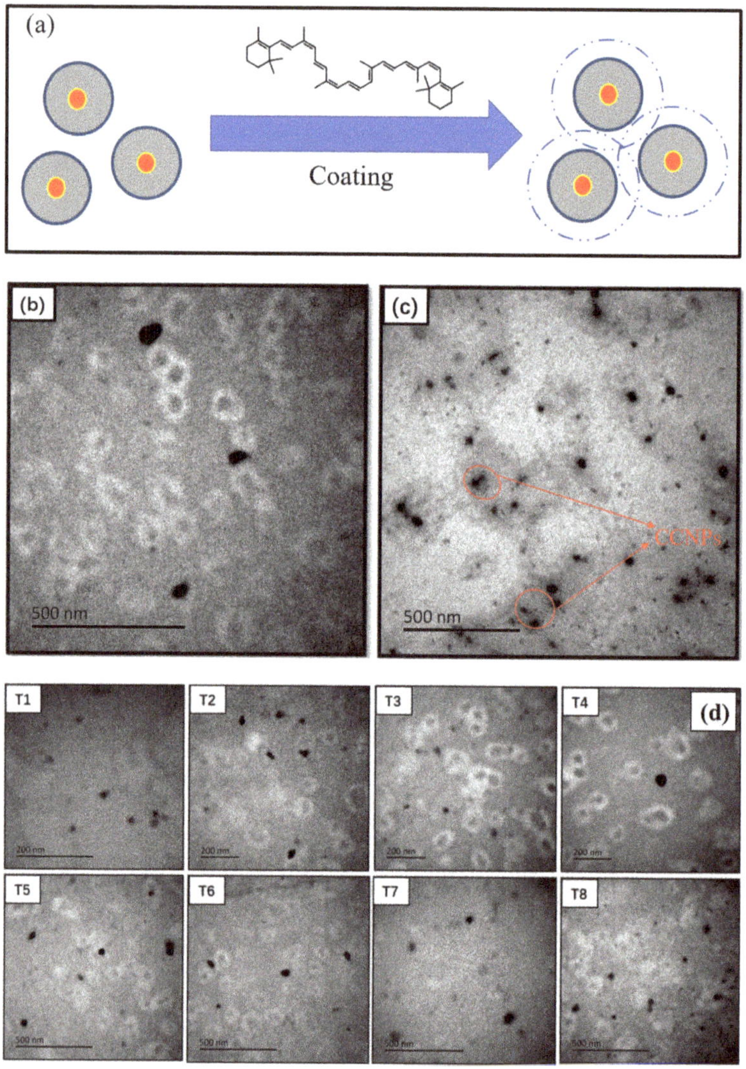

Figure 1. Mechanism process of carotenoid-coated chitosan nanoparticles (CCNPs) (**a**); and the representative TEM images of bare chitosan nanoparticles (CNPs) (**b**), and CNPs generated by ionic gelation at a reaction temperature of 20 °C, a TPP solution concentration of 10 mg/mL, and a pH of 6 (i.e., T2) coated with β-carotene (CCNPs) (**c**); and TEM analysis of CNPs cross-linked by ionotropic gelation with TPP. (Note: The treatments of T1–T8 are shown in Table 1). (**d**) Note, T1, 20 °C, 10 mg/mL TPP, pH = 5; T2, 20 °C, 10 mg/mL TPP, pH = 6; T3, 20 °C, 20 mg/mL TPP, pH = 5; T4, 20 °C, 20 mg/mL TPP, pH = 6; T5, 40 °C, 10 mg/mL TPP, pH = 5; T6, 40 °C, 10 mg/mL TPP, pH = 6; T7, 40 °C, 20 mg/mL TPP, pH = 5; and T8, 40 °C, 20 mg/mL TPP, pH = 6.

Table 1. Treatment information for comparison experiment.

Treatments	T/°C	TPP (mg/mL)	pH
T1	20	10	5
T2	20	10	6
T3	20	20	5
T4	20	20	6
T5	40	10	5
T6	40	10	6
T7	40	20	5
T8	40	20	6

2.3. PAHs Extraction and Analysis

The extraction and purification procedure for PAHs was based on that reported by Gao et al. [2]. Specifically, the chopped plant samples were extracted with an extraction agent (acetone: dichloromethane = 2:1, v/v) using an ultrasonic water bath for 30 min, and this step was repeated three times. The combined extracts were passed through a silica gel column (silica gel 3 g, anhydrous sodium sulfate 3 g) and eluted with 10 mL of 1:1 (v/v) dichloromethane and n-hexane. The filtrate was passed through a rotary evaporator (RE-25A; Yarong Biochemical Instrument Factory, Shanghai, China), exchanged with 2 mL of methanol, filtered through a 0.45 μm Teflon membrane, and transferred into a 2 mL sample vial. PAH concentrations were measured by HPLC (UltiMate 3000 HPLC; Thermo Co., Ltd., Waltham, MA, USA). The HPLC conditions were as follows: the pump model was LPG-3400 SDN, the UV detector model was VWD-3100, and the column was a 4.6 mm × 150 mm C18 column with a temperature of 30 °C. The mobile phase was methanol/water (80/20, v/v) at a flow rate of 1.0 mL/min. The injection volumes for phenanthrene and pyrene were set to 10 and 40, respectively, and the UV detection wavelength of phenanthrene was 254 nm, and that of pyrene was 234 nm. The peak areas were quantified by the external standard method.

2.4. Bioconcentration Factor (BCF) and Translocation Factor (TF)

The ability of CCNPs to reduce PAHs accumulation is by measuring the bioconcentration factor (BCF) (1) and translocation factor (TF) (2), defined as the ratio of PAH concentration in plant shoots to roots and the ratio of PAH concentration in plant roots to soils, respectively [33]; and the results are presented in Table 2.

$$\mathrm{BCF} = \frac{\text{PAHs concentration in plants}}{\text{PAHs concentration in sediment}} \tag{1}$$

and

$$\mathrm{TF} = \frac{\text{PAHs concentration in shoot}}{\text{PAHs concentration in root}} \tag{2}$$

Table 2. Bioconcentration factors (BCFs) and transfer factor (TFs) of PAHs in the shoots and roots of spinach.

Treatment	Bioconcentration Factor (BCF)		Transfer Factor (TF)	
	BCF_{PYR}	BCF_{PHE}	TF_{PYR}	TF_{PHE}
T1	0.390	0.198	0.011	0.063
T2	0.088	0.035	0.007	0.079
T3	0.329	0.380	0.166	0.040
T4	0.180	0.165	0.085	0.026
T5	0.188	0.158	0.032	0.034
T6	0.567	0.084	0.009	0.064
T7	0.200	0.191	0.268	0.061
T8	0.226	0.037	0.049	0.037

2.5. Statistics

Statistical analysis was performed using the Excel (Microsoft, Redmond, WA, USA) (version 2019), IBM SPSS Statistics 25.0 (IBM, Armonk, NY, USA). Sampling and chemical analyses were examined in triplicate to decrease the experimental errors and to increase the experimental reproducibility. The confidence of the data generated in the present investigations was analyzed by standard statistical methods to determine the mean values and standard deviation (S.D.). Descriptive Statistics were applied to assess the normality of the distribution, and the test data meets the normality. The differences among the treatments were analyzed by one-way ANOVA (LSD test).

3. Results

3.1. The CCNPs Characteristics

The sizes of CNPs were around 260–682 nm; the smallest CNPs occurred with a reaction temperature at 20 °C, a TPP concentration of 10 mg/mL, and the pH 6 (T2); and the largest were prepared with the reaction conditions of 40 °C, pH 6, and 10 mg/mL of TPP (Figure 1b,c). Based on the preparation of the coating NPs, *β*-carotene and ribonucleic acid (RNA) had similar functional groups as the hydrophobic effect, such as the nitrogenous bases, ribose, and phosphate groups of RNA and long carbon chain of *β*-carotene. Due to the similar chemical functions, it is considered that *β*-carotene can be coated on the CNPs like RNA (Figure 1a) [34], and CNPs are produced with *β*-carotene finally for the spinach growth experiment.

3.2. The Phenotype and Growth of the Spinach

After four weeks' treatment, we found that the spinach presented stress in the treatment of combined PAHs of PHE and PYR with no exogenous addition (CK), and the leaves went curly; the spinach presented growth limitations when treated with T3, T4, and T7 in the combined PAH contaminations, and those were shorter than that in the CK (Figure 2a). The spinach grew the least in the T7 treatment; the leaves were shriveled after the treatment with the CCNPs of different temperatures and pH, but the TPP concentration was focused at 20 mg/mL. The spinach grew the most in T1, T2, T5, and T8; the spinach growth was the best in T2, for which the CCNPs were prepared with a reaction temperature of 20 °C, 10 mg/mL TPP, and pH 6. The leaves in the PAHs treatment of PAH and PYR were larger and stronger when compared with the spinach treated with only PAHs (Figure 2a).

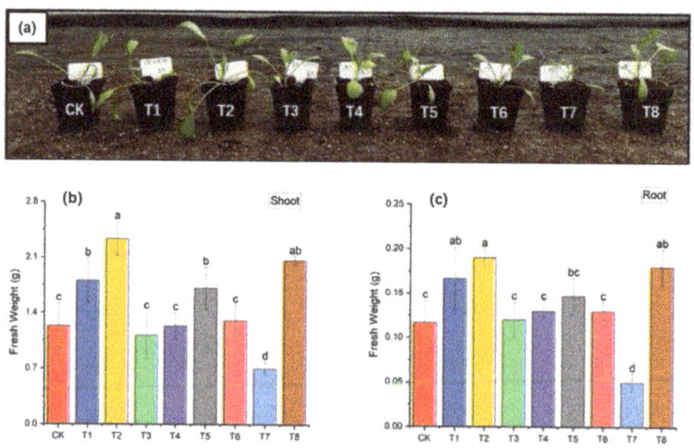

Figure 2. The phenotypes of spinach in the screening experiment of PAH-contaminated soil after 40 days (**a**), and the fresh weight of spinach shoots and roots after 40 days under CCNPs (**b**,**c**), respectively (Note, the error bars indicate the standard deviation (n = 3); each sample with a different letter above indicates statistically significant differences with at least $p < 0.05$; CK was the blank control group, which grew naturally without reagent or contamination. The T1–T8 treatments are shown in Table 1).

It was recorded that the whole fresh weight of the spinach presented as the highest at 2.52 g in the treatment of 20 °C, pH 6, and the TPP concentration of 10 mg/mL (T2) with a shoot of 2.34 g and a root of 0.18 g; and the smallest, of 0.75 g, was recorded in the treatment of 40 °C, pH 5, and 20 mg/mL TPP (T7), with the shoot weighing 0.70 g and the root weighing 0.05 g (Figure 2b,c). In addition, the spinach's fresh weight of the shoot and root was 1.235 g and 0.11 g in the PAHs only treatment, respectively; and the shoot's fresh weights were 1.12, 1.24, and 0.70 g in the T3, T4, and T7 treatments, which were lower than that in the CK after 4 weeks. The shoot's fresh weight presented an increasing trend in the T2, T5, T6, and T8 treatments (Figure 2b). Despite the spinach root's fresh weight being lower than that in the CK of 0.06 g, the other root's fresh weights presented an increased response after the CCNPs were added; the fresh weight reached 0.12 to 0.19 g (seven kinds of CCNPs), respectively (Figure 2c).

3.3. *The PAHs Concentration in Spinach*

After 4 weeks of PAH treatment, the roots were the major location for PYR and PHE accumulation (Figure 3). In the treatment of CCNPs with a reaction temperature of 40 °C, 10 mg/mL TPP, and the solution with pH 6 (T6), the spinach root accumulated the highest ($p < 0.05$) PYR of 8.22 mg kg^{-1}, while the spinach root only accumulated 3.88, 0.97, 2.15, 1.30, 1.59, 1.29, and 2.08 mg kg^{-1} when treated with the other CCNPs; the spinach root in T6 presented the highest at 8.22 mg kg^{-1} among all treatments (Figure 3b). The PYR accumulated the lowest at 0.041 mg kg^{-1} in treatment T2, and the PYR increased to the highest of 0.42 and 0.47 mg kg^{-1} when treated with T3 and T7 after 4 weeks, respectively (Figure 3a). PHE accumulated more in the spinach shoots after 4 weeks of PAHs treatment, it reached 0.13 mg kg^{-1}, and the spinach shoot had similar contents of 0.10 and 0.11 mg kg^{-1} when the spinach was treated with CCNPs of the 20 mg/mL TPP (T3 and T7), respectively; The PHE was lowest at 0.01 mg kg^{-1} when the spinach was treated with CCNPs of 40 °C, pH 6, and the TPP concentration of 20 mg/mL (T8) (Figure 3c). PHE accumulated more in the roots when the spinach was treated with CCNPs of 20 °C, pH 5, and 20 mg/mL. For the TPP concentration of T3, when compared with the spinach root in the CK with 1.83 mg kg^{-1}, the PHE accumulated the least in the spinach root treated with the CCNPs of 20 °C, pH 6, and 10 mg/mL TPP (T2) and 40 °C, pH 6, and 20 mg/mL TPP (T8); the PHE concentrations are 0.32 and 0.36 mg kg^{-1}, respectively (Figure 3d).

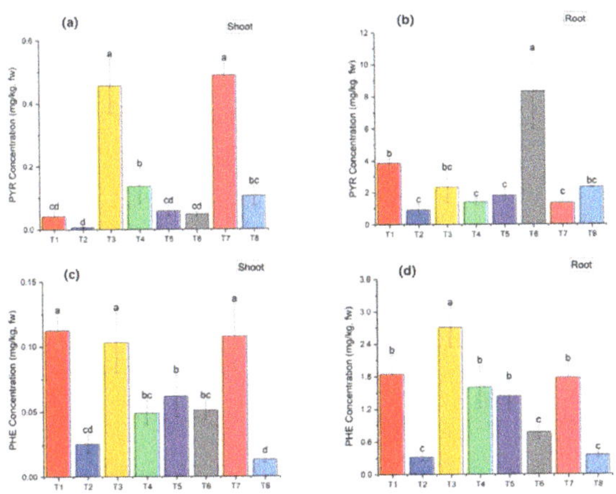

Figure 3. Concentrations of pyrene (PYR) and phenanthrene (PHE) in the shoots (**a**,**c**) and roots (**b**,**d**) of spinach (Note: the error bars indicate the standard deviation (n = 3); each sample with a different letter above indicates statistically significant differences with at least $p < 0.05$; CK was the blank control group, which grew naturally without reagent or contamination. The treatments T1–T8 are shown in Table 1).

3.4. BCFs and TFs

The BCFs and TFs of PYR and PAH are calculated in Table 2. It is shown that the PYR presented higher BCF in most treatments than PHE in the spinach, the BCF_{PHE} is higher than BCF_{PYR} only in T3, and BCF_{PYR} is highest, at 0.567, in T6 and lowest, at 0.088, in T2. Meanwhile, the TF presented different trends than BCF in the spinach under PAH treatments. As shown in Table 2, the TF_{PYR}s are higher than the TF_{PHE}s with values of 0.166, 0.085, 0.268, and 0.049 in T3, T4, T7, and T8, respectively; and in other treatments, the TF_{PYR}s are lower than the TF_{PHE}s, with the values of 0.063, 0.079, 0.034, and 0.064 in T1, T2, T5, and T6, respectively.

4. Discussion

4.1. Preparation of CCNPs and PAHs Accumulation

The synthesis of CCNPs involved varying reaction conditions, resulting in particles of different sizes. The sizes of CNPs ranged from approximately 260 to 682 nm, with the smallest size observed at a reaction temperature of 20 °C, TPP concentration of 10 mg/mL, and pH 6. On the other hand, the largest size was obtained under the conditions of 40 °C, pH 6, and 10 mg/mL TPP (Figure 1). Among the CCNP treatments, the one prepared at 40 °C, 10 mg/mL TPP, and pH 6 (T6) showed the highest accumulation of PYR in spinach roots, with a concentration of 8.22 mg kg^{-1}, significantly higher ($p < 0.05$) than the control treatment without CCNPs (CK). The other CCNP treatments resulted in varying levels of PYR accumulation in spinach roots, ranging from 0.97 to 2.15 mg kg^{-1}, depending on the specific CCNP treatment.

In subsequent experiments, CCNPs were used to investigate the transport blocking of polycyclic aromatic hydrocarbons (PAHs), specifically PHE and PYR, in spinach plants. After four weeks of treatment, the concentration of PAHs in the spinach was assessed, revealing significant accumulation in the roots. Interestingly, the spinach shoots contained higher concentrations of PHE compared to the roots. The CCNP treatments at TPP concentrations of 20 mg/kg (T3 and T7) resulted in shoot PHE concentrations ranging from 0.10 to 0.11 mg kg^{-1}. The lowest shoot PHE accumulation of 0.01 mg kg^{-1} was observed in the treatment with CCNPs prepared at 40 °C, pH 6, and 20 mg/mL TPP concentration (T8). In contrast, the accumulation of PAHs in the spinach roots varied depending on the specific CCNP treatment. For example, in the treatment with CCNPs prepared at 20 °C, TPP concentration of 5 mg/mL, and pH 6 (T3), the root PHE accumulation was higher compared to the CK. Conversely, the treatments with CCNPs prepared at 20 °C, pH 6, and 10 mg/mL TPP (T2), as well as 40 °C, pH 6, and 20 mg/mL TPP (T8), resulted in the lowest root PHE accumulation, with concentrations of 0.32 and 0.36 mg kg^{-1}, respectively. The data indicate that smaller CCNPs facilitated reduced PAH accumulation and transfer in spinach. Furthermore, no significant differences were observed in the concentration of PYR in shoots between samples T5 (0.058 mg kg^{-1}) and T6 (0.048 mg kg^{-1}), as well as in the concentration of PHE between samples T5 (0.052 mg kg^{-1}) and T6 (0.050 mg kg^{-1}). Similarly, there were no significant differences in the concentration of PYR in roots among samples T4 (1.3 mg kg^{-1}), T5 (1.59 mg kg^{-1}), and T7 (1.29 mg kg^{-1}), as well as in the concentration of PHE among samples T4 (1.61 mg kg^{-1}), T5 (1.53 mg kg^{-1}), and T7 (1.80 mg kg^{-1}). These findings suggest that reaction conditions may differentially influence root and shoot responses at various levels.

Commonly, CNPs are sensitive to temperature and pH [35]. Varying pH can affect the size of CNPs and probe the states of water in CNP hydrogels [36], while smaller CNPs have high water imbibing capability, minimal invasiveness, porous networks, and can mold perfectly into an irregular defect [37]. Temperature is another key factor for CNP synthesis; CNPs formed at high temperatures may have remaining associations, confirmed by their spontaneous recovery after breakup at low temperatures [35]. Moreover, chitosan treated at 25 °C possessed similar or weaker antibacterial activity compared to those at 4 °C, which can influence the CNP coating with β-carotene. In our study, we selected 20 °C as a reasonable temperature for CCNP synthesis and 20 °C with pH 6 as the ideal condition for CCNP coating with β-carotene. TPP serves as a polymerization agent for NPs formation, and lower TPP concentrations were found to result in smaller CNP sizes. Our results match well with the previous report that the coated CNPs with nano-size can have biological effects on the crops via carrying materials to the target cells in plants [27]. With their nano-size, CCNPs exhibited increased adsorption capacity for PAHs. The preparation of CCNPs influenced the accumulation of PAHs in spinach plants, as specific conditions, such as reaction temperature, TPP concentration, and pH, played a role in determining the extent of PAH accumulation in the roots and shoots of the spinach plants. These findings highlight the importance of understanding the interaction between CCNPs and PAHs to assess their potential impact on plant health and food safety.

4.2. CCNPs and PAHs Transfer

The interaction between CCNPs and polycyclic aromatic hydrocarbons (PAHs) in the environment is crucial as it can influence the transfer of these contaminants within ecosystems. In this study, we investigated the transfer of CCNPs and PAHs, focusing on their movement within spinach plants. The synthesis of CCNPs resulted in particles of varying sizes, influenced by reaction conditions such as temperature, TPP, and pH [36]. The resulting CCNPs ranged in size from approximately 260 to 682 nm, with the smallest size observed under specific conditions (20 °C, 10 mg/mL TPP, and pH 6) and the largest size obtained under different conditions (40 °C, 10 mg/mL TPP, and pH 6) (Figure 1).

We then analyzed the transfer and accumulation of PAHs, specifically PHE and PYR, within the spinach plants. After four weeks of treatment, we observed significant PAH accumulation in the roots of the spinach plants. Notably, the treatment with CCNPs prepared at 40 °C, TPP concentration of 10 mg/mL, and pH 6 (T6) resulted in the highest accumulation of PYR in spinach roots, with a concentration of 8.22 mg kg^{-1}, significantly higher ($p < 0.05$) than the control treatment without CCNPs (CK). The other CCNP treatments led to varying levels of PYR accumulation in spinach roots, ranging from 0.97 to 2.15 mg kg^{-1}, depending on the specific CCNP treatment.

It was reported that coated CNPs have stronger adsorption potential than organic containments, such as paraquat, 4-nitrophenol, methyl orange, cango red, etc. [38]. A similar synthesized process was reported where the CNPs when coated with oxide metals and oligo, performed better at absorbing the PAHs in the environment [39,40]. Our results agree with the previous study; the CCNPs reduce the PAH transfer from soil to root. Regarding PHE accumulation, the spinach shoots exhibited higher concentrations of this PAH compared to the roots. After the four-week treatment period, the spinach shoots accumulated PHE concentrations ranging from 0.10 to 0.11 mg kg^{-1} in the treatments with CCNPs prepared at a TPP concentration of 20 mg/mL (T3 and T7), respectively. The lowest shoot PHE accumulation of 0.01 mg kg^{-1} was observed in the treatment with CCNPs prepared at 40 °C, pH 6, and 20 mg/mL TPP concentration (T8). In contrast, the accumulation of PAHs in the spinach roots varied depending on the specific CCNP treatment. For example, in the treatment with CCNPs prepared at 20 °C, TPP concentration of 5 mg/mL, and pH 6 (T3), the root PHE accumulation was higher compared to the control treatment (CK). Conversely, the treatments with CCNPs prepared at 20 °C, pH 6, and a TPP concentration of 10 mg/mL (T2), as well as 40 °C, pH 6, and 20 mg/mL TPP (T8), resulted in the lowest root PHE accumulation, with concentrations of 0.32 and 0.36 mg kg^{-1}, respectively.

The result indicates that the exogenous application of CCNPs can significantly reduce the movement of PAHs from the environment to the roots and shoots of spinach plants. CCNPs are known for their role in chemical transfer, and we found that their application increased resistance to PAH contamination. Understanding the interaction between CCNPs and PAHs is crucial for evaluating their potential impact on plant health and food safety.

5. Conclusions

These findings demonstrate that the reduction of PAHs transfer and accumulation within spinach plants by CCNPs is influenced by various factors, including the physicochemical properties of the nanoparticles and the specific exposure conditions. Particularly, CCNPs exhibit better performance at room temperature and neutral pH, making them more suitable as carriers for β-carotene in plants compared to conditions of high temperature and acidic pH. Under appropriate reaction conditions, the synthesized CCNPs show significant protective effects on plants by efficiently providing and releasing β-carotene in PAH-contaminated environments. The decreased uptake and translocation of PAHs within plants can have significant implications for food safety and environmental health.

Furthermore, the main materials used in CCNP synthesis are derived from natural and biologically harmless sources, indicating minimal risk in consuming them. Additionally, their sustainability is enhanced by the abundance of the precursor materials used in their synthesis. However, further research is necessary to fully comprehend the mechanisms of transfer and potential risks associated with the interaction between CCNPs and PAHs in plant systems. This knowledge will aid in understanding the sustainable agricultural benefits of CCNPs and their potential applications in environmental remediation.

Author Contributions: Conceptualization, J.H. and Y.S.; Methodology, J.H.; Formal analysis, M.C.; Investigation, J.Z., M.C., R.T. and Y.Y.; Resources, J.H.; Data curation, J.Z.; Writing—original draft, M.C. and Y.S.; Writing—review & editing, Y.S. All authors have read and agreed to the published version of the manuscript.

Funding: The present work was carried out with the financial support of Agricultural Innovation Project (CX(22)3133) of Jiangsu Academy of Agricultural Sciences; Yu Shen thanks the Chinese Scholarship Council (CSC) supporting his study overseas.

Institutional Review Board Statement: Not applicable.

Informed Consent Statement: Not applicable.

Data Availability Statement: Not applicable.

Conflicts of Interest: The authors declare that they have no known competing financial interests or personal relationships that could have appeared to influence the work reported in this paper.

References

1. Li, Q.; Zhang, Y.; Xie, Z.; Zhen, Y.; Hu, W.; Dong, H. Polycyclic aromatic hydrocarbon-based organic semiconductors: Ring-closing synthesis and optoelectronic properties. *J. Mater. Chem. C* **2022**, *10*, 2411–2430. [CrossRef]
2. Wang, Y.; Ding, L.; Shi, Q.; Liu, S.; Qian, L.; Yu, Z.; Wang, H.; Lei, J.; Gao, Z.; Long, H.; et al. Volatile organic compounds (VOC) emissions control in iron ore sintering process: Recent progress and future development. *Chem. Eng. J.* **2022**, *448*, 137601. [CrossRef]
3. Vijayanand, M.; Ramakrishnan, A.; Subramanian, R.; Issac, P.K.; Nasr, M.; Khoo, K.S.; Rajagopal, R.; Greff, B.; Wan Azelee, N.I.; Jeon, B.H.; et al. Polyaromatic hydrocarbons (PAHs) in the water environment: A review on toxicity, microbial biodegradation, systematic biological advancements, and environmental fate. *Env. Res.* **2023**, *227*, 115716. [CrossRef] [PubMed]
4. Souza, M.C.O.; Rocha, B.A.; Adeyemi, J.A.; Nadal, M.; Domingo, J.L.; Barbosa, F., Jr. Legacy and emerging pollutants in Latin America: A critical review of occurrence and levels in environmental and food samples. *Sci. Total Environ.* **2022**, *848*, 157774. [CrossRef]
5. Janneh, M.; Qu, C.; Zhang, Y.; Xing, X.; Nkwazema, O.; Nyihirani, F.; Qi, S. Distribution, sources, and ecological risk assessment of polycyclic aromatic hydrocarbons in agricultural and dumpsite soils in Sierra Leone. *RSC Adv.* **2023**, *13*, 7102–7116. [CrossRef]
6. Han, L.; Bai, J.; Gao, Z.; Wang, W.; Wang, D.; Cui, B.; Liu, X. Polycyclic aromatic hydrocarbons (PAHs) in surface soils from reclaimed and ditch wetlands along a 100-year chronosequence of reclamation in a Chinese estuary: Occurrence, sources, and risk assessment. *Agric. Ecosyst. Environ.* **2019**, *286*, 106648. [CrossRef]
7. Wu, F.; Tian, K.; Wang, J.; Bao, H.; Luo, W.; Zhang, H.; Hong, H. Accumulation and translocation of phenanthrene, anthracene and pyrene in winter wheat affected by soil water content. *Ecotoxicol. Environ. Saf.* **2019**, *183*, 109567. [CrossRef]
8. Molina, L.; Segura, A. Biochemical and metabolic plant responses toward polycyclic aromatic hydrocarbons and heavy metals present in atmospheric pollution. *Plants* **2021**, *11*, 2305. [CrossRef]
9. Hu, J.; Chen, J.; Wang, W.; Zhu, L. Mechanism of growth inhibition mediated by disorder of chlorophyll metabolism in rice (*Oryza sativa*) under the stress of three polycyclic aromatic hydrocarbons. *Chemosphere* **2023**, *329*, 138554. [CrossRef]

10. Fismes, J.; Perrin-Ganier, C.; Empereur-Bissonnet, P.; Morel, J.L. Soil-to-root transfer and translocation of polycyclic aromatic hydrocarbons by vegetables grown on industrial contaminated soils. *J. Env. Qual.* **2002**, *31*, 1649–1656. [CrossRef]
11. Schwab, A.P.; Dermody, C.L. Pathways of polycyclic aromatic hydrocarbons assimilation by plants growing in contaminated soils. *Adv. Agron.* **2021**, *169*, 193–250.
12. Shen, Y.; He, F.; Zhu, J.; Zhang, H.; Wang, J.; Wang, H.; Zhan, X. Proton-coupled cotransporter involves phenanthrene xylem loading in roots. *Sci. Total Env.* **2021**, *773*, 145637. [CrossRef] [PubMed]
13. Tao, S.; Jiao, X.C.; Chen, S.H.; Liu, W.X.; Coveney, R.M., Jr.; Zhu, L.Z.; Luo, Y.M. Accumulation and distribution of polycyclic aromatic hydrocarbons in rice (*Oryza sativa*). *Env. Pollut.* **2006**, *140*, 406–415. [CrossRef] [PubMed]
14. Kang, F.; Chen, D.; Gao, Y.; Zhang, Y. Distribution of polycyclic aromatic hydrocarbons in subcellular root tissues of ryegrass (*Lolium multiflorum* Lam.). *BMC Plant Biol.* **2010**, *10*, 210. [CrossRef] [PubMed]
15. Gessner, I.; Neundorf, I. Nanoparticles Modified with Cell-Penetrating Peptides: Conjugation Mechanisms, Physicochemical Properties, and Application in Cancer Diagnosis and Therapy. *Int. J. Mol. Sci.* **2020**, *21*, 2536. [CrossRef] [PubMed]
16. Jia, H.R.; Zhu, Y.X.; Duan, Q.Y.; Chen, Z.; Wu, F.G. Nanomaterials meet zebrafish: Toxicity evaluation and drug delivery applications. *J. Control Release* **2019**, *311–312*, 301–318. [CrossRef]
17. Nair, R.; Varghese, S.H.; Nair, B.G.; Maekawa, T.; Yoshida, Y.; Kumar, D.S. Nanoparticulate material delivery to plants. *Plant Sci.* **2010**, *179*, 154–163. [CrossRef]
18. Ma, J.; Zhang, S.; Liu, J.; Liu, F.; Du, F.; Li, M.; Chen, A.T.; Bao, Y.; Suh, H.W.; Avery, J.; et al. Targeted Drug Delivery to Stroke via Chemotactic Recruitment of Nanoparticles Coated with Membrane of Engineered Neural Stem Cells. *Small* **2019**, *15*, e1902011. [CrossRef]
19. Kalam, M.A. Development of chitosan nanoparticles coated with hyaluronic acid for topical ocular delivery of dexamethasone. *Int. J. Biol. Macromol.* **2016**, *89*, 127–136. [CrossRef]
20. Lopes, M.; Shrestha, N.; Correia, A.; Shahbazi, M.A.; Sarmento, B.; Hirvonen, J.; Veiga, F.; Seica, R.; Ribeiro, A.; Santos, H.A. Dual chitosan/albumin-coated alginate/dextran sulfate nanoparticles for enhanced oral delivery of insulin. *J. Control Release* **2016**, *232*, 29–41. [CrossRef]
21. Tong, T.; Wang, L.; You, X.; Wu, J. Nano and microscale delivery platforms for enhanced oral peptide/protein bioavailability. *Biomater. Sci.* **2020**, *8*, 5804–5823. [CrossRef]
22. Faqir, Y.; Ma, J.; Chai, Y. Chitosan in modern agriculture production. *Plant Soil. Environ.* **2021**, *67*, 679–699. [CrossRef]
23. Bernkop-Schnürch, A.; Dünnhaupt, S. Chitosan-based drug delivery systems. *Eur. J. Pharm. Biopharm.* **2012**, *81*, 463–469. [CrossRef] [PubMed]
24. Boutrif, E. Institutions involved in food safety: Food and agriculture organization of the United Nations (FAO). *Encycl. Food Saf.* **2014**, *4*, 354–358.
25. Yan, N.; Chen, X. Don't waste seafood waste. *Nature* **2015**, *524*, 155–157. [CrossRef]
26. Şenol, Z.M.; Şimşek, S. Insights into effective adsorption of lead ions from aqueous solutions by using chitosan-bentonite composite beads. *J. Polym. Environ.* **2022**, *30*, 3677–3687. [CrossRef]
27. Buchman, J.T.; Elmer, W.H.; Ma, C.; Landy, K.M.; White, J.C.; Haynes, C.L. Chitosan-coated mesoporous silica nanoparticle treatment of *Citrullus lanatus* (Watermelon): Enhanced fungal disease suppression and modulated expression of stress-related genes. *ACS Sustain. Chem. Eng.* **2019**, *7*, 19649–19659. [CrossRef]
28. Grillo, R.; Pereira, A.E.S.; Nishisaka, C.S.; De Lima, R.; Oehlke, K.; Greiner, R.; Fraceto, L. Chitosan/tripolyphosphate nanoparticles loaded with paraquat herbicide: An environmentally safer alternative for weed control. *J. Hazard. Mater.* **2014**, *278*, 163171. [CrossRef]
29. Shen, Y.; Li, J.; Gu, R.; Yue, L.; Wang, H.; Zhan, X.; Xing, B. Carotenoid and superoxide dismutase are the most effective antioxidants participating in ROS scavenging in phenanthrene accumulated wheat leaf. *Chemosphere* **2018**, *197*, 513–525. [CrossRef]
30. Shen, Y.; Li, J.; Shi, S.; Gu, R.; Zhan, X. Application of carotenoid to alleviate the oxidative stress caused by phenanthrene in wheat. *Environ. Sci. Pollut. Res.* **2019**, *26*, 3593–3602. [CrossRef]
31. Stutz, H.; Bresgen, N.; Eckl, P.M. Analytical tools for the analysis of β-carotene and its degradation products. *Free. Radic. Res.* **2015**, *49*, 650–680. [CrossRef] [PubMed]
32. Iammarino, M.; Di Taranto, A.; Cristino, M. Monitoring of nitrites and nitrates levels in leafy vegetables (spinach and lettuce): A contribution to risk assessment. *J. Sci. Food Agric.* **2014**, *94*, 773–778. [CrossRef] [PubMed]
33. Sushkova, S.; Minkina, T.; Tarigholizadeh, S. PAHs accumulation in soil-plant system of *Phragmites australis* Cav. in soil under long-term chemical contamination. *Eurasian J. Soil Sci.* **2020**, *9*, 242253. [CrossRef]
34. Ragelle, H.; Riva, R.; Vandermeulen, G.; Naeye, B.; Pourcelle, V.; Le Duff, C.S.; D'Haese, C.; Nysten, B.; Braeckmans, K.; De Smedt, S.C.; et al. Chitosan nanoparticles for siRNA delivery: Optimizing formulation to increase stability and efficiency. *J. Control. Release* **2014**, *176*, 54–63. [CrossRef]
35. Qu, X.; Wirsen, A.; Albertsson, A.C. Novel pH-sensitive chitosan hydrogels: Swelling behavior and states of water. *Polymer* **2000**, *41*, 4589–4598. [CrossRef]
36. Lavanya, K.; Chandran, S.V.; Chandran, K.; Selvamurugan, N. Temperature-and pH-responsive chitosan-based injectable hydrogels for bone tissue engineering. *Mater. Sci. Eng. C* **2020**, *111*, 110862. [CrossRef]
37. Sreekumar, S.; Goycoolea, F.M.; Moerschbacher, B.M.; Rivera-Rodriguez, G.R. Parameters influencing the size of chitosan-TPP nano-and microparticles. *Sci. Rep.* **2018**, *8*, 4695. [CrossRef]

38. Ali, F.; Khan, S.B.; Kamal, T.; Anwar, Y.; Alamry, K.A.; Asiri, A.M. Anti-bacterial chitosan/zinc phthalocyanine fibers supported metallic and bimetallic nanoparticles for the removal of organic pollutants. *Carbohydr. Polym.* **2017**, *173*, 676–689. [CrossRef]
39. Rani, M.; Shanker, U. Metal oxide-chitosan based nanocomposites for efficient degradation of carcinogenic PAHs. *J. Environ. Chem. Eng.* **2020**, *8*, 103810. [CrossRef]
40. Rodrigues, P.R.; Nascimento, L.E.S.; Godoy, H.T.; Vieira, R.P. Improving chitosan performance in the simultaneous adsorption of multiple polycyclic aromatic hydrocarbons by oligo (β-pinene) incorporation. *Carbohydr. Polym.* **2023**, *302*, 120379. [CrossRef]

coatings

Article

N-Rich Algal Sludge Biochar for Peroxymonosulfate Activation toward Sulfadiazine Removal

Chao Liu [1,2,†], Zhenxiang Chen [3,†], Ruiqin Kang [1], Jing Wang [1], Qingwei Lu [1], Tao Wang [1], Dayong Tian [1,*], Ying Xu [2,*], Zhan Wang [4] and Huiping Ding [5]

1 College of Chemistry and Environmental Engineering, Anyang Institute of Technology, Anyang 455000, China
2 College of Chemistry and Chemical Engineering, Henan University, Kaifeng 475004, China
3 Jiangsu Xingzhou Ecological Environment Technology Co., Ltd., Nanjing 210004, China
4 Beijing Key Laboratory for Green Catalysis and Separation, Department of Chemistry and Chemical Engineering, Beijing University of Technology, Beijing 100124, China
5 College of Chemistry and Chemical Engineering, Henan Chemical Technician College, Kaifeng 475002, China
* Correspondence: tiandayong@163.com (D.T.); hdccxu@126.com (Y.X.)
† These authors contributed equally to this work.

Abstract: The fabrication of a green, high activity and low-cost carbon-based catalyst capable of activating new oxidant (peroxymonosulfate, PMS) for contaminants abatement is needed. In this research, we prepared novel N-doped biochars via one-step pyrolysis of algal sludge without external nitrogen sources. The obtained ASBC800 possessed the largest specific surface area (S_{BET} = 145.596 $m^2\ g^{-1}$) and thus it displayed the best catalytic performance, as revealed by the effective elimination of sulfadiazine (SDZ, >95% within 70 min) with 0.2 g L^{-1} ASBC800 and 0.5 mM PMS. Both radical species (e.g., $SO_4^{\bullet -}$, and $^{\bullet}OH$), and nonradical regime ($^1O_2$ and electron-transfer) contributed to SDZ oxidation, in which ASBC800 played essential roles in activating PMS, accumulating SDZ, and regulating electron shuttle from SDZ to ASBC800-PMS*. Overall, this work not only provides a novel strategy for the synthesis of N-rich and cost-effective biochar but also promotes the development and application of carbon-based functional materials in environmental remediation.

Keywords: peroxymonosulfate; pre-adsorption; oxidation; electron-transfer; N doping

Citation: Liu, C.; Chen, Z.; Kang, R.; Wang, J.; Lu, Q.; Wang, T.; Tian, D.; Xu, Y.; Wang, Z.; Ding, H. N-Rich Algal Sludge Biochar for Peroxymonosulfate Activation toward Sulfadiazine Removal. *Coatings* **2023**, *13*, 431. https://doi.org/10.3390/coatings13020431

Academic Editor: Ioannis V. Yentekakis

Received: 19 October 2022
Revised: 29 November 2022
Accepted: 1 December 2022
Published: 14 February 2023

1. Introduction

Recently, peroxymonosulfate-based advanced oxidation processes (PMS−AOPs) have received increasing attention for pollutants removal [1–3]. Various reactive oxygen species (ROS) can be produced in this oxidation process, including sulfate radical ($SO_4^{\bullet -}$), hydroxyl radical ($^{\bullet}OH$) and singlet oxygen ($^1O_2$), which possess much higher redox potentials than their parent oxidants (PMS, 1.75–1.82 V) [4]. In addition, a novel mediated electron transfer regime may occur between target contaminants and oxidants without the generation of ROS [5–7].

In order to produce ROS, it is of great importance to break the O–O bond of the PMS. Metal-free carbon-based catalysts have recently aroused widespread attention owing to their environmental protection and easy preparation [8]. Carbonaceous materials (e.g., activated carbon [9], reduced graphene oxide [10], porous carbon [11], carbon nanotubes [12] and nanodiamonds [13]) have been proven to be potential green catalytic materials for PMS activation due to their abundant oxygen-containing functional groups, edge defects, and porous structures. In addition to the abovementioned carbon-based catalysts, biochar is deemed a promising candidate for catalysis [14]. Biochar is a black solid obtained via high-temperature pyrolysis and often used as an excellent adsorbent [15,16]. Compared with other catalytic materials such as metal-based catalysts (e.g., Co, Fe, Cu, Mn), biochar has the advantages of easy preparation, a wide variety of raw precursors and low cost [17,18]. However, the catalytic activity of pristine biochar is often unsatisfactory owing to its limited

S_{BET} and undeveloped pore structure. Thus, several modification methods (e.g., metal doping and heteroatomic doping) have been applied to enhance the catalytic performance of pristine biochar [19,20]. Among others, N doping has been deemed a popular tool to boost the catalytic activity of pristine biochar [21]. N doping can enhance the interaction between catalysts and oxidants through changing the electrical potential on the carbon surface [22]. Nevertheless, the addition of external nitrogen sources (urea, thiourea, melamine, NH_4NO_3, NH_4Cl, $NH_3 \cdot H_2O$, etc.) is needed when preparing the N-doped biochar, which increases the preparation cost [21,23,24]. For instance, Wu et al. [23] prepared N-doped biochar by using pharmaceutical sludge and urea solid as the feedstocks, which may increase the preparation cost. Therefore, it is of great importance to develop N-doped biochars without exogenous nitrogen dopants, which may promote the development and application of biochars in environmental remediation. Interestingly, algae, as it contains a large amount of protein, can be directly transferred to N-doped biochars without using extra nitrogenous chemical reagents [25–27]. It was reported that N-abundant Taihu blue algae biomass could be used as the feedstock for N self-doping porous carbon preparation [28]. Thus, it could be speculated that the algal sludge obtained by the mechanical pressure filtration of algal slurries might be a promising precursor for N-doped biochars preparation. Notably, it is still a big challenge to effectively dispose of algal sludge and thus translating the algal sludge into valuable biochars may be an alternative solution. As one of the most widely used antimicrobial compounds in animal husbandry, relatively high sulfadiazine (SDZ) residues were detected in the environment [29]. Therefore, it is urgent that green and efficient methods for SDZ-contaminated water purification are developed.

Consequently, this study aimed to prepare N-rich biochars from algal sludge for PMS activation. The physicochemical properties of the N-doped biochars along with the adsorptive and oxidative degradation of SDZ were systematically investigated. Moreover, the activation mechanism of PMS by N-doped biochar was analyzed in detail. This study will provide a low-cost and efficient strategy for N-doped biochars preparation and establish an integrated oxidation process for SDZ-contaminated water cleanup.

2. Materials and Methods

2.1. Chemicals and Regents

SDZ (Table S1) and PMS were obtained from Aladdin, China. All other reagents were analytical grade (Text S1).

2.2. Preparation and Characterization of the ASBCx

The algal sludge was collected from a sewage plant in Wuxi, China. Firstly, the algal sludge was dried at 80 °C and then mechanically grinded and sieved to 0.15 mm. The obtained powdered sample was placed in an oven and heated to a required temperature (400, 500, 600 and 800 °C) at 5 °C min^{-1} and held for 2 h. The obtained black solids were then ground into fine powders and named as ASBCx, where x represented the annealing temperature (400, 500, 600 and 800 °C). Detailed information on various characterizations and electrochemical experiments is given in Text S2 and Text S3, respectively.

2.3. Adsorption Studies

Experiments on SDZ adsorption were performed in a 100 mL bottle containing 50 mL of SDZ solution (5 mg L^{-1}) and 0.2 g L^{-1} of ASBCx. The mixture was stirred at ambient temperature with a magnetic stirrer. At regular intervals (2, 5, 10, 20, 40, 60, 120 min), liquid samples were taken from the mixture and filtered with 0.2 mm of filter to remove the solid catalyst for analysis. Then, the adsorption behavior of SDZ on ASBCx was further studied by adsorption kinetics (Text S4).

2.4. Catalytic Degradation of SDZ

SDZ, as a model pollutant, was selected as a degradative substrate to evaluate the catalytic activity of ASBCx. The SDZ degradation was conducted in a glass beaker (100 mL) containing 50 mL of SDZ solution (5 mg L^{-1}). First, 10 mg of ASBCx was dispersed in a SDZ solution and stirred for 30 min to reach adsorption equilibrium. Then, 0.5 mM of oxidants were injected into the suspension to initiate catalytic reaction. At regular intervals (2, 5, 10, 20, 30, 40 min), 0.5 mL of solution was withdrawn, filtered with 0.2 mm of filter to remove the ASBCx, and quenched with 0.5 mL of methanol for analysis. More details of degradation experiments and analytical methods are provided in Text S5 and Text S6, respectively.

3. Results and Discussion

3.1. Characterizations

The SEM images of the biochars prepared under 400 and 800 °C are shown in Figure 1a,b, respectively. As depicted in Figure 1a, ASBC400 exhibited a massive and plane structure, which had a relatively smooth surface. In contrast, ASBC800 (Figure 1b) underwent a dramatic change and a large number of irregular defect structures appeared on the surface, which might be the active sites for pollutant adsorption. In addition, compared with the TEM image of ASBC400 (Figure 1c), it could be observed that plenty of nano-spaces/channels were formed in ASBC800 (Figure 1d), which was able to expose abundant active sites and beneficial for nanoconfinement effects, boosting the PMS activation and pollutants adsorption.

Figure 1. SEM images of ASBC400 (**a**), ASBC800 (**b**), and TEM images of ASBC400 (**c**) and ASBC800 (**d**).

As depicted in Figure 2, N_2 adsorption/desorption isotherms were characterized, which presented type IV, and the hysteresis loop occurred at the medium pressure end. These phenomena were due to N_2 condensation and accumulation in the porous channel, demonstrating that ASBCx were porous materials [21]. Moreover, the pore size distribution of ASBCx is listed in Table 1, which was mainly between 6 and 10 nm, and belonged to microporous and mesoporous. With the increase in pyrolysis temperature, the S_{BET} increased from 84.017 to 145.596 m^2 g^{-1}. Comparing the results, it could be seen that the higher pyrolysis temperature endowed ASBC800 with a larger S_{BET}. Thus, the ASBC800 with larger S_{BET} may have more active sites for pollutants and oxidants adsorption.

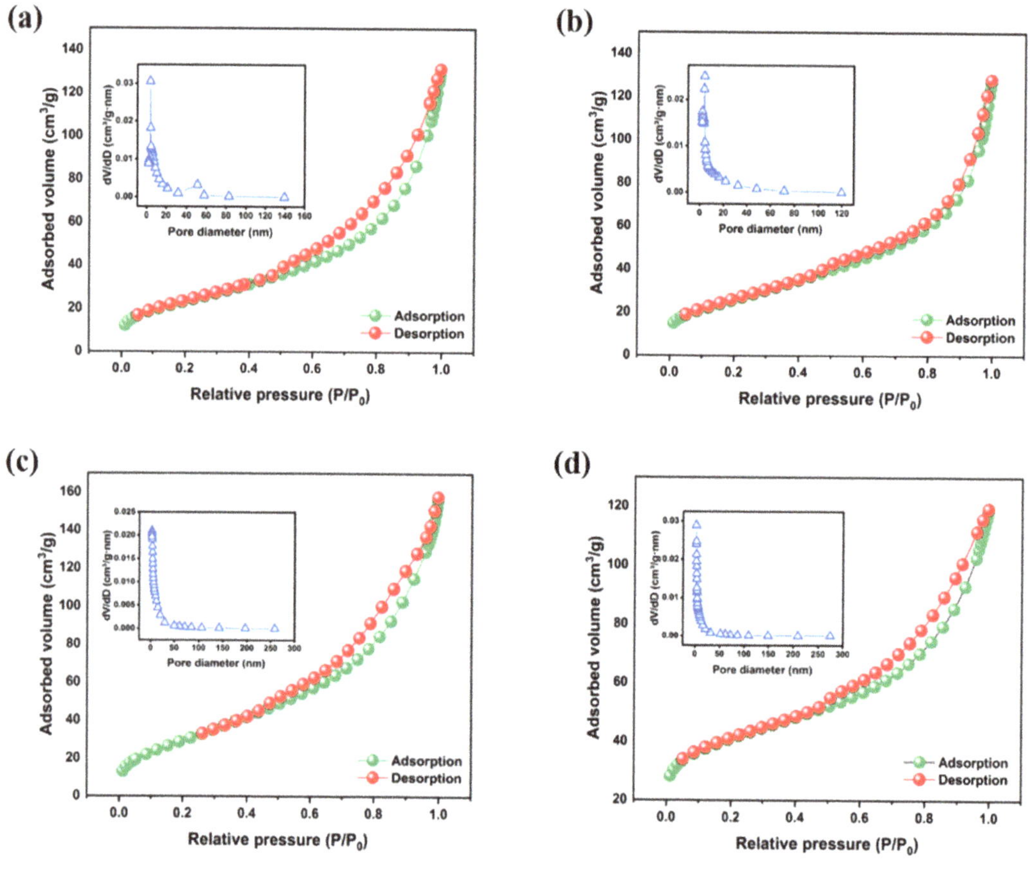

Figure 2. BET and pore size distribution of ASBC400 (**a**), ASBC500 (**b**), ASBC600 (c), and ASBC800 (**d**).

Table 1. Basic information of ASBCx.

Samples	S_{BET} (m^2 g^{-1})	Pore Volume (cm^3 g^{-1})	Average Pore Diameter (nm)
ASBC400	84.017	0.194	9.756
ASBC500	93.210	0.188	7.447
ASBC600	110.686	0.232	7.636
ASBC800	145.596	0.179	6.220

As depicted in Figure 3a, the Raman spectra of ASBCx showed two representative bands at 1343 cm^{-1} (D bands) and 1588 (G bands) cm^{-1}, which originated from the defects and disorder of carbon atomic crystals and the graphitic structures, respectively. The intensity ratios of D to G (I_D/I_G) clearly indicated the defective degree of the ASBCx [30]. Therefore, the higher I_D/I_G value of ASBC800 (0.972) than ASBC600 (0.969), ASBC500 (0.957), and ASBC400 (0.860) suggested higher defected carbon and it could be concluded that more new structural defects would be created under higher pyrolysis temperature. FTIR spectroscopy in the wavenumber range of 4000 to 500 cm^{-1} was conducted to investigate evolution of the surface of ASBCx. As displayed in Figure 3b, the band at 3428 cm^{-1} could be allotted to –OH stretching vibrations, whereas the peak at 2928 cm^{-1} was assigned to the C–H bond stretching vibrations [19]. Moreover, the peaks at 1470 cm^{-1} (C–C), 1592 cm^{-1} (C=C), and 1064 cm^{-1} (C–O) were identified [19]. Notably, the peak at 1174 cm^{-1} on ASBCx was owing to the formation of C–N bonds, which might be beneficial for oxygen reduction reaction (ORR) kinetics, boosting the ORR process [31]. In addition, the peak at 560 cm^{-1} was the Fe–O characteristic vibration mode of ASBCx [32]. XPS spectra (Figure 3c) confirmed ASBCx were composed of C, N, and O elements. Moreover, a signal appeared at 708.8 eV representing Fe 2p was observed in the XPS measurement spectra of ASBCx [33], which was due to the addition of iron–containing reagent in the process of algal sludge pressure filtration.

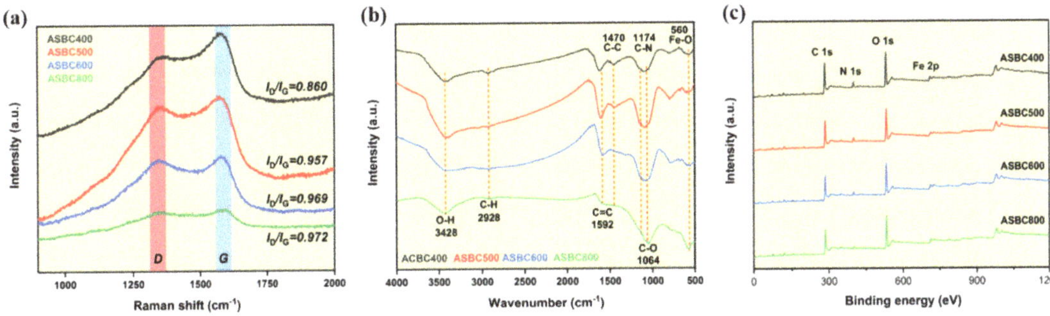

Figure 3. Raman spectra (**a**), FTIR spectra (**b**), and XPS spectra (**c**) of ASBCx.

The C 1s XPS spectra (Figure S1) were divided into four sections at: 284.6, 285.3, 286.4, and 533 eV, indicating C=C (sp^2-C), C–C (sp^3-C), C–O, and C=O, respectively. The O 1s spectra (Figure S2) were also divided into four sections: –C=O (531.4), –O–C=O (532.5), C–O (533.7), and H_2O (535.0). The Fe 2p core levels of ASBCx were analyzed, and Fe 2p was divided into two pairs of doublets and two satellite peaks (Figure S3). The peaks at 710.6 and 723.8 eV belonged to Fe^{2+} species, whereas the peaks at 713.6 and 726.8 eV belonged to Fe^{3+} species [34]. Additionally, the other two peaks at 717.2 and 732.8 eV originated from two satellites. The N 1s spectra of ASBCx were fitted into different components, including pyridinic N (N in 6-atom ring, 397.9 eV), pyrrolic N (N in 5-atom ring, 399.5 eV) and graphitic N (N in graphitic carbon plane, 401.1 eV). The content of the N configurations also exhibited correlation with the pyrolysis temperature (Figure 4). The percentage of graphitic N significantly increased from 15.62% to 34.82% when the pyrolysis temperature increased from 400 to 800 °C. It was well accepted that the graphitic N could be more reactive in PMS activation than other N configurations [35,36]. Thus, the increased graphitic N in ASBC800 implied its best performance for PMS activation and contaminants removal.

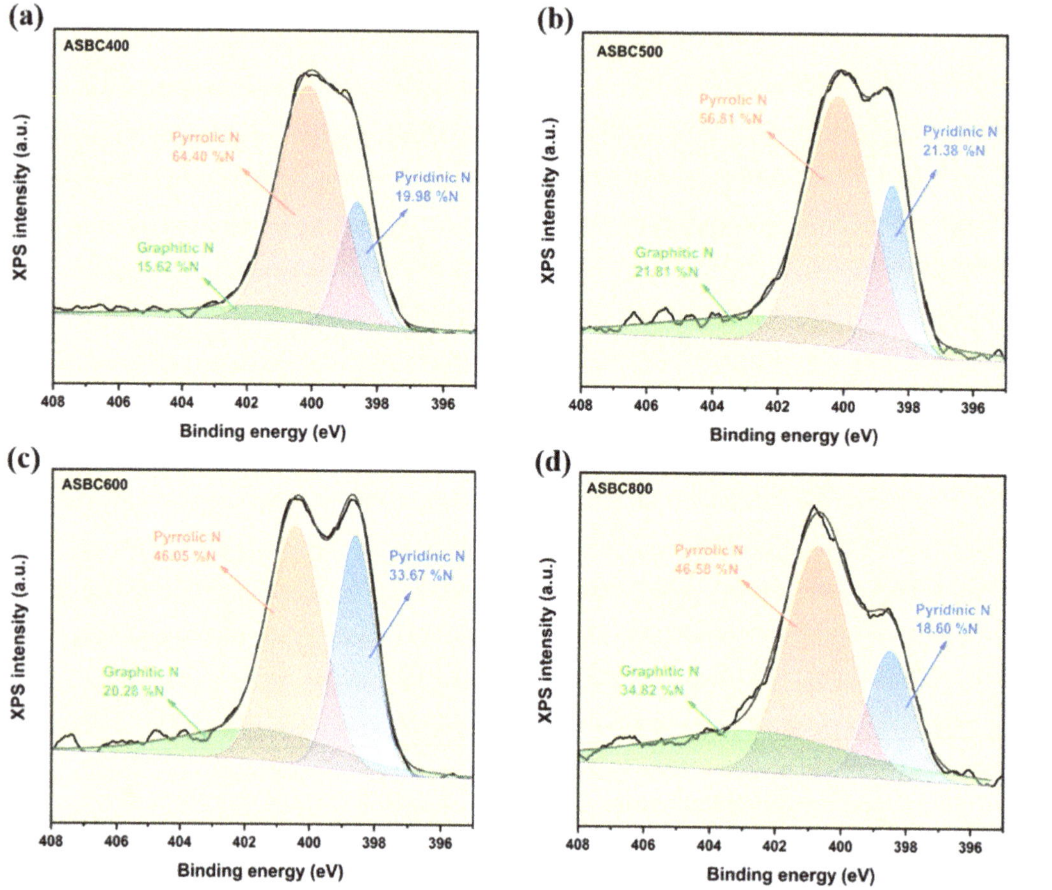

Figure 4. N 1s high-resolution spectrum of ASBC400 (**a**), ASBC500 (**b**), ASBC600 (**c**), and ASBC800 (**d**).

3.2. Adsorptive Experiment

As shown in Figure 5a,b, the SDZ molecule could be rapidly transferred to the ASBCx surface via adsorption. The better correlation coefficients (Table 2, $R^2 > 0.967$) were obtained by fitting with a pseudo-second-order kinetic model, indicating that chemisorption might dominate the adsorption process. More specifically, ASBC800 had the highest q_e value of 12.532 mg g^{-1}, followed with ASBC600 (1.175 mg g^{-1}), ASBC500 (0.993 mg g^{-1}), and ASBC400 (0.482 mg g^{-1}), suggesting the adsorption affinity of the catalyst could be boosted under a higher pyrolysis temperature. Additionally, as abovementioned in Table 1, the largest S_{BET} would be obtained under 800 °C, which would be beneficial for contaminant adsorption.

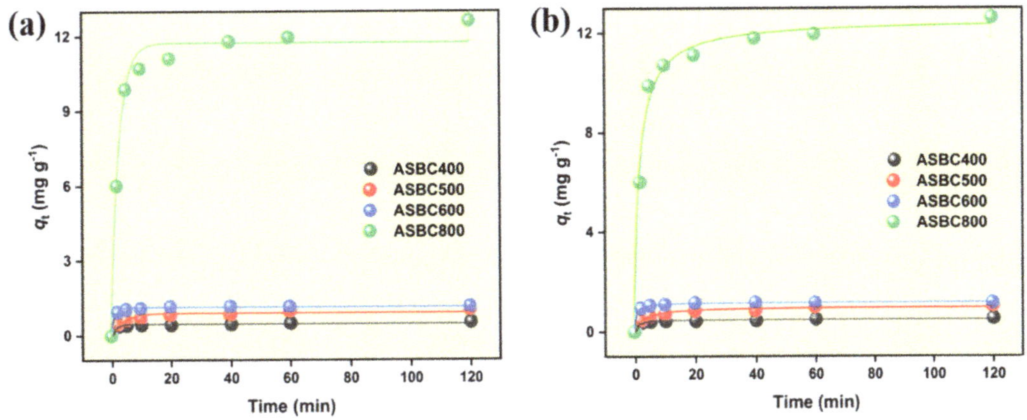

Figure 5. Pseudo-first-order (**a**) and pseudo-second-order (**b**) adsorption kinetic modelling results.

Table 2. Adsorption kinetic parameters.

Biochar	Pseudo First-Order			Pseudo Second-Order		
	q_e	k_1	R^2	q_e	k_2	R^2
ASBC400	0.461	0.721	0.941	0.482	2.696	0.967
ASBC500	0.915	0.239	0.912	0.993	0.435	0.969
ASBC600	1.144	0.881	0.991	1.175	1.816	0.999
ASBC800	11.763	0.352	0.985	12.532	0.043	0.990

3.3. Catalytic Degradation of SDZ

In order to evaluate the performance of ASBCx activated PMS, SDZ was selected as the model contaminant. As depicted in Figure S4, only 23.05% of the SDZ could be oxidized by PMS alone. For ASBC400, ASBC500 and ASBC600, negligible adsorption of SDZ was obtained (<3%), whereas ASBC800 could adsorb 37.4% of SDZ, which may be owing to the larger S_{BET} caused by higher pyrolysis temperature. As for the catalytic performance, a complete SDZ degradation removal was achieved in the ASBC800/PMS system in 40 min, while removals of 28.7%, 30.2% and 35.0% were obtained in 40 min in ASBC400/PMS, ASBC500/PMS, and ASBC600/PMS systems (Figure 6a), respectively. In addition, compared to the biochars prepared at a low temperature (k_{obs} = 0.00425–0.00587 min^{-1}), ASBC800 gave rise to an approximately 10-fold enhancement of reaction rate constant (0.0586 min^{-1}). As depicted in Figure 6b, a good correlation (R^2 = 0.866) between S_{BET} and q_e was observed, indicating that the larger S_{BET} of ASBC800 endowed it with stronger adsorptive capacity towards SDZ, and thus enhanced SDZ removal. Similarly, a good linear relationship was observed between k_{app} and q_e (R^2 = 0.999), suggesting that the adsorption of SDZ was beneficial for the subsequent degradation of SDZ (Figure 6c). Therefore, it could be concluded that the S_{BET} of ASBCx played a vital role both in SDZ adsorption and degradation. Notably, the ASBC800/PMS system attained a comparatively good SDZ removal rate and even surpassed some state-of-the-art metal/metal-free based oxidation systems (e.g., BC/PMS, Fe_3O_4/PMS, Cu_2O/PMS, and MoS_2/PMS) according to the higher SDZ normalized removal rate, faster kinetics, and lower oxidant dosage (Figure 6d).

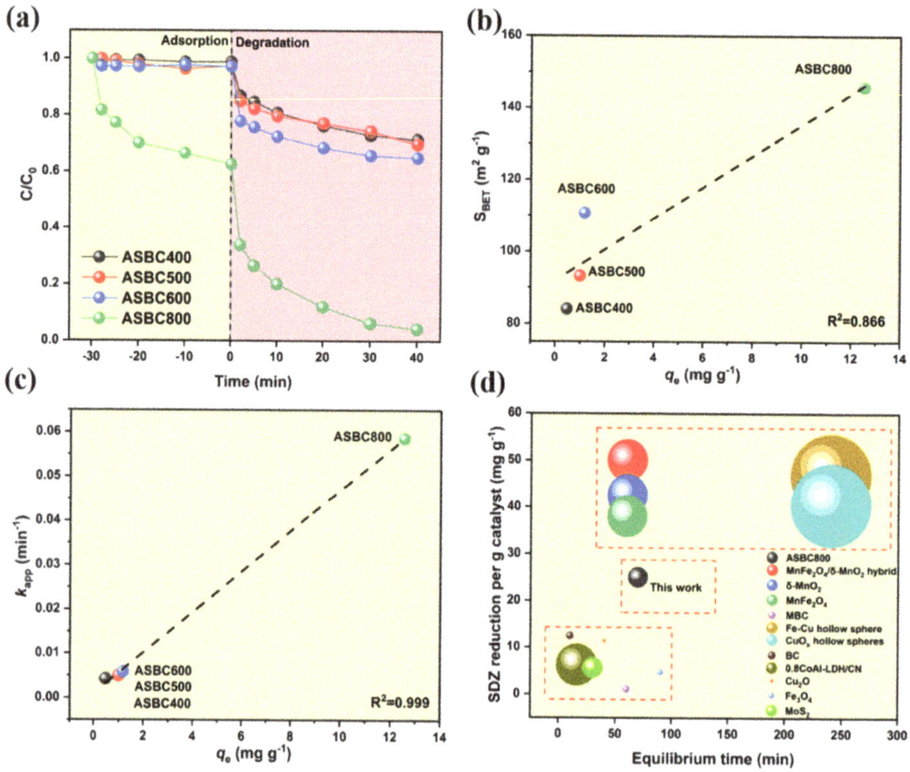

Figure 6. SDZ elimination in the PMS/ASBCx system (**a**), the correlation between S_{BET} and q_e (**b**), the correlation between k_{app} and q_e (**c**), and the comparisons between the ASBC800/PMS system and other PMS-based oxidation processes for SDZ removal (**d**). (The diameters of the balls were the ratios of PMS/SDZ).

3.4. Stability of the ASBC800/PMS System

In order to explore the practical application prospects of the ASBC800/PMS system, investigating the stability of the oxidation process was of great significance. First of all, the effect of ASBC800 dosage on SDZ degradation was explored (Figure 7a). As the ASBC800 dosage increased from 5 to 15 mg, the SDZ could be removed more quickly. This was because a higher dosage of ASBC800 could increase adsorptive sites for the PMS and SDZ, which could enhance the interaction among catalysts, oxidants and target contaminants. The effects of organic matter and inorganic anions (e.g., humic acid (HA), Cl^-, $H_2PO_4^-$) were further confirmed. HA, as a representative NOM, is often used as the model compound to explore the effects of NOM on the oxidation processes. HA with abundant electrons can function as a scavenger to consume the generated ROS, and thus obviously depress the oxidation process. More specifically, HA can attach to the surface of catalysts through π–π stacking, hydrophobic or other interactions and cover the active sites, and thus hinder the adsorption of pollutants. On the other hand, HA may compete with the target pollutants for ROS, inhibiting organic contaminants degradation. Notably, the ASBC800/PMS system could exhibit strong tolerance to HA (Figure 7b). Interestingly, Cl^- remarkably accelerated the SDZ oxidation process (Figure 7c). This was probably because Cl^- reacted with PMS, which further produced other species such as $Cl^\bullet$ or HOCl, and thus synergistically degraded SDZ (Equations (1)–(11)). In addition, $H_2PO_4^-$ showed weak influences on SDZ oxidation processes (Figure 7d), which might be due to the fact that

$H_2PO_4^-$ can convert $SO_4^{\bullet-}$ and $^{\bullet}OH$ to $HPO_4^{\bullet-}$ that have a weaker oxidizing capacity, (Equations (12) and (13)), and thus slightly suppress the oxidation process.

$$SO_4^{\bullet-} + Cl^- \rightarrow SO_4^{2-} + Cl^{\bullet} \quad (1)$$

$$HO^- + Cl^{\bullet} \rightarrow ClOH^{\bullet-} \quad (2)$$

$$Cl^{\bullet} + Cl^{\bullet} \rightarrow Cl_2 \quad (3)$$

$$Cl^{\bullet} + Cl^- \rightarrow Cl_2^{\bullet-} \quad (4)$$

$$Cl_2^{\bullet-} + Cl_2^{\bullet-} \rightarrow Cl_2 + 2Cl^- \quad (5)$$

$$Cl_2^{\bullet-} + H_2O \rightarrow ClOH^- + H^+ + Cl^- \quad (6)$$

$$ClOH^{\bullet-} + H^+ \rightarrow Cl^{\bullet} + H_2O \quad (7)$$

$$R^{\bullet} + Cl_2^{\bullet-} \rightarrow R - Cl + Cl^- \quad (8)$$

$$Cl^- + HSO_5^- \rightarrow SO_4^{2-} + HOCl \quad (9)$$

$$2Cl^- + HSO_5^- + H^+ \rightarrow SO_4^{2-} + Cl_2 + H_2O \quad (10)$$

$$R - H + HOCl \rightarrow R - Cl + H_2O \quad (11)$$

$$HPO_4^{2-} + SO_4^{\bullet-} \rightarrow SO_4^{2-} + HPO_4^{\bullet-} \quad (12)$$

$$HPO_4^{2-} + HO^{\bullet} \rightarrow OH^- + HPO_4^{\bullet-} \quad (13)$$

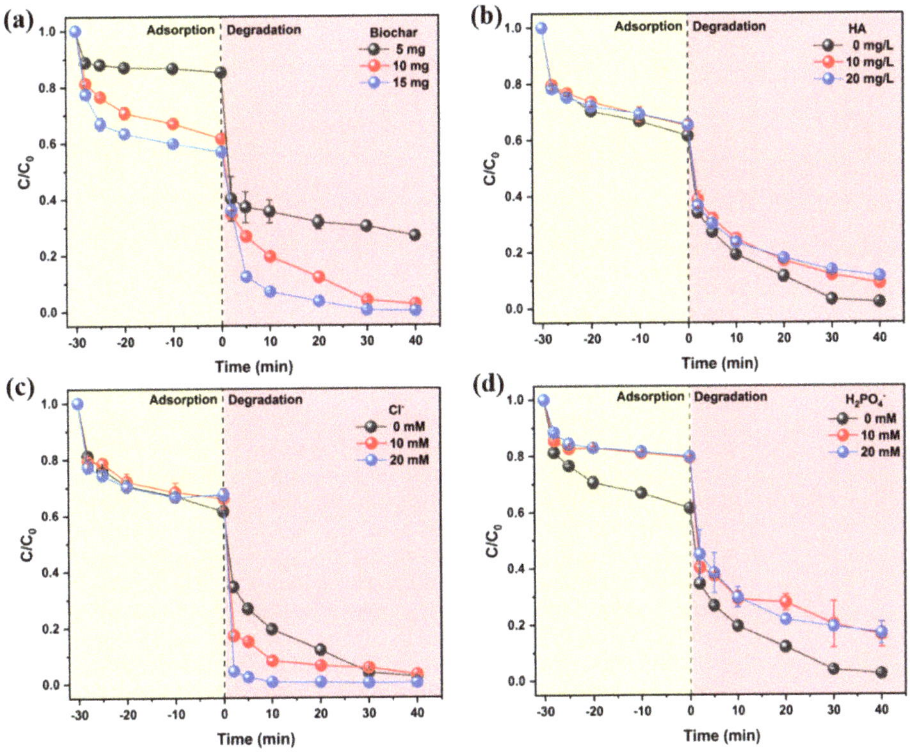

Figure 7. Effects of ASBC800 dosage (**a**), HA (**b**), Cl^- (**c**), $H_2PO_4^-$ (**d**) on SDZ removal.

3.5. Activation Mechanism

3.5.1. Identification of Reactive Oxygen Species

As evidenced in Figure 6a, the removal of SDZ was significantly boosted in the ASBC800/PMS system, suggesting the formation of reactive oxygen species (ROS). On the one hand, $^{\bullet}OH$ and $SO_4^{\bullet-}$ could function as two primary ROS, which often dominate the radical-based oxidation process. Quenching experiments were taken firstly for elucidating the generated ROS in the ASBC800/PMS system. If radicals played vital roles in such oxidation system, the addition of methanol (MeOH) and tertiary butanol (TBA) would obviously inhibit the degradation of SDZ (Table S3). As depicted in Figure 8a, a moderate inhibition of SDZ elimination was obtained, indicating that $^{\bullet}OH$ and $SO_4^{\bullet-}$ were produced in the ASBC800-mediated PMS activation process. Moreover, we monitored the leaching of Fe^{2+} during the oxidation process (Figure S5), and the leached Fe^{2+} content was 0.23 mg L^{-1}, which met the requirements of surface water environmental quality standards (GB3838-2002, 0.3 mg L^{-1}). In addition, the SDZ degradation experiments were also conducted in the PMS/Fe^{2+} (0.23 mg L^{-1}) system, in which the SDZ could be removed by the homogeneous reaction system (Figure S6). Notably, both high concentrations of MeOH and TBA could not completely suppress the SDZ removal, and thus other ROS might also participate in the oxidation reaction.

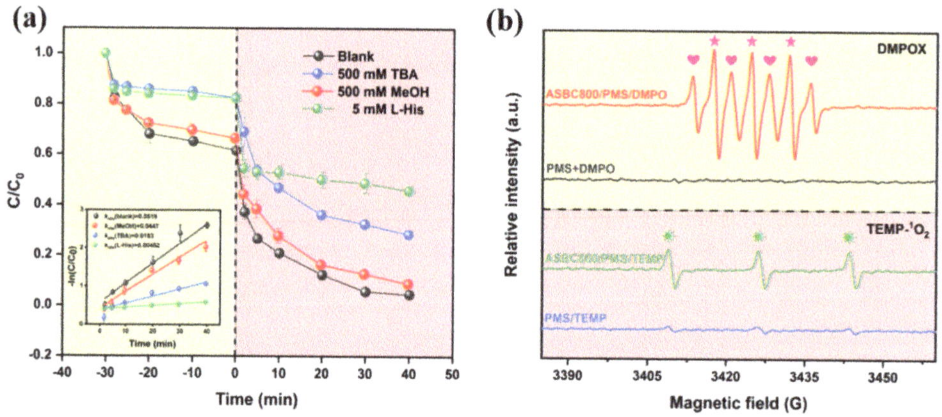

Figure 8. Effects of scavengers on SDZ degradation in PMS/ASBC800 system (**a**) and the result of EPR characterizations (**b**). (5,5-dimethylpyrroline-N-oxide: (DMPO), DMPO-adducts: (DMPOX), 2,2,6,6-tetramethyl-4-piperidinol: (TEMP), TEMP-adducts: (TEMP-$^1O_2$)).

$^1O_2$, as the selective ROS, could also oxidize contaminants, especially the electron-rich compounds. To verify the generation of $^1O_2$ in the ASBC800/PMS system, *L*-histidine (*L*-his) was chosen for the quenching experiments. As shown in Figure 8a, the oxidizing capability of the ASBC800/PMS system was effectively inhibited after adding 5 mM of *L*-his, indicating that $^1O_2$ was also generated and obligated to the elimination of SDZ. In addition, D_2O was able to accelerate the $^1O_2$-dominated oxidation process because the lifetime of $^1O_2$ in D_2O (55 μs) was longer than that in H_2O (4.2 μs). As shown in Figure 7, SDZ removal was accelerated in D_2O, suggesting that $^1O_2$ was the contributor. The above inference could be further verified by EPR tests. As shown in Figure 8b, both DMPO-OH and DMPO-SO_4 signals were detected in the ASBC800/PMS system, and the signal of TEMP-$^1O_2$ was also observed in the ASBC800/PMS system, confirming that ASBC800 was capable of generating ROS via PMS activation.

3.5.2. Electrochemical Analysis

Apart from common reactive species produced in the oxidation system, electron-transfer regime was further investigated by electrochemical tests (Text S6). Firstly, electrochemical impedance spectroscopy (EIS) was measured to evaluate the conductivity of the biochars prepared under different temperatures. As depicted in Figure 9a, the semicircle diameter of ASBC800 was smallest, which suggested that ASBC800 possessed the best conductivity among other biochars and thus could serve as the favorable medium for electron transfer. Secondly, a noticeable increase in current was seen from the LSV curves with the injection of PMS (Figure 9b), demonstrating the formation of metastable ASBC800-PMS complexes (ASBC800-PMS*). Notably, another remarkable increase in current was triggered upon the injection of SDZ, manifesting the formation of current flow from SDZ to ASBC800-PMS*. In addition, an I-t plot was further conducted to verify this conclusion. As evidenced in Figure 9c, an obvious current jump was detected with sequential injection of PMS and SDZ, which was the solid evidence proving the occurrence of electron transfer. Overall, both radicals (e.g., $SO_4^{\bullet-}$, and $^{\bullet}OH$), and nonradical pathways ($^1O_2$ and electron-transfer) were responsible for SDZ oxidation, in which the leached Fe^{2+} was beneficial for the generation of radicals and ASBC800 played essential roles in activating PMS, accumulating SDZ, and regulating electron shuttle from SDZ to ASBC800-PMS*.

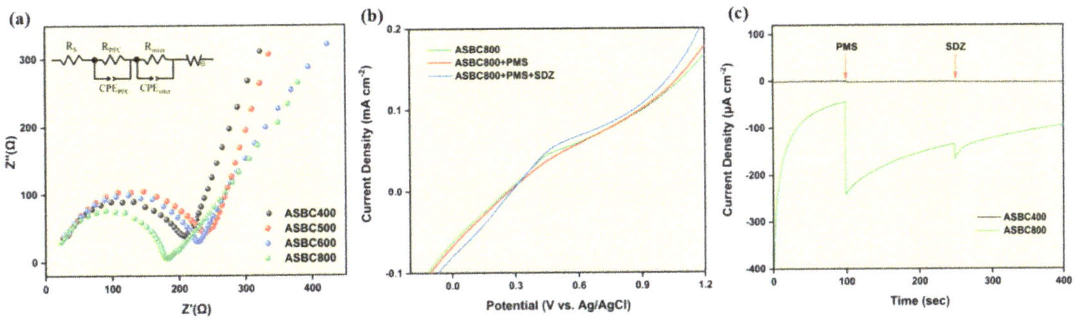

Figure 9. EIS analysis of ASBC800 (**a**), LSV under different conditions (**b**), and I-t curves (**c**).

3.6. Durability and Reusability

To examine the practical application potential of ASBC800 in eliminating SDZ-contaminated water, SDZ degradation experiments were performed in real river water and wastewater. As displayed in Figure 10a, unremarkable deterioration of SDZ removal was observed in river water, whereas a moderate deterioration occurred in wastewater, which might be owing to the ultra-high concentration compounds in the wastewater. Nevertheless, over 79% of SDZ was successfully degraded in river water and wastewater, indicating that ASBC800 has an excellent resistance. In addition, cycle experiments were performed to evaluate the stability of the ASBC800/PMS system. Unfortunately, we noticed a remarkable deterioration of SDZ degradation in the second run (Figure 10b). We supposed that the depletion was induced by the retarded adsorption capacity (from 38.7% of the first run to 22.75% of the second run). This was probably due to the adsorbed SDZ being not completely degraded, and thus only limited active sites could be provided for the SDZ adsorption. To verify this hypothesis, we used acetonitrile to wash the residual SDZ on the surface of ASBC800 in order to regenerate the catalyst. As expected, the adsorption efficiency of SDZ was significantly recovered from 22.75% to 28.60%, and thus the total removal rate of SDZ was also improved.

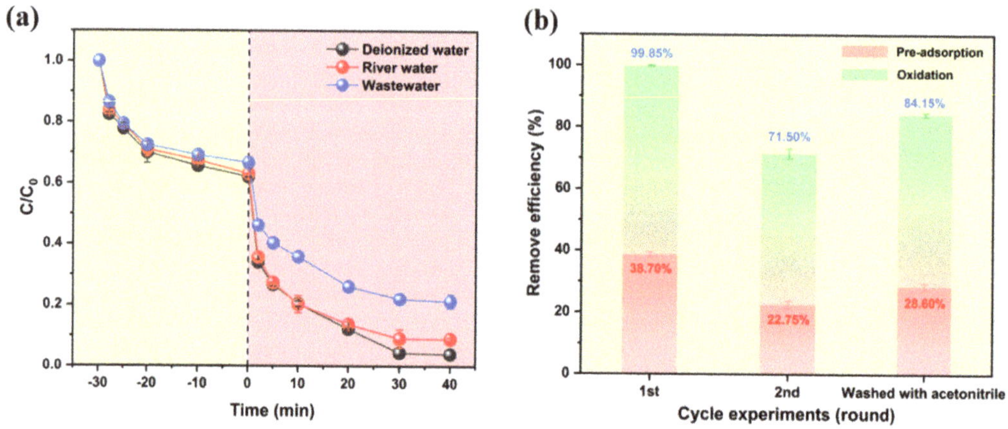

Figure 10. Elimination of SDZ in real aquatic systems (**a**) and the regeneration of ASBC800 (**b**).

4. Conclusions

In conclusion, N-rich and cost-effective biochars were fabricated by algal sludge, which were further used to catalyze PMS for SDZ degradation. To be specific, the ASBC800/PMS system exhibited excellent performance for SDZ removal (>95% within 70 min), which even surpass many state-of-the-art metal/metal-free based oxidation systems. Based on the quenching experiments, solvent exchange experiments and EPR characterizations, both radical pathways (i.e., $^{\bullet}OH$ and $SO_4^{\bullet-}$) and nonradical pathways (i.e., $^1O_2$ and electron-transfer) were demonstrated to be responsible for SDZ removal. Notably, during the electron transfer process, PMS molecules were first adsorbed on the ASBC800 surface to form the surface-bound ASBC800-PMS* complexes, which could subsequently degrade the co-adsorbed SDZ. Owing to the selectivity of non-radical mechanisms, the ASBC800/PMS system could maintain outstanding SDZ removal efficiency even in real wastewater, indicating that ASBC800 possesses promising prospects for practical application. Overall, this study might provide a new insight into N-rich and environment friendly biochar preparation from algal sludge and deepen the insight into mechanisms of PMS activation with carbon-based functional materials.

Supplementary Materials: The following supporting information can be downloaded at: https://www.mdpi.com/article/10.3390/coatings13020431/s1, Figure S1: Deconvoluted C 1s XPS spectra of the ASBCx samples; Figure S2: Deconvoluted O 1s XPS spectra of the ASBCx samples; Figure S3: Deconvoluted Fe 2p XPS spectra of the ASBCx samples; Figure S4: Removal of SDZ by PMS alone; Figure S5: The leached Fe^{2+} during PMS activation; Figure S6: Removal of SDZ by PMS alone and Fe^{2+}/PMS; Figure S7: Removal of SDZ by solvent exchange experiment; Table S1: The physicochemical properties of SDZ; Table S2: Catalytic performance of reported Fenton-like catalysts for PMS activation [37–44]; Table S3: Second-order rate constants for reactions of the MeOH and TBA with different radicals; Text S1: Chemicals and regents; Text S2: Characterizations of the biochars; Text S3: Details of electrochemical experiment; Text S4: Adsorption kinetics; Text S5: Details of degradation experiments.

Author Contributions: C.L.: writing—original draft, formal analysis, investigation, and visualization. Z.C.: writing—original draft, formal analysis, investigation, and supervision. R.K.: investigation and formal analysis. H.D.: writing—reviewing and editing, investigation. Q.L.: writing—reviewing and editing, investigation. T.W.: writing—reviewing and editing, investigation. J.W.: writing—reviewing and editing, investigation. D.T.: conceptualization, methodology. Y.X.: formal analysis, resources, writing—review and editing. Z.W.: visualization, supervision. All authors have read and agreed to the published version of the manuscript.

Funding: This work has been supported by the Research Project of Science and Technology of the Anyang City (2022C01NY016), the Research Project of Science and Technology of the Henan Province (222102320274), PhD research startup foundation of Anyang Institute of Technology (BSJ2021035), Postdoc research startup foundation of Anyang Institute of Technology (BHJ2022005), and Major science and technology projects of Anyang (201928).

Institutional Review Board Statement: All authors agreed to publish this manuscript in Environmental Science and Pollution Research.

Informed Consent Statement: All authors read and approved the final manuscript.

Data Availability Statement: The data used and/or analyzed during the current study are available from the corresponding author on reasonable request.

Conflicts of Interest: The authors declare no competing interests.

References

1. Miao, J.; Zhu, Y.; Lang, J.; Zhang, J.; Cheng, S.; Zhou, B.; Zhang, L.; Alvarez, P.; Long, M. Spin-State-Dependent Peroxymonosulfate Activation of Single-Atom M–N Moieties via a Radical-Free Pathway. *ACS Catal.* **2021**, *11*, 9569–9577. [CrossRef]
2. Xiao, C.; Hu, Y.; Li, Q.; Liu, J.; Li, X.; Shi, Y.; Chen, Y.; Cheng, J. Carbon-doped defect MoS_2 co-catalytic Fe^{3+}/peroxymonosulfate process for efficient sulfadiazine degradation: Accelerating Fe^{3+}/Fe^{2+} cycle and $^1O_2$ dominated oxidation. *Sci. Total Environ.* **2023**, *858*, 159587. [CrossRef] [PubMed]
3. Wang, L.; Jiang, S.; Huang, J.; Jiang, H. Oxygen-doped biochar for the activation of ferrate for the highly efficient degradation of sulfadiazine with a distinct pathway. *J. Environ. Chem. Eng.* **2022**, *10*, 108537. [CrossRef]
4. Yu, J.; Feng, H.; Tang, L.; Pang, Y.; Zeng, G.; Lu, Y.; Dong, H.; Wang, J.; Liu, Y.; Feng, C.; et al. Metal-free carbon materials for persulfate-based advanced oxidation process: Microstructure, property and tailoring. *Prog. Mater. Sci.* **2020**, *111*, 100654. [CrossRef]
5. Huang, K.; Zhang, H. Direct Electron-Transfer-Based Peroxymonosulfate Activation by Iron-Doped Manganese Oxide (delta-MnO_2) and the Development of Galvanic Oxidation Processes (GOPs). *Environ. Sci. Technol.* **2019**, *53*, 12610–12620. [CrossRef] [PubMed]
6. Hu, P.; Su, H.; Chen, Z.; Yu, C.; Li, Q.; Zhou, B.; Alvarez, P.; Long, M. Selective Degradation of Organic Pollutants Using an Efficient Metal-Free Catalyst Derived from Carbonized Polypyrrole via Peroxymonosulfate Activation. *Environ. Sci. Technol.* **2017**, *51*, 11288–11296. [CrossRef] [PubMed]
7. Miao, J.; Geng, W.; Alvarez, P.; Long, M. 2D N-Doped Porous Carbon Derived from Polydopamine-Coated Graphitic Carbon Nitride for Efficient Nonradical Activation of Peroxymonosulfate. *Environ. Sci. Technol.* **2020**, *54*, 8473–8481. [CrossRef]
8. Duan, X.; Sun, H.; Wang, S. Metal-Free Carbocatalysis in Advanced Oxidation Reactions. *Acc. Chem. Res.* **2018**, *51*, 678–687. [CrossRef]
9. Cheng, X.; Li, P.; Zhou, W.; Wu, D.; Luo, C.; Liu, W.; Ren, Z.; Liang, H. Effect of peroxymonosulfate oxidation activated by powdered activated carbon for mitigating ultrafiltration membrane fouling caused by different natural organic matter fractions. *Chemosphere* **2019**, *221*, 812–823. [CrossRef]
10. Fu, C.; Sun, G.; Wang, C.; Wei, B.; Ran, G.; Song, Q. Fabrication of nitrogen-doped graphene nanosheets anchored with carbon nanotubes for the degradation of tetracycline in saline water. *Environ. Res.* **2022**, *206*, 112242. [CrossRef]
11. Huo, X.; Zhou, P.; Zhang, J.; Liu, Y.; Cheng, X.; Liu, Y.; Li, W.; Zhang, Y. N, S-Doped porous carbons for persulfate activation to remove tetracycline: Nonradical mechanism. *J. Hazard. Mater.* **2020**, *391*, 122055. [CrossRef] [PubMed]
12. Xiong, W.; Wang, Z.; He, S.; Hao, F.; Yang, Y.; Lv, Y.; Zhang, W.; Liu, P.; Luo, H. Nitrogen-doped carbon nanotubes as a highly active metal-free catalyst for nitrobenzene hydrogenation. *Appl. Catal. B Environ.* **2020**, *260*, 118105. [CrossRef]
13. Yang, B.; Kang, H.; Ko, Y.; Woo, H.; Gim, G.; Choi, J.; Kim, J.; Cho, K.; Kim, E.; Lee, S.; et al. Persulfate activation by nanodiamond-derived carbon onions: Effect of phase transformation of the inner diamond core on reaction kinetics and mechanisms. *Appl. Catal. B Environ.* **2021**, *293*, 120205. [CrossRef]
14. Ye, S.; Zeng, G.; Tan, X.; Wu, H.; Liang, J.; Song, B.; Tang, N.; Zhang, P.; Yang, Y.; Chen, Q.; et al. Nitrogen-doped biochar fiber with graphitization from Boehmeria nivea for promoted peroxymonosulfate activation and non-radical degradation pathways with enhancing electron transfer. *Appl. Catal. B Environ.* **2020**, *269*, 118850. [CrossRef]
15. Chen, T.; Zhou, Z.; Han, R.; Meng, R.; Wang, H.; Lu, W. Adsorption of cadmium by biochar derived from municipal sewage sludge: Impact factors and adsorption mechanism. *Chemosphere* **2015**, *134*, 286–293. [CrossRef]
16. Chen, Y.; Lin, Y.; Ho, S.; Zhou, Y.; Ren, N. Highly efficient adsorption of dyes by biochar derived from pigments-extracted macroalgae pyrolyzed at different temperature. *Bioresour. Technol.* **2018**, *259*, 104–110. [CrossRef]
17. Yu, J.; Tang, L.; Pang, Y.; Zeng, G.; Feng, H.; Zou, J.; Wang, J.; Feng, C.; Zhu, X.; Ouyang, X.; et al. Hierarchical porous biochar from shrimp shell for persulfate activation: A two-electron transfer path and key impact factors. *Appl. Catal. B Environ.* **2020**, *260*, 118160. [CrossRef]

18. Wang, H.; Guo, W.; Liu, B.; Si, Q.; Luo, H.; Zhao, Q.; Ren, N. Sludge-derived biochar as efficient persulfate activators: Sulfurization-induced electronic structure modulation and disparate nonradical mechanisms. *Appl. Catal. B Environ.* **2020**, *279*, 119361. [CrossRef]
19. Mian, M.; Liu, G.; Fu, B.; Song, Y. Facile synthesis of sludge-derived MnOx-N-biochar as an efficient catalyst for peroxymonosulfate activation. *Appl. Catal. B Environ.* **2019**, *255*, 117765. [CrossRef]
20. Zaeni, J.; Lim, W.; Wang, Z.; Ding, D.; Chua, Y.; Ng, S.; Oh, W. In situ nitrogen functionalization of biochar via one-pot synthesis for catalytic peroxymonosulfate activation: Characteristics and performance studies. *Sep. Purif. Technol.* **2020**, *241*, 116702. [CrossRef]
21. Mian, M.; Liu, G.; Zhou, H. Preparation of N-doped biochar from sewage sludge and melamine for peroxymonosulfate activation: N-functionality and catalytic mechanisms. *Sci. Total Environ.* **2020**, *744*, 140862. [CrossRef] [PubMed]
22. Zhang, P.; Yang, Y.; Duan, X.; Liu, Y.; Wang, S. Density Functional Theory Calculations for Insight into the Heterocatalyst Reactivity and Mechanism in Persulfate-Based Advanced Oxidation Reactions. *ACS Catal.* **2021**, *11*, 11129–11159. [CrossRef]
23. Wu, Q.; Zhang, Y.; Liu, H.; Liu, H.; Tao, J.; Cui, M.; Zheng, Z.; Wen, D.; Zhan, X. FexN produced in pharmaceutical sludge biochar by endogenous Fe and exogenous N doping to enhance peroxymonosulfate activation for levofloxacin degradation. *Water Res.* **2022**, *224*, 119022. [CrossRef]
24. Ren, F.; Zhu, W.; Zhao, J.; Liu, H.; Zhang, X.; Zhang, H.; Zhu, H.; Peng, Y.; Wang, B. Nitrogen-doped graphene oxide aerogel anchored with spinel $CoFe_2O_4$ nanoparticles for rapid degradation of tetracycline. *Sep. Purif. Technol.* **2020**, *241*, 116690. [CrossRef]
25. Ouasfi, N.; Zbair, M.; Bouzikri, S.; Anfar, Z.; Bensitel, M.; Ait Ahsaine, H.; Sabbar, E.; Khamliche, L. Selected pharmaceuticals removal using algae derived porous carbon: Experimental, modeling and DFT theoretical insights. *RSC Adv.* **2019**, *9*, 9792–9808. [CrossRef] [PubMed]
26. Ji, R.; Wu, Y.; Bian, Y.; Song, Y.; Sun, Q.; Jiang, X.; Zhang, L.; Han, J.; Cheng, H. Nitrogen-doped porous biochar derived from marine algae for efficient solid-phase microextraction of chlorobenzenes from aqueous solution. *J. Hazard. Mater.* **2021**, *407*, 124785. [CrossRef] [PubMed]
27. Wang, H.; Wang, H.; Zhao, H.; Yan, Q. Adsorption and Fenton-like removal of chelated nickel from Zn-Ni alloy electroplating wastewater using activated biochar composite derived from Taihu blue algae. *Chem. Eng. J.* **2020**, *379*, 122372. [CrossRef]
28. Wang, H.; Wang, H.; Liu, G.; Yan, Q. In-situ pyrolysis of Taihu blue algae biomass as appealing porous carbon adsorbent for CO_2 capture: Role of the intrinsic N. *Sci. Total Environ.* **2021**, *771*, 145424. [CrossRef]
29. Gaballah, M.; Guo, J.; Hassanein, A.; Sobhi, M.; Zheng, Y.; Philbert, M.; Li, B.; Sun, H.; Dong, R. Removal performance and inhibitory effects of combined tetracycline, oxytetracycline, sulfadiazine, and norfloxacin on anaerobic digestion process treating swine manure. *Sci. Total Environ.* **2023**, *857*, 159536. [CrossRef]
30. Yu, J.; Tang, L.; Pang, Y.; Zeng, G.; Wang, J.; Deng, Y.; Liu, Y.; Feng, H.; Chen, S.; Ren, X. Magnetic nitrogen-doped sludge-derived biochar catalysts for persulfate activation: Internal electron transfer mechanism. *Chem. Eng. J.* **2019**, *364*, 146–159. [CrossRef]
31. Zhong, K.; Li, M.; Yang, Y.; Zhang, H.; Zhang, B.; Tang, J.; Yan, J.; Su, M.; Yang, Z. Nitrogen-doped biochar derived from watermelon rind as oxygen reduction catalyst in air cathode microbial fuel cells. *Appl. Energ.* **2019**, *242*, 516–525. [CrossRef]
32. Zhu, K.; Bin, Q.; Shen, Y.; Huang, J.; He, D.; Chen, W. In-situ formed N-doped bamboo-like carbon nanotubes encapsulated with Fe nanoparticles supported by biochar as highly efficient catalyst for activation of persulfate (PS) toward degradation of organic pollutants. *Chem. Eng. J.* **2020**, *402*, 126090. [CrossRef]
33. Li, X.; Jia, Y.; Zhou, M.; Su, X.; Sun, J. High-efficiency degradation of organic pollutants with Fe, N co-doped biochar catalysts via persulfate activation. *J. Hazard. Mater.* **2020**, *397*, 122764. [CrossRef] [PubMed]
34. Cheng, X.; Dou, S.; Qin, G.; Wang, B.; Yan, P.; Isimjan, T.; Yang, X. Rational design of highly selective nitrogen-doped Fe_2O_3-CNTs catalyst towards H_2O_2 generation in alkaline media. *Int. J. Hydrogen Energ.* **2020**, *45*, 6128–6137. [CrossRef]
35. Qi, Y.; Ge, B.; Zhang, Y.; Jiang, B.; Wang, C.; Akram, M.; Xu, X. Three-dimensional porous graphene-like biochar derived from Enteromorpha as a persulfate activator for sulfamethoxazole degradation: Role of graphitic N and radicals transformation. *J. Hazard. Mater.* **2020**, *399*, 123039. [CrossRef]
36. Ho, S.; Chen, Y.; Li, R.; Zhang, C.; Ge, Y.; Cao, G.; Ma, M.; Duan, X.; Wang, S.; Ren, N. N-doped graphitic biochars from C-phycocyanin extracted Spirulina residue for catalytic persulfate activation toward nonradical disinfection and organic oxidation. *Water Res.* **2019**, *159*, 77–86. [CrossRef]
37. Dong, F.; Yan, L.; Huang, S.; Liang, J.; Zhang, W.; Yao, X.; Chen, X.; Qian, W.; Guo, P.; Kong, L.; et al. Removal of antibiotics sulfadiazine by a biochar based material activated persulfate oxidation system: Performance, products and mechanism. *Process Saf. Environ.* **2021**, *157*, 411–419. [CrossRef]
38. Feng, Y.; Wu, D.; Deng, Y.; Zhang, T.; Shih, K. Sulfate Radical-Mediated Degradation of Sulfadiazine by $CuFeO_2$ Rhombohedral Crystal-Catalyzed Peroxymonosulfate: Synergistic Effects and Mechanisms. *Environ. Sci. Technol.* **2016**, *50*, 3119–3127. [CrossRef]
39. Jiang, Z.; Li, Y.; Zhou, Y.; Liu, X.; Wang, C.; Lan, Y.; Li, Y. Co_3O_4-MnO_2 nanoparticles moored on biochar as a catalyst for activation of peroxymonosulfate to efficiently degrade sulfonamide antibiotics. *Sep. Purif. Technol.* **2021**, *281*, 119935. [CrossRef]
40. Li, Y.; Feng, Y.; Yang, B.; Yang, Z.; Shih, K. Activation of peroxymonosulfate by molybdenum disulfide-mediated traces of Fe(III) for sulfadiazine degradation. *Chemosphere* **2021**, *283*, 131212. [CrossRef]
41. Liu, T.; Wu, K.; Wang, M.; Jing, C.; Chen, Y.; Yang, S.; Jin, P. Performance and mechanisms of sulfadiazine removal using persulfate activated by Fe_3O_4@CuOx hollow spheres. *Chemosphere* **2021**, *262*, 127845. [CrossRef] [PubMed]

42. Tan, C.; Lu, X.; Cui, X.; Jian, X.; Hu, Z.; Dong, Y.; Liu, X.; Huang, J.; Deng, L. Novel activation of peroxymonosulfate by an easily recyclable VC@Fe_3O_4 nanoparticles for enhanced degradation of sulfadiazine. *Chem. Eng. J.* **2019**, *363*, 318–328. [CrossRef]
43. Zeng, H.; Zhang, H.; Deng, L.; Shi, Z. Peroxymonosulfate-assisted photocatalytic degradation of sulfadiazine using self-assembled multi-layered CoAl-LDH/g-C_3N_4 heterostructures: Performance, mechanism and eco-toxicity evaluation. *J. Water Process Eng.* **2019**, *33*, 101084. [CrossRef]
44. Zhu, L.; Shi, Z.; Deng, L.; Duan, Y. Efficient degradation of sulfadiazine using magnetically recoverable $MnFe_2O_4$/δ-MnO_2 hybrid as a heterogeneous catalyst of peroxymonosulfate. *Colloid Surface A* **2021**, *609*, 125637. [CrossRef]

coatings

Article

Sedum Plumbizincicola Derived Functional Carbon for Activation of Peroxymonosulfate to Eliminate Bisphenol A: Performance and Reaction Mechanisms

Chao Liu [1,2,†], Zhenxiang Chen [3,†], Ruiqin Kang [1], Yongsheng Niu [1,*], Wenhui Su [1], Xiaolong Wang [1], Dayong Tian [1] and Ying Xu [2,*]

1 Department of Environmental Engineering, College of Chemistry and Environmental Engineering, Anyang Institute of Technology, Anyang 455000, China
2 College of Chemistry and Chemical Engineering, Henan University, Kaifeng 475004, China
3 Jiangsu Xingzhou Ecological Environment Technology Co., Ltd., Nanjing 210004, China
* Correspondence: nys2205@163.com (Y.N.); hdccxu@126.com (Y.X.)
† These authors contributed equally to this work.

Abstract: Carbon-based functional materials are deemed to be excellent candidates to adsorb contaminants from wastewater, yet their catalytic roles in advanced oxidation processes (AOPs) are still ambiguous. Therefore, four functional carbons (SPFCx) were fabricated in this study under various pyrolysis temperatures by using Sedum plumbizincicola (SP) residues (a kind of phytoremediation plant) as the precursors. Notably, SPFC800 exhibited the best adsorption capacity (q_e = 26.081 mg g^{-1}) toward bisphenol A (BPA) due to its having the largest specific surface area (121.57 m^2 g^{-1}). By injecting peroxymonosulfate (PMS, 5.0 mM), BPA (10 mg L^{-1}) could be completely removed within 70 min. More importantly, the BPA removal was stable and effective even in complex wastewater. Interestingly, radicals played minor roles in the SPFC800/PMS system, while nonradical mechanisms (i.e., $^1O_2$ and electron-transfer regime) were responsible for the BPA elimination, which was verified by quenching tests, solvent exchange experiments ($H_2O_2 \rightarrow D_2O$), and electrochemical experiments. Overall, this work may provide a facile and green method for BPA contaminated-wastewater purification and promote the application of AOPs in environmental remediation.

Keywords: peroxymonosulfate; pre-adsorption; degradation; nonradical oxidation

Citation: Liu, C.; Chen, Z.; Kang, R.; Niu, Y.; Su, W.; Wang, X.; Tian, D.; Xu, Y. Sedum Plumbizincicola Derived Functional Carbon for Activation of Peroxymonosulfate to Eliminate Bisphenol A: Performance and Reaction Mechanisms. *Coatings* **2022**, *12*, 1892. https://doi.org/10.3390/coatings12121892

Academic Editor: Alexandru Enesca

Received: 27 October 2022
Accepted: 28 November 2022
Published: 5 December 2022

1. Introduction

Phytoremediation technology is deemed to be a green environmental protection technology for soil remediation [1–3]. Notably, the core principle of this method is to use hyperaccumulators (e.g., pteris vittata, violabaoshaensis, and ellshohzia splendens) to concentrate the metals from the contaminated soil and then harvest the plants to remove heavy metal pollutants [4,5]. As a result, large amounts of plant-based wastes containing heavy metals are produced during the phytoremediation processes. Notably, these harvests often contain very high concentrations of heavy metals, which may become new sources of pollution upon improper disposal. Therefore, the safe disposal of the harvest is deemed to be an important issue [6,7]. Up until now, several methods (e.g., liquid extraction, landfill, composting treatment, and thermal treatment) have been developed to treat the harvest in the previous studies. Although liquid-phase extraction can effectively extract heavy metals from the plants, the generated extraction liquid needs to be treated, which may cause high treatment costs. In addition, the landfill may take up a lot of land resources, and the ultra-high concentrations of heavy metals in the harvest may migrate into the leachate, which may cause secondary pollution to the soil and groundwater. Therefore, developing an effective, green, and promising technology for the reuse of hyperaccumulator residues is urgent.

Recently, thermal treatment technology has been widely recognized. In particular, thermal treatment can prevent metal leaching, which is conducive to harmless disposal. Notably, value-added functional carbon (FC) can be obtained by pyrolyzing these plant residues under oxygen-limited conditions, which may further be used for water decontamination [8–11]. To our satisfaction, the inherent heavy metals can be converted into metal oxide nanoparticles, which can reduce the release of metallic pollutants [12]. Meanwhile, metal species can tailor the structures of carbon materials [13–16]. Recently, FC has been successfully applied for contaminants degradation in advanced oxidation processes (AOPs) by coupling with various oxidants (e.g., hydrogen peroxide (H_2O_2), periodate (PI), peroxymonosulfate (PMS), and peracetic acid (PAA)) [17–21]. As a result, the specific roles of carbon materials in AOPs were tentatively investigated in previous studies [22–25]. For instance, peroxydisulfate (PDS) could effectively be activated to produce reactive oxygen species (ROS, e.g., $SO_4^{\bullet-}$, $O_2^{\bullet-}$ and $^1O_2$) with the assistance of FC prepared by peanut shells, corn straw, and water hyacinth [26–30]. Despite many attempts, the mechanism of interaction between target pollutants, oxidants, and carbon materials in heterogeneous AOPs remained unclear [31–33].

Due to the large scale of bisphenol A (BPA) utilization, BPA contamination can be found in various environmental compartments (e.g., air, water, and soil), which may cause disease by disrupting the human endocrine system [34]. To this end, several methods (e.g., adsorption, flocculation, and biological treatment) are used for BPA removal [35–38]. Notably, AOPs have been considered as promising methods for BPA elimination in recent years. For example, a series of photocatalysts were designed and applied for the photocatalytic degradation of BPA [34]. However, the constant input of energy may increase the cost of the technology; thus carbon-catalyzed AOPs are deemed to be the promising candidates for pollutants removal.

In this study, Sedum plumbizincicola, which is often used to concentrate heavy metals from contaminated soil, was used for the FC preparation. BPA, which is widely used in the manufacturing industry, was selected as the target pollutant. First, the apparent morphology of the FC was characterized. After that, the adsorptive behaviors of BPA on FC were analyzed by fitting with the kinetic models. Then, the BPA removal was investigated in several AOPs (i.e., FC/PMS, FC/PDS, and FC/H_2O_2 systems). Furthermore, the BPA degradation mechanism was systematically explored by quenching tests, solvent exchange tests, and electrochemical experiments. Additionally, the catalytic activity and stability of the FC were also investigated in real water samples. Overall, this work not only provides a facile and low-cost method for the preparation of functional materials toward environmental remediation but also deepens the understanding of carbon-catalyzed AOPs.

2. Materials and Methods

2.1. Materials

The BPA (≥99%) was purchased from Aladdin. D_2O (99.9%) was purchased from Macklin, Nanjing, China. Quenching agents (e.g., methanol (MeOH, ≥99.9%), L-histidine (*L-his*, 99%), Furfuryl alcohol (FFA, 98%) and *Tert*-butanol (TBA, 99%)) were obtained from Aladdin, Nanjing, China. Other chemicals were at least analytical grade and used without further purification. Ultrapure water (18.25 MΩ) was produced by Spring-S60i + PALL system. In addition, a multifunctional crusher (LINGSUM, 1000C) was used for grinding Sedum plumbizincicola.

2.2. Methods

2.2.1. Preparation and Characterization of FC

Sedum plumbizincicola, a kind of hyperaccumulator collected in Yunnan, China, was used for the FC preparation. Firstly, sedum plumbizincicola was washed with pure water and then dried at 60 °C for 12 h. After that, the dried sedum plumbizincicola was ground to powder, transferred into a crucible, and carbonized in the muffle furnace at the required temperature (i.e., 500 °C, 600 °C, 700 °C, and 800 °C) for 2 h. Then, the black solid block was ground to desired size by using a 100-mesh sieve, washed with the pure water, and dried at 60 °C (about 4 h) until constant weight. The final products were designated as SPFCx, in which x was the annealing temperature (i.e., 500 °C, 600 °C, 700 °C, and 800 °C). In addition, Field emission scanning electron microscopy (SEM, Fei Quanta 400 FEG, Hillsboro, OR, USA) was used to determine the morphologies of SPFCx. The specific surface area (S_{BET}) was calculated according to the Brunauer-Emmett-Teller equation (Micromeritics ASAP 2020, Micromeritics Instrument Corporation, Norcross, GA, USA).

2.2.2. Adsorption

Adsorption was carried out in several conical flasks (100 mL) that contained 50 mL of BPA solution (10 mg L^{-1}) and 0.2 g L^{-1} of SPFCx. At predetermined time intervals, samples were collected, quenched by MeOH and then analyzed by liquid chromatography. The adsorption behaviors of BPA on SPFCx were determined by kinetic studies. Two kinetic models (i.e., pseudo-first and pseudo-second order kinetic models) were selected to analyze the adsorptive behaviors of BPA on SPFCx:

$$\text{Pseudo first} - \text{order kinetic model}: q_t = q_e\left(1 - e^{-k_1 t}\right) \quad (1)$$

$$\text{Pseudo second} - \text{order kinetic model}: q_t = \frac{k_2 q_e^2 t}{1 + k_2 q_e t} \quad (2)$$

where, q_e and q_t are the adsorption amounts of BPA on SPFCx at equilibrium and time t (mg g^{-1}), respectively. k_1 (min^{-1}) and k_2 (g mg^{-1} min^{-1}) are the corresponding adsorption rate constants, respectively.

2.2.3. Catalytic Degradation of BPA

BPA, an endocrine disruptor, was selected as the target contaminant to assess the catalytic performance of SPFCx. The experiments were conducted as follows: 10 mg SPFCx was first added into the BPA solution (10 mg L^{-1}) and stirred for 40 min to achieve adsorption equilibrium. Then, 5 mM oxidants (e.g., PMS, PDS, and H_2O_2) were injected into the mixture to initiate the catalytic reaction. At predetermined time intervals, 0.5 mL solution was withdrawn, filtered with a 0.2 mm filter to remove the solid catalyst, and quenched with 0.5 mL methanol for analysis.

The elimination of BPA was also tested in real water matrixes (i.e., pure water (PW), river water (RW) from Taihu in Wuxi, chemical wastewater (WW) collected from a sewage treatment plant in Anyang, and tap water (TW)).

2.2.4. Analytical Methods

The concentration of BPA could be detected by Agilent 1260 High Performance Liquid Chromatography (HPLC, Agilent, Beijing, China) equipped with a reversed-phase C18 column (5 μm, 4.6 × 250 mm). The mobile phase contained water and methanol (15/85, v/v). The flow rate was 1.0 mL min^{-1} and the injection volume was 20 μL. Electrochemical experiments were performed on a CHI660E electrochemical workstation and the details of the analytical methods were provided in the Supplementary Materials Text S1. In addition, the $^1O_2$ in SPFC800/PMS systems was identified via electron paramagnetic resonance (EPR, Bruker A320 spectrometer, Bruker, Beijing, China) analysis.

3. Results and Discussion

3.1. Characterizations

In order to evaluate the apparent morphology of the prepared SPFCx, SEM images were conducted. As depicted in Figure 1, the particle size of SPFC500 (Figure 1a) was relatively larger than that of the SPFC800 (Figure 1b), indicating that higher pyrolysis temperature might be beneficial for the complete decomposition of the precursors and the finer particle size may endow SPFC800 with larger S_{BET} and more active sites. It was reported that higher surface areas and more adsorptive sites were beneficial for removing contaminants [39–42].

Figure 1. SEM images of SPFC500 (**a**) and SPFC800 (**b**).

In order to further verify the enlargement of surface area with the increasing pyrolysis temperature, the nitrogen adsorption method was carried out to measure the S_{BET} of SPFCx. The structure information of SPFCx was summarized in Table 1. The results demonstrated that with the increase of annealing temperature (from 500 to 800 °C), the S_{BET} significantly increased from 36.09 to 121.57 $m^2\ g^{-1}$, indicating more adsorptive sites were created, which may be beneficial for accumulating the BPA from the solution. The largest S_{BET} of SPFC800 might be owing to the complete decomposition of contents (e.g., fatty acid esters) in precursors. Moreover, the pore volume of SPFCx also increased (from 0.0277 to 0.0847 $cm^3\ g^{-1}$) with increasing pyrolysis temperature, which might be beneficial for producing more active sites. The average pore diameters of SPFCx were between 3.651 and 3.977 nm, suggesting that SPFCx could produce abundant mesoporous pores. Notably, much more meso/micro pores might be formed during higher pyrolysis temperature as evidenced by the smaller pore diameter. The well-interconnected porous structures may contain plenty of nano-spaces/channels, which might be beneficial for the formation of nanoconfinement effects, boosting BPA removal [43,44].

Table 1. The basic information of SPFCx.

Samples	S_{BET} ($m^2\ g^{-1}$)	Pore Volume ($cm^3\ g^{-1}$)	Average Pore Diameter (nm)
SPFC500	36.09	0.0277	3.997
SPFC600	59.88	0.0592	3.764
SPFC700	78.29	0.0631	3.799
SPFC800	121.57	0.0847	3.651

3.2. Adsorption

As shown in Figure 2a,b, BPA molecule could quickly accumulate on SPFCx surface within 30 min.

Notably, better correlation coefficients (Table 2, $R^2 > 0.982$) were obtained by fitting with the pseudo-second-order kinetic model, suggesting the adsorption process might be chemisorption. More specifically, SPFC800 exhibited the highest q_e value of 26.081 mg g^{-1}, followed with SPFC700 (21.158 mg g^{-1}), SPFC600 (18.619 mg g^{-1}), and SPFC500 (18.308 mg g^{-1}), indicating the pyrolysis temperature was beneficial for enhancing the adsorption affinity. Thus, it could be speculated that a larger S_{BET} would be obtained under high pyrolysis temperature, which would be beneficial for contaminant adsorption. As depicted in Figure 2c, a good linear relationship between q_e and S_{BET} ($R^2 = 0.941$) was observed, further confirming that the adsorption of BPA was determined by the specific surface area of SPFCx.

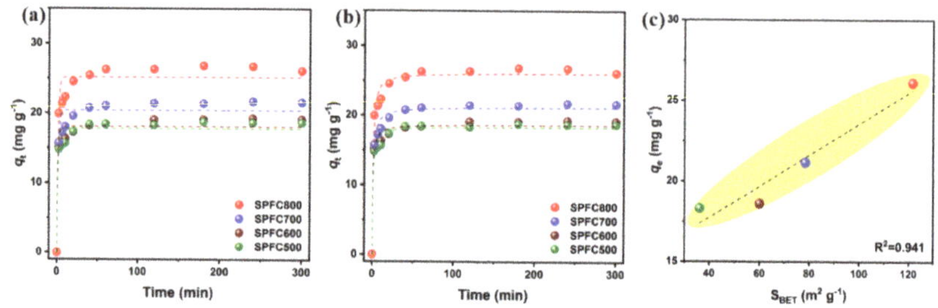

Figure 2. Pseudo first (**a**); pseudo second (**b**) order kinetic models; and the correlation analysis between q_e and S_{BET} (**c**). (Experimental conditions: [BPA] = 10 mg L^{-1}, and [SPFCx] = 0.2 g L^{-1}).

Table 2. The parameters obtained by fitting with adsorption kinetic equation.

Biochar	Pseudo First-Order			Pseudo Second-Order		
	q_e	k_1	R^2	q_e	k_2	R^2
SPFC500	17.771 (0.408)	0.862	0.951	18.308 (0.302)	0.0847	0.982
SPFC600	18.084 (0.411)	0.876	0.951	18.619 (0.307)	0.0855	0.982
SPFC700	20.399 (0.509)	0.674	0.949	21.158 (0.323)	0.0521	0.985
SPFC800	25.184 (0.624)	0.724	0.944	26.081 (0.418)	0.0461	0.983

3.3. Catalytic Degradation Performance

Although H_2O_2 was a traditional and effective oxidant that was usually utilized in Fenton and/or Fenton-Like processes, it could hardly oxidize the BPA even with the help of SPFCx (Figure 3a). The removal of BPA in such an oxidation process was almost attributed to the adsorption by SPFCx. However, a moderate synergistic effect was observed between BPA adsorption and BPA oxidation in SPFCx/PDS systems (Figure 3b). Notably, BPA was more rapidly removed by SPFCx/PMS systems (especially SPFC800), which might be attributed to the non-symmetrical structure of PMS, and thus it was easier to be decomposed (Figure 3c). Due to the highest BPA removal rates in SPFC800/PMS systems, the subsequent BPA degradation experiments would be performed in the SPFC800-induced systems (Figure 3d).

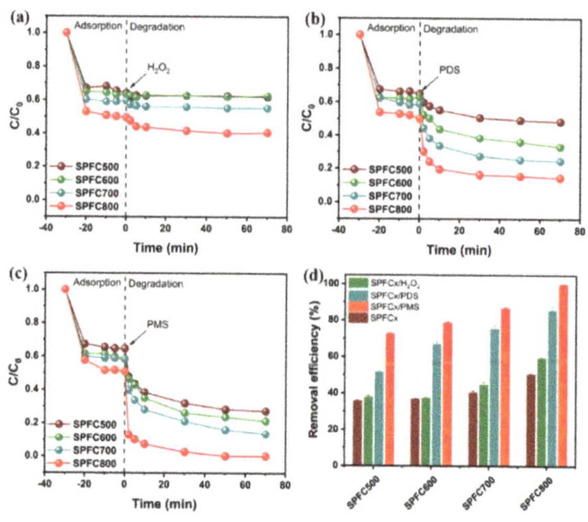

Figure 3. Catalytic degradation of BPA in SPFCx/H_2O_2 (**a**); SPFCx/PDS (**b**); SPFCx/PMS (**c**); systems, respectively, and the removal efficiencies of BPA in various oxidation process (**d**). (Experimental conditions: [BPA] = 10 mg L^{-1}, [Oxidant] = 5 mM, and [SPFCx] = 0.2 g L^{-1}).

3.4. Catalytic Mechanism

In order to reveal why BPA could be quickly and effectively removed in SPCF800/PMS system, quenching experiments were conducted first to demonstrate the involved ROS for BPA removal. Apparently, with the injection of MeOH (Figure 4a, from 500 to 2000 mM), the removal efficiency of BPA was only slightly inhibited, indicating that BPA degradation was not dependent on the radicals (i.e., SO_4^{-} and $\cdot OH$). Furthermore, a similar phenomenon was observed by using TBA (quencher for $\cdot OH$, from 100 to 1000 mM) as the quenching agent, confirming the minor role of $\cdot OH$ in BPA elimination (Figure 4b). Nevertheless, both *L-his* and FFA (classic quenchers for $^1O_2$) could significantly inhibit the BPA removal, which suggested that $^1O_2$ might be obligated to the BPA removal (Figure 4c,d). In order to further confirm the existence of $^1O_2$ in SPFC800/PMS system, BPA degradation was conducted in D_2O, which could prolong the half-life of $^1O_2$. As shown in the inset of Figure 4d, BPA removal was boosted in D_2O, demonstrating the vital roles of $^1O_2$ in BPA elimination.

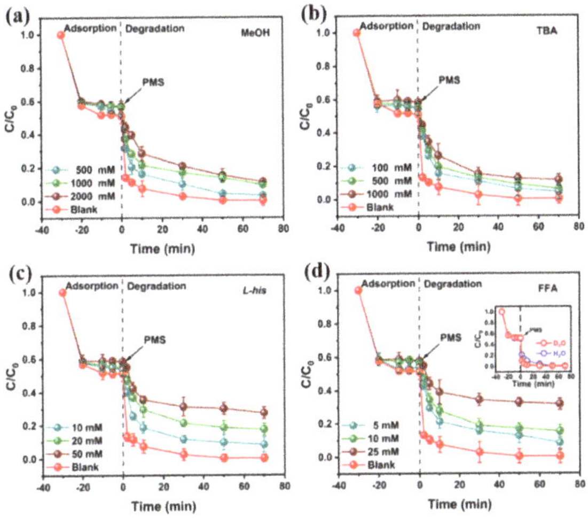

Figure 4. The effects of several quenching agents (i.e., MeOH (**a**); TBA (**b**); *L-his* (**c**); and FFA (**d**)) on BPA removal in SPFC800/PMS system and the solvent exchange experiments (inset of d). (Experimental conditions: [BPA] = 10 mg L^{-1}, [Oxidant] = 5 mM, [SPFCx] = 0.2 g L^{-1}, [MeOH] = 0–2000 mM, [TBA] = 0–1000 mM, [*L-his*] = 0–50 mM, and [FFA] = 0–25 mM).

The EPR test was further conducted to verify the generation of $^1O_2$ in SPFC800/PMS system by using TEMP as the spin-trapping agent (Figure 5). A distinctive triplet signal was detected upon the injection of SPFC800 into the PMS solution, which indicated the existence of $^1O_2$, matching well with the quenching results.

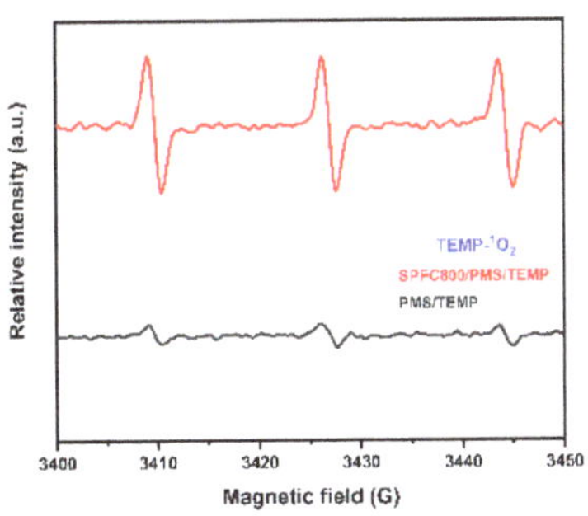

Figure 5. Electron paramagnetic resonance (EPR) spectra of PMS activation by SPFC800: TEMP-$^1O_2$.

In addition, electrochemical experiments were performed to verify the occurrence of electron transfer regime. The EIS results could be used to reveal the resistance of the catalyst. As shown in Figure 6a, SPFC800 possessed a smaller semicircle diameter than SPFC500, indicating the better conductivity of SBC800, and thus electrons could be easier to transfer on SPFC800 surface than SPFC500. The LSV and I-T curves of SPFC800 were also carried out. As shown in Figure 6b, the current density increased upon the injection of PMS, suggesting the rapid interaction between PMS and SPFC800. Furthermore, another obviously increased current density was detected after adding BPA, which might be owing to the electron transfer among BPA (electron donor), PMS (electron acceptor), and SPFC800 (electric conductor). Similarly, as illustrated in Figure 6c, the current responses showed obvious variations upon the injection of PMS and BPA, revealing the vital role of electron transfer mechanism. The potential of the SPFCx surface was characterized to reveal the electron transfer regime in depth. It was well accepted that the PMS could first combine with the catalyst to form the catalyst-PMS complexes (C-PMS*) with higher oxidation capacity. Notably, the C-PMS* was formed first and the potential of C-PMS* significantly increased with the higher dosage of PMS, revealing why BPA generally could be more effectively degraded with higher dosage of oxidants. According to the abovementioned analysis, the reaction mechanism between PMS, BPA, and SPFC800 was elucidated. Nonradicals were the dominant species in the catalytic process, and the SPFC800 could not only directly activate PMS to produce $^1O_2$ but accumulate PMS to form the C-PMS*, leading to effective BPA removal [24,45].

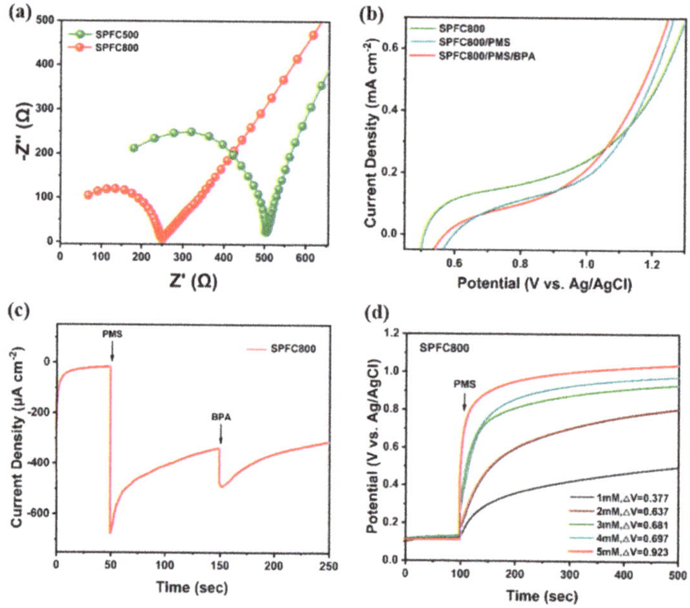

Figure 6. Electrochemical impedance spectroscopy (EIS) of SPFC500 and SPFC800 (**a**); linear sweep voltammetry (LSV) of SPFC800 with/without PMS and BPA (**b**); I-T curves after adding PMS and BPA (**c**); and the change of SPFC800 potential with injection of different concentrations of PMS (**d**).

3.5. Degradation in Real Water

BPA degradation experiments were further conducted in SPFC800/PMS system by adding several interfering substances (i.e., HA, Cl^-, and HCO_3^-). As shown in Figure 7a, the removal of BPA was slightly inhibited by adding 10 mM of HA, which might be owing to that HA could consume PMS and/or the ROS. Similarly, with the injection of HCO_3^-, the BPA degradation was inhibited, which could be due to that the ROS generated in PMS/SPFC800 system were converted into $CO_3^{\bullet-}$ with lower oxidation capacity (Equations (3) and (4)). Interestingly, BPA removal was slightly accelerated by adding 10 mM of Cl^-, which was attributed to that Cl^- may react with PMS and/or the ROS to form $Cl^•$, synergistically eliminating BPA (Equations (5)–(8)).

$$HO^• + HCO_3^- \rightarrow H_2O + CO_3^{•-} \quad (3)$$

$$SO_4^{-•} + HCO_3^- \rightarrow SO_4^{2-} + HCO_3^-, k = 9.1 \times 10^6\ \mathrm{M^{-1}s^{-1}} \quad (4)$$

$$SO_4^{-•} + Cl^- \rightarrow SO_4^{2-} + Cl^•, k = 2.0 \times 10^8\ \mathrm{M^{-1}s^{-1}} \quad (5)$$

$$HO^• + Cl^• \rightarrow ClOH^{-•} \quad (6)$$

$$ClOH^{-•} + H^+ \rightarrow Cl^• + H_2O \quad (7)$$

$$Cl^• + Cl^- \rightarrow Cl_2^{-•}, k = (0.8 - 2.1) \times 10^{10}\ \mathrm{M^{-1}s^{-1}} \quad (8)$$

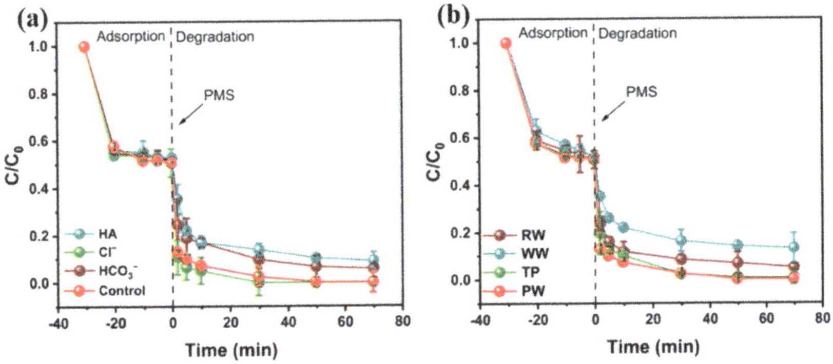

Figure 7. Elimination of BPA in SPFC800/PMS system after adding interfering substances ((**a**), i.e., HA, Cl^-, and HCO_3^-) and degradation of BPA in actual water samples; ((**b**), i.e., RW, WW, TW, and PW). (Experimental conditions: [BPA] = 10 mg L^{-1}, [Oxidant] = 5 mM, [SPFCx] = 0.2 g L^{-1}, [HA] = 5 mg L^{-1}, and [Cl^{-1}] = [HCO_3^-] = 5 mM).

To further evaluate the feasibility of the SPFC800 for BPA elimination, the BPA degradation experiments were carried out in different water matrixes (Table 3). As shown in Figure 7b, BPA removal did not change in TW and was slightly inhibited in RW, which was due to that RW often contained a lot of impurities, consuming the ROS. In addition, a moderate deterioration was observed in WW, which could be owing to the ultra-high concentrations of coexisting contaminants, which could compete with BPA for the ROS.

Table 3. The basic information of the real waters.

Water Sample	pH	COD_{Cr} (mg L^{-1})	Cl^- (mg L^{-1})	NH_4^+-N (mg L^{-1})	PO_4^{3-} (mg L^{-1})
PW	7.2 ± 0.03	<10	0.47	0.09	0.02
TW	7.13 ± 0.03	<10	12.6	1.37	0.138
RW	7.08 ± 0.03	<10	8.52	2.14	0.577
WW	7.44 ± 0.03	572 ± 20	1388	4.11	2.06

4. Conclusions

In this study, a series of sedum plumbizincicola derived functional carbon (SPFCx) catalysts were successfully fabricated. The SPFCx could effectively and rapidly activate PMS for BPA removal. Notably, the catalytic performance was highly decided by the pyrolysis conditions. Among others, SPFC800 exhibited the best catalytic activity because of its highest surface area and pore volumes. In addition, the non-radical pathway (i.e., $^1O_2$ and electron transfer regime) rather than the radical pathway dominated the BPA degradation. More importantly, the BPA elimination (>80%) was highly stable and effective even in complex wastewater, demonstrating that the SPFC800/PMS system was promising for BPA-contaminated water remediation. Overall, this work may promote the application of AOPs in environmental remediation.

Supplementary Materials: The following supporting information can be downloaded at: https://www.mdpi.com/article/10.3390/coatings12121892/s1, Supporting Information Text S1: Details of electrochemical experiment.

Author Contributions: C.L.: writing—original draft, formal analysis, investigation, and visualization. Z.C.: writing—original draft, formal analysis, investigation, and supervision. R.K.: writing—reviewing and editing, investigation. Y.N.: writing—reviewing and editing, investigation. W.S.: writing—reviewing and editing, investigation. X.W.: writing—reviewing and editing, investigation. D.T.: conceptualization, methodology. Y.X.: formal analysis, resources, writing—review and editing. All authors have read and agreed to the published version of the manuscript.

Funding: This work has been supported by the Research Project of Science and Technology of the Anyang City (2022C01NY016), the Research Project of Science and Technology of the Henan Province (222102320274), PhD research startup foundation of Anyang Institute of Technology (BSJ2021035), Postdoc research startup foundation of Anyang Institute of Technology (BHJ2022005), Postdoc research startup foundation of the Henan Province (202103099), and Major science and technology projects of Anyang (201928).

Institutional Review Board Statement: This article does not contain any studies with human participants or animals performed by any of the authors.

Informed Consent Statement: Not applicable.

Data Availability Statement: Not applicable.

Conflicts of Interest: The authors declare no conflict of interest.

References

1. Oladoye, P.O.; Olowe, O.M.; Asemoloye, M.D. Phytoremediation technology and food security impacts of heavy metal contaminated soils: A review of literature. *Chemosphere* **2022**, *288*, 132555. [CrossRef]
2. Yin, Z.; Yu, J.; Han, X.; Wang, H.; Yang, Q.; Pan, H.; Lou, Y.; Zhuge, Y. A novel phytoremediation technology for polluted cadmium soil: Salix integra treated with spermidine and activated carbon. *Chemosphere* **2022**, *306*, 135582. [CrossRef]
3. Sarma, H.; Islam, N.F.; Prasad, R.; Prasad, M.N.V.; Ma, L.Q.; Rinklebe, J. Enhancing phytoremediation of hazardous metal(loid)s using genome engineering CRISPR-Cas9 technology. *J. Hazard. Mater.* **2021**, *414*, 125493. [CrossRef]
4. Fadhile Almansoory, A.; Abu Hasan, H.; Idris, M.; Sheikh Abdullah, S.R.; Anuar, N. Potential application of a biosurfactant in phytoremediation technology for treatment of gasoline-contaminated soil. *Ecol. Eng.* **2015**, *84*, 113–120. [CrossRef]
5. Zhang, J.; Cao, X.; Yao, Z.; Lin, Q.; Yan, B.; Cui, X.; He, Z.; Yang, X.; Wang, C.H.; Chen, G. Phytoremediation of Cd-contaminated farmland soil via various Sedum alfredii-oilseed rape cropping systems: Efficiency comparison and cost-benefit analysis. *J. Hazard. Mater.* **2021**, *419*, 126489. [CrossRef]
6. Cui, X.; Zhang, J.; Wang, X.; Pan, M.; Lin, Q.; Khan, K.Y.; Yan, B.; Li, T.; He, Z.; Yang, X.; et al. A review on the thermal treatment of heavy metal hyperaccumulator: Fates of heavy metals and generation of products. *J. Hazard. Mater.* **2021**, *405*, 123832. [CrossRef]
7. Guo, X.; Zhang, S.; Luo, J.; Pan, M.; Du, Y.; Liang, Y.; Li, T. Integrated glycolysis and pyrolysis process for multiple utilization and cadmium collection of hyperaccumulator Sedum alfredii. *J. Hazard. Mater.* **2022**, *422*, 126859. [CrossRef]
8. Huo, X.; Zhou, P.; Zhang, J.; Liu, Y.; Cheng, X.; Liu, Y.; Li, W.; Zhang, Y. N, S-Doped porous carbons for persulfate activation to remove tetracycline: Nonradical mechanism. *J. Hazard. Mater.* **2020**, *391*, 122055. [CrossRef]
9. Pei, X.; Peng, X.; Jia, X.; Wong, P.K. N-doped biochar from sewage sludge for catalytic peroxydisulfate activation toward sulfadiazine: Efficiency, mechanism, and stability. *J. Hazard. Mater.* **2021**, *419*, 126446. [CrossRef]
10. Xiao, P.; Yi, X.; Wu, M.; Wang, X.; Zhu, S.; Gao, B.; Liu, Y.; Zhou, H. Catalytic performance and periodate activation mechanism of anaerobic sewage sludge-derived biochar. *J. Hazard. Mater.* **2021**, *424*, 127692. [CrossRef]
11. Yu, J.; Feng, H.; Tang, L.; Pang, Y.; Zeng, G.; Lu, Y.; Dong, H.; Wang, J.; Liu, Y.; Feng, C.; et al. Metal-free carbon materials for persulfate-based advanced oxidation process: Microstructure, property and tailoring. *Prog. Mater. Sci.* **2020**, *111*, 100654. [CrossRef]
12. Yao, Y.; Gao, B.; Chen, J.; Yang, L. Engineered biochar reclaiming phosphate from aqueous solutions: Mechanisms and potential application as a slow-release fertilizer. *Environ. Sci. Technol.* **2013**, *47*, 8700–8708. [CrossRef] [PubMed]

13. Xiao, X.; Chen, B.; Chen, Z.; Zhu, L.; Schnoor, J.L. Insight into Multiple and Multilevel Structures of Biochars and Their Potential Environmental Applications: A Critical Review. *Environ. Sci. Technol.* **2018**, *52*, 5027–5047. [CrossRef] [PubMed]
14. Yao, B.; Luo, Z.; Du, S.; Yang, J.; Zhi, D.; Zhou, Y. Magnetic MgFe2O4/biochar derived from pomelo peel as a persulfate activator for levofloxacin degradation: Effects and mechanistic consideration. *Bioresour. Technol.* **2022**, *346*, 126547. [CrossRef] [PubMed]
15. Liu, J.; Liu, H.; Yang, X.; Jia, X.; Cai, M.; Bao, Y. Preparation of Si-Mn/biochar composite and discussions about characterizations, advances in application and adsorption mechanisms. *Chemosphere* **2021**, *281*, 130946. [CrossRef]
16. Long, Y.; Dai, J.; Zhao, S.; Huang, S.; Zhang, Z. Metal-organic framework-derived magnetic carbon for efficient decontamination of organic pollutants via periodate activation: Surface atomic structure and mechanistic considerations. *J. Hazard. Mater.* **2021**, *424*, 126786. [CrossRef]
17. Zhang, X.; Miao, X.; Xiang, W.; Zhang, J.; Cao, C.; Wang, H.; Hu, X.; Gao, B. Ball milling biochar with ammonia hydroxide or hydrogen peroxide enhances its adsorption of phenyl volatile organic compounds (VOCs). *J. Hazard. Mater.* **2021**, *403*, 123540. [CrossRef]
18. Hassan, M.F.; Sabri, M.A.; Fazal, H.; Hafeez, A.; Shezad, N.; Hussain, M. Recent trends in activated carbon fibers production from various precursors and applications—A comparative review. *J. Anal. Appl. Pyrolysis* **2020**, *145*, 104715. [CrossRef]
19. Wang, X.; Liu, Y.; Zhu, L.; Li, Y.; Wang, K.; Qiu, K.; Tippayawong, N.; Aggarangsi, P.; Reubroycharoen, P.; Wang, S. Biomass derived N-doped biochar as efficient catalyst supports for CO_2 methanation. *J. CO2 Util.* **2019**, *34*, 733–741. [CrossRef]
20. Feng, D.; Lü, J.; Guo, S.; Li, J. Biochar enhanced the degradation of organic pollutants through a Fenton process using trace aqueous iron. *J. Environ. Chem. Eng.* **2021**, *9*, 104677. [CrossRef]
21. Anfar, Z.; Ait El Fakir, A.; Ait Ahsaine, H.; Zbair, M.; Farsad, S.; Morlet-Savary, F.; Jada, A.; El Alem, N. Nitrogen doped graphitic porous carbon from almond shells as an efficient persulfate activator for organic compound degradation. *New J. Chem.* **2020**, *44*, 9391–9401. [CrossRef]
22. Dou, J.; Cheng, J.; Lu, Z.; Tian, Z.; Xu, J.; He, Y. Biochar co-doped with nitrogen and boron switching the free radical based peroxydisulfate activation into the electron-transfer dominated nonradical process. *Appl. Catal. B Environ.* **2022**, *301*, 120832. [CrossRef]
23. He, L.; Yang, C.; Ding, J.; Lu, M.-Y.; Chen, C.-X.; Wang, G.-Y.; Jiang, J.-Q.; Ding, L.; Liu, G.-S.; Ren, N.-Q.; et al. Fe, N-doped carbonaceous catalyst activating periodate for micropollutant removal: Significant role of electron transfer. *Appl. Catal. B Environ.* **2022**, *303*, 120880. [CrossRef]
24. Wang, H.; Guo, W.; Liu, B.; Si, Q.; Luo, H.; Zhao, Q.; Ren, N. Sludge-derived biochar as efficient persulfate activators: Sulfurization-induced electronic structure modulation and disparate nonradical mechanisms. *Appl. Catal. B Environ.* **2020**, *279*, 119361. [CrossRef]
25. Xiong, W.; Wang, Z.; He, S.; Hao, F.; Yang, Y.; Lv, Y.; Zhang, W.; Liu, P.; Luo, H.a. Nitrogen-doped carbon nanotubes as a highly active metal-free catalyst for nitrobenzene hydrogenation. *Appl. Catal. B Environ.* **2020**, *260*, 118105. [CrossRef]
26. Feng, Z.; Zhou, B.; Yuan, R.; Li, H.; He, P.; Wang, F.; Chen, Z.; Chen, H. Biochar derived from different crop straws as persulfate activator for the degradation of sulfadiazine: Influence of biomass types and systemic cause analysis. *Chem. Eng. J.* **2022**, *440*, 135669. [CrossRef]
27. Lian, F.; Cui, G.; Liu, Z.; Duo, L.; Zhang, G.; Xing, B. One-step synthesis of a novel N-doped microporous biochar derived from crop straws with high dye adsorption capacity. *J. Environ. Manag.* **2016**, *176*, 61–68. [CrossRef]
28. Wang, Y.P.; Liu, Y.L.; Tian, S.Q.; Yang, J.J.; Wang, L.; Ma, J. Straw biochar enhanced removal of heavy metal by ferrate. *J. Hazard. Mater.* **2021**, *416*, 126128. [CrossRef]
29. Wang, H.; Xu, J.; Sheng, L. Preparation of straw biochar and application of constructed wetland in China: A review. *J. Clean. Prod.* **2020**, *273*, 123131. [CrossRef]
30. Duan, R.; Ma, S.; Xu, S.; Wang, B.; He, M.; Li, G.; Fu, H.; Zhao, P. Soybean straw biochar activating peroxydisulfate to simultaneously eliminate tetracycline and tetracycline resistance bacteria: Insights on the mechanism. *Water Res* **2022**, *218*, 118489. [CrossRef] [PubMed]
31. Miao, J.; Geng, W.; Alvarez, P.J.J.; Long, M. 2D N-Doped Porous Carbon Derived from Polydopamine-Coated Graphitic Carbon Nitride for Efficient Nonradical Activation of Peroxymonosulfate. *Environ. Sci. Technol.* **2020**, *54*, 8473–8481. [CrossRef]
32. Minakata, D.; Kamath, D.; Maetzold, S. Mechanistic Insight into the Reactivity of Chlorine-Derived Radicals in the Aqueous-Phase UV-Chlorine Advanced Oxidation Process: Quantum Mechanical Calculations. *Environ. Sci. Technol.* **2017**, *51*, 6918–6926. [CrossRef] [PubMed]
33. Peng, J.; Zhou, P.; Zhou, H.; Liu, W.; Zhang, H.; Zhou, C.; Lai, L.; Ao, Z.; Su, S.; Lai, B. Insights into the Electron-Transfer Mechanism of Permanganate Activation by Graphite for Enhanced Oxidation of Sulfamethoxazole. *Environ. Sci. Technol.* **2021**, *55*, 9189–9198. [CrossRef] [PubMed]
34. Reddy, P.V.L.; Kim, K.H.; Kavitha, B.; Kumar, V.; Raza, N.; Kalagara, S. Photocatalytic degradation of bisphenol A in aqueous media: A review. *J. Environ. Manag.* **2018**, *213*, 189–205. [CrossRef]
35. Moradi, F.G.M. Application of peroxymonosulfate and its activation methods for degradation of environmental organic pollutants: Review. *Chem. Eng. J.* **2017**, *310*, 41–62. [CrossRef]
36. Lu, C.S.; Tsai, H.Y.; Shaya, J.; Golovko, V.B.; Wang, S.Y.; Liu, W.J.; Chen, C.C. Degradation of sulfamethoxazole in water by AgNbO3 photocatalyst mediated by persulfate. *RSC Adv.* **2022**, *12*, 29709–29718. [CrossRef]

37. Zhang, Y.; Cui, W.; An, W.; Liu, L.; Liang, Y.; Zhu, Y. Combination of photoelectrocatalysis and adsorption for removal of bisphenol A over TiO2-graphene hydrogel with 3D network structure. *Appl. Catal. B Environ.* **2018**, *221*, 36–46. [CrossRef]
38. Wang, K.; Qiu, L.; Zhu, J.; Sun, Q.; Qu, W.; Yu, Y.; Zhao, Z.; Yu, Y.; Shao, G. Environmental contaminant BPA causes intestinal damage by disrupting cellular repair and injury homeostasis in vivo and in vitro. *Biomed. Pharm.* **2021**, *137*, 111270. [CrossRef] [PubMed]
39. Huang, D.; Zhang, Q.; Zhang, C.; Wang, R.; Deng, R.; Luo, H.; Li, T.; Li, J.; Chen, S.; Liu, C. Mn doped magnetic biochar as persulfate activator for the degradation of tetracycline. *Chem. Eng. J.* **2020**, *391*, 123532. [CrossRef]
40. Liu, B.; Guo, W.; Wang, H.; Si, Q.; Zhao, Q.; Luo, H.; Ren, N. B-doped graphitic porous biochar with enhanced surface affinity and electron transfer for efficient peroxydisulfate activation. *Chem. Eng. J.* **2020**, *396*, 125119. [CrossRef]
41. Qu, S.; Yuan, Y.; Yang, X.; Xu, H.; Mohamed, A.K.; Zhang, J.; Zhao, C.; Liu, L.; Wang, B.; Wang, X.; et al. Carbon defects in biochar facilitated nitrogen doping: The significant role of pyridinic nitrogen in peroxymonosulfate activation and ciprofloxacin degradation. *Chem. Eng. J.* **2022**, *441*, 135864. [CrossRef]
42. Zhu, K.; Wang, X.; Chen, D.; Ren, W.; Lin, H.; Zhang, H. Wood-based biochar as an excellent activator of peroxydisulfate for Acid Orange 7 decolorization. *Chemosphere* **2019**, *231*, 32–40. [CrossRef] [PubMed]
43. Qian, J.; Gao, X.; Pan, B. Nanoconfinement-Mediated Water Treatment: From Fundamental to Application. *Environ. Sci. Technol.* **2020**, *54*, 8509–8526. [CrossRef] [PubMed]
44. Teng, Y.; Liu, E.; Ding, R.; Liu, K.; Liu, R.; Wang, L.; Yang, Z.; Jiang, H. Bean dregs-based activated carbon/copper ion supercapacitors. *Electrochim. Acta* **2016**, *194*, 394–404. [CrossRef]
45. Chen, Y.-d.; Duan, X.; Zhang, C.; Wang, S.; Ren, N.-Q.; Ho, S.-H. Graphitic biochar catalysts from anaerobic digestion sludge for nonradical degradation of micropollutants and disinfection. *Chem. Eng. J.* **2020**, *384*, 123244. [CrossRef]

coatings

MDPI

Article

A Novel Magnetic Fluorescent Fe_3O_4@ZnS@MPS Nanosensor for Highly Sensitive Determination and Removal of Ag^+

Yan Gao [1], Xin Chen [2], Ping Xu [2], Jie Chen [1], Shihua Yu [2], Zhigang Liu [1,*] and Xiaodan Zeng [1,*]

1 Center of Characterization and Analysis, Jilin Institute of Chemical Technology, Jilin 132013, China; gyaxz@126.com (Y.G.); jiechendr@163.com (J.C.)
2 School of Chemical and Pharmaceutical Engineering, Jilin Institute of Chemical Technology, Jilin 132013, China; chenxin92@jlict.edu.cn (X.C.); xuping@jlict.edu.cn (P.X.); ysh@jlict.edu.cn (S.Y.)
* Correspondence: lzg@jlict.edu.cn (Z.L.); jiangzxd@jlict.edu.cn (X.Z.)

Abstract: A novel magnetic fluorescent nanoprobe (Fe_3O_4@ZnS@MPS(MFNPs)) was synthesized, which recognized and cooperated with Ag^+ ions, and a rapid method for detecting Ag^+ was established in solution. It was found by fluorescence spectroscopy analysis that the MFNPs could detect Ag^+ in PBS solution and, upon addition of Ag^+ ions, the fluorescence (FL) of MFNPs could be quenched significantly. The sensor has a low limit of detection (LOD) of 7.04 μM for Ag^+. The results showed that MFNPs were extremely specific and sensitive for the quantitative detection of Ag^+ over a wide pH range. Then, the recognition mechanism between MFNPs and guest Ag^+ was explored via measures of infrared spectroscopy and electron microscopy. It was speculated that the oxygen atoms in the sulfonic acid group cooperated with Ag^+ to form a synergistic complexation. The assay was successfully used to determine the content of Ag+ in real samples.

Keywords: magnetic fluorescent nanosensor; selective recognition; silver ion; 3-sulfhydryl-1-propane sodium (MPS)

Citation: Gao, Y.; Chen, X.; Xu, P.; Chen, J.; Yu, S.; Liu, Z.; Zeng, X. A Novel Magnetic Fluorescent Fe_3O_4@ZnS@MPS Nanosensor for Highly Sensitive Determination and Removal of Ag^+. *Coatings* **2023**, *13*, 1557. https://doi.org/10.3390/coatings13091557

Academic Editor: M. Shaheer Akhtar

Received: 24 July 2023
Revised: 23 August 2023
Accepted: 29 August 2023
Published: 6 September 2023

1. Introduction

Silver has always attracted attention due to its unique chemical properties of strong corrosion resistance and high antioxidant capacity. Because of its rarity and high gloss, silver is widely used in the production of daily necessities [1,2]. A large number of widespread uses have a relatively significant impact on the environment. Therefore, it is necessary to adopt appropriate detection methods to analyze it. Excessive human exposure to silver is likely to cause silver poisoning, growth retardation, etc., and too much silver hurts the eyes and skin [3,4]. Thus, it is of great significance to investigate an effective method for detection of Ag^+.

Silver ions can be detected by a variety of methods, such as fast but unstable electrochemical detection and accurate but expensive flame atomic absorption spectrometry [5–7]. In addition to these techniques, the extraction method using molecular receptors or chelating ligands is also widely used in the detection of Ag(I). Compared with conventional assays, fluorescence detection has greater advantages, such as rapid reaction, simple operation, low cost, and high sensitivity [8–10].

However, these fluorescent probes also cause further pollution to the environment when they are used to detect Ag^+. Compared with these fluorescent probes, ZnS QDs have unique advantages, such as a narrow and symmetric emission spectrum, excellent light stability, and resistance to photobleaching [11]. Therefore, the development of multifunctional fluorescence chemosensors for sensing of heavy metals, that are friendly to the environment, is increasingly attracting attention [12].

In recent years, magnetic nanomaterials have attracted significant attention in the academic field due to their excellent magnetic reaction ability, high biocompatibility, and good stability [13–15]. Among these, superparamagnetic iron oxide (Fe_3O_4) is particularly versatile due to its unique high coercivity, easily functional modification, excellent controllable magnetic responsiveness, which can be manipulated by external magnetic fields, and controllable size and surface [16–18]. This enables researchers in various fields, such as chemistry, biology, medicine, and materials science, to use MFNPs to construct multifunctional nanoprobes and conduct significant research in sewage treatment, biological imaging, drug delivery, and other fields [19,20].

However, due to the strong magnetic dipole attraction between the particles, Fe_3O_4 nanoparticles tend to aggregate. Therefore, stabilizers such as organic matter, semiconductors, and oxides with specific functional groups are often applied to the nanoparticles' surface in order to improve stability. Using a variety of biocompatible polymers to functionalize and modify the surface of Fe_3O_4 nanoparticles to provide new functions is the focus of current research. When the stability of Fe_3O_4 nanoparticles is guaranteed, the introduction of fluorescent substances is conducive to the separation and transfer of detection substances, especially heavy metals in water, thus expanding the application field of fluorescent probes. Aggregation-induced sedimentation technology can efficiently and rapidly remove Ag^+ from water samples and can avoid secondary pollution [21].

In this work, a novel environmentally friendly magnetic fluorescent nanosensor (Fe_3O_4@ZnS@MPS(MFNPs)) modified with 3-sulfhydryl-1-propane sodium for coinstantaneous detection and removal of Ag^+ from water samples is reported (Scheme 1). The results showed that MFNPs achieved highly specific recognition and were extremely sensitive for the quantitative detection of Ag^+ over a wide pH range. The sensor has a low limit of detection (LOD) of 7.04 μM for Ag^+. The optimal adsorption capacity of MFNPs was calculated to be about 395.79 mg/g and the optimal adsorption rate capacity of MFNPs was calculated to be about 98%. This work provides a new method for the synthesis of magnetic fluorescent nanosensors, exploiting their application not only for the detection of Ag^+, but also other heavy metal ions, with the multiple functions of enrichment, detection, and separation.

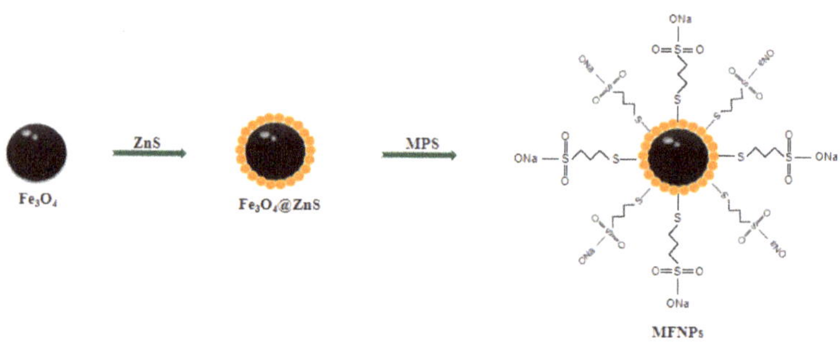

Scheme 1. Fabrication of MFNPs.

2. Materials and Methods

2.1. Preparation of Fe_3O_4 Magnetic Nanoparticle

The magnetic Fe_3O_4 nanoparticle was synthesized via the solvothermal method and prepared as follows: $FeCl_3 \cdot 6H_2O$ (2.7 g) was dissolved in 60 mL ethylene glycol in a water bath and stirred until fully dissolved. Sodium acetate (7.2 g) and surfactant (0.5 g) were added and stirred for 30 min. The solution was moved to a Polytetrafluoroethylene kettle, which was sealed tightly, and the hydrothermal reaction was carried out at high temperature. The reaction was then allowed to end and to cool naturally. The solution was moved to a Teflon kettle, which was sealed, followed by a hydrothermal reaction at a high temperature and natural cooling at the end of the reaction. The black precipitate was collected after washing several times, and the dispersive solution was prepared by adding water for further modification.

2.2. Preparation of Fe_3O_4@ZnS

The chemical coprecipitation method was employed to synthesize Fe_3O_4@ZnS core/shell nanocomposites. First, 10 mL of Fe_3O_4 was diluted to 100 mL and the solution was kept neutral. Then, 2.5 mmol of $Zn(Ac)_2 \cdot 2H_2O$ and $Na_2S \cdot 9H_2O$ were dissolved in turn under the condition of water bath stirring, and were stirred for 6 h. The item was attracted by a magnetic force and then underwent a thorough cleansing process with ultrapure water and ethanol. The Fe_3O_4@ZnS nanoprobe was completed and water was added to make a dispersion for further modification [22].

2.3. Preparation of Fe_3O_4@ZnS@MPS

A quantity of 5 mL of Fe_3O_4@ZnS ethanol solution was placed in a round-bottom flask. Then, 5 mL 3-sulfhydryl-1-propane sodium (MPS) (89.105 mg, 0.1 mmol/L) ethanol solution was added. The mixture was then stirred in a water bath at 40 °C for 4 h in the dark. After the item was attracted by a magnetic force, it underwent a thorough cleansing process with ultrapure water and ethanol. Water was added to the prepared Fe_3O_4@ZnS@MPS microspheres to make a dispersion solution.

2.4. Instrumentation

The surface topography of the microspheres was observed via a scanning electron microscope (SEM) (Quanta 200, FEI, Hillsboro, OR, USA) (EHT = 5.00 kV) and a transmission electron microscope (TEM) (JEM 2100, JEOL, Tokyo, Japan) (Accelerating Voltage = 200 kV). Fourier transform-infrared (FTIR) spectra were obtained from a spectrometer (Thermo Scientific Nicolet IS50, Thermofisher, Waltham, MA, USA). X-ray diffraction (XRD) images were obtained using an X-ray diffractometer (D8FOCUS, Bruker, Berlin, German). The measurement was conducted using CuKα radiation (λ = 1.5406 Å) in the range of $2\theta = 20°–80°$ at a scan speed of 2°/min. XPS spectra data were obtained on an X-ray photoelectron spectrometer (Krayos AXIS Ultra DLDX, Shimadzu, Tokyo, Japan) (Analyzer Mode: Constant Analyzer Energy: Pass Energy 30.0 eV; Energy Step Size: 0.100 eV). The magnetism (VSM) was measured using a vibrating sample magnetometer (7404 vibrating sample magnetometer, LakeShore, Carson, CA, USA). The thermogravimetric (TG) curves were obtained using a thermalgravimetric analyzer (Discovery SDT650 synchronous TGA-DTA Instrument, TA, Los Angeles, CA, USA). Fluorescence spectra were measured on a fluorescence spectrophotometer (FluoroMax-plus instrument, Horiba, Austin, TX, USA) with an excitation of (370) nm. The concentration of Ag^+ in solution was determined by atomic absorption spectrometry (AAS) (ICP-OES:Thermo Fisher iCAP 7400, Thermofisher, USA).

2.5. Materials

Other chemicals used in this study, such as $FeCl_3 \cdot 6H_2O$, $CH_3COONa \cdot 3H_2O$, $Zn(ac)_2$, Na_2S, polyethylene glycol 2000, 3-sulfhydryl-1-propane sodium, Co^{2+}, Pb^{2+}, Ni^{2+}, Hg^{2+}, Al^{3+}, Cu^{2+}, Zn^{2+}, Cd^{2+}, Fe^{3+}, Fe^{2+}, K^+, Ca^{2+}, and NaCl, were purchased from Shanghai Aladdin Bio-Chem Technology Co., Ltd. (Shanghai, China).

2.6. Procedure for the Fluorescent Detection of Ag^+

Certain volumes of Ag^+ solutions were added to 0.4 mg of the Fe_3O_4@ZnS@MPS suspension, whose final volume was tuned to 5 mL by PBS buffer solution. After a 5 min reaction, the emission spectra were measured at an excitation wavelength of 370 nm. The selectivity performance of Fe_3O_4@ZnS@MPS assay for the detection of Ag^+ was carried out by measuring its fluorescence response in the presence of various common ions (Co^{2+}, Pb^{2+}, Ni^{2+}, Hg^{2+}, Al^{3+}, Cu^{2+}, Zn^{2+}, Cd^{2+}, Fe^{3+}, Fe^{2+}, K^+, Ca^{2+}, Na^+).

2.7. Removal and Adoption Capacity Performance of Ag^+

The removal and adoption capacity performance of Ag^+ testing was determined in a centrifuge tube, for different concentrations of $AgNO_3$, after shaking well.

The removal activities can be evaluated by the following equation:

$$R = \frac{C_0 - C_1}{C_0} \times 100\% \quad (1)$$

C_1 is the final concentration of Ag^+, and C_0 is the initial concentration of Ag^+ ions. The concentration of Ag^+ ions was analyzed via AAS.

The adsorption amount of Ag^+ (Q mg/g) can be calculated according to the following equation:

$$Q_e = \frac{(C_0 - C_1)V}{m} \quad (2)$$

Q_e is the adsorption capacity (mg·g^{-1}), C_0 and C_e are the initial and equilibrium concentrations of Ag^+ (mg·L^{-1}), respectively, V is the volume of the metal ion solution (L), and m is the Fe_3O_4@ZnS@MPS mass (g).

3. Results and Discussion

3.1. Characterization of MFNS

3.1.1. Morphology Analysis

The morphology and particle size of Fe_3O_4 MNPs and Fe_3O_4@ZnS@MPS core-shell nanocomposites were investigated using SEM and TEM.

Figure 1a,c show the SEM and TEM images of Fe_3O_4 MNPs, from the commercial Fe_3O_4 with a diameter distribution from 150 to 250 nm, which reveal that Fe_3O_4 MNPs have a regular and uniform distribution. In addition, the SEM image of Fe_3O_4@ZnS@MPS is shown in Figure 1b. Compared with Figure 1a,b, it can be seen that ZnS particles are accumulated on the surface of Fe_3O_4 after the modification of the thiol group, and the surface of Fe_3O_4 MNPs becomes irregular and rough. In addition, TEM images (Figure 1d) of Fe_3O_4@ZnS@MPS with a diameter distribution from 200 to 300 nm were obtained, which show that ZnS particles modified with sulfhydryl groups have been deposited on the surface of Fe_3O_4 MNPs [23].

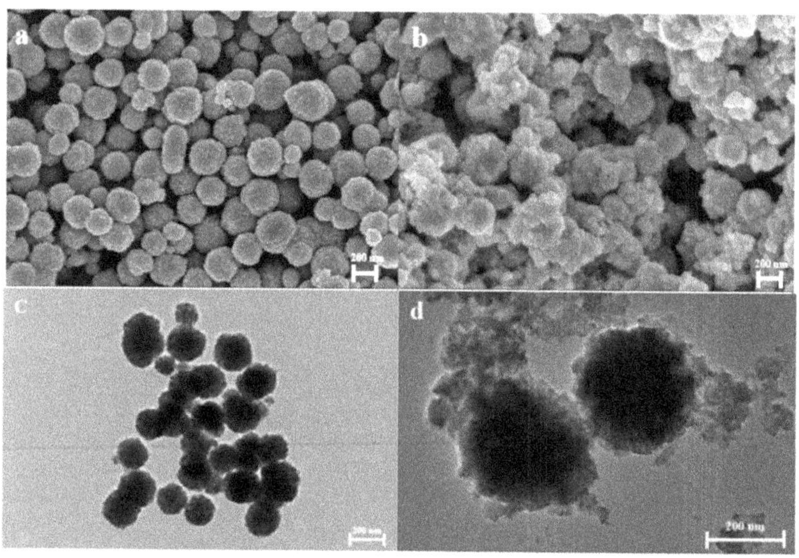

Figure 1. SEM and TEM images of Fe_3O_4 (**a**,**c**), Fe_3O_4@ZnS@MPS (**b**,**d**).

3.1.2. XRD Analysis

Figure 2 shows the XRD patterns of Fe_3O_4, Fe_3O_4@ZnS, and MFNPs. In the XRD pattern, several diffraction peaks around 2θ = 30.1°, 35.6°, 43.1°, 53.7°, 57.1°, and 62.8° were observed for Fe_3O_4@ZnS@MPS, which could be assigned to the (220), (311), (400), (422), (511), and (440) planes of the cubic spinel crystal structure of Fe_3O_4, respectively. There were also three new diffraction peaks at around 2θ = 28.9°, 47.7°, and 57.0°, which could confirm the existence of ZnS crystals in the composites [24]. The position of the diffraction peak did not change when the quantum dots and sulfhydryl groups were modified, indicating that the Fe_3O_4 magnetic core did not undergo chemical or structural changes during the coating process. Figure 2 clearly shows that the spectral peaks are basically the same before and after functionalization, indicating that the crystal structure of Fe_3O_4 does not change during different functionalization processes [25,26].

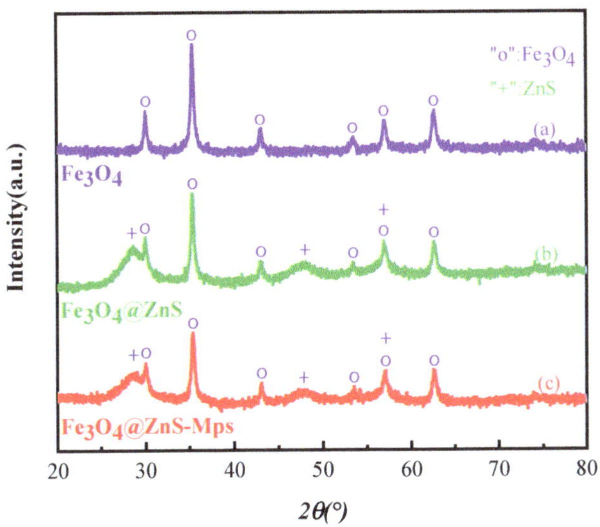

Figure 2. XRD patterns of (**a**) Fe_3O_4, (**b**) Fe_3O_4@ZnS, and (**c**) Fe_3O_4@ZnS@MPS nanocomposites.

3.1.3. FT−IR Analysis

Figure 3 shows the infrared (IR) spectra of Fe_3O_4, Fe_3O_4@ZnS, Fe_3O_4@ZnS@MPS, and MPS. It can be seen from the infrared spectra of the magnetic microspheres that the strong absorption peak at 584 cm^{-1} is related to the stretching vibration of the Fe–O bond. For Fe_3O_4 @ZnS magnetic fluorescent nanoparticles, the peaks corresponding to the stretching vibration of Fe–O and Zn–S appeared at 580 cm^{-1} and 627 cm^{-1}. The stretching vibration of -SH in MPS corresponds to a peak at 2555 cm^{-1}. For MFNP nanoparticles, the Zn–S stretching vibration peak was obviously masked after sulfhydryl modification at 1018 cm^{-1}, and a new C–H peak appeared at 2922 cm^{-1} [27]. The corresponding peak of the new S–O stretching vibration was 1042 cm^{-1}, while the peak at 1178 cm^{-1} was assigned to the O=S=O symmetric stretching vibration. This can prove that the sulfhydryl group is bound to the surface of the microspheres and the ligand is successfully modified on the surface of the microspheres [28–31].

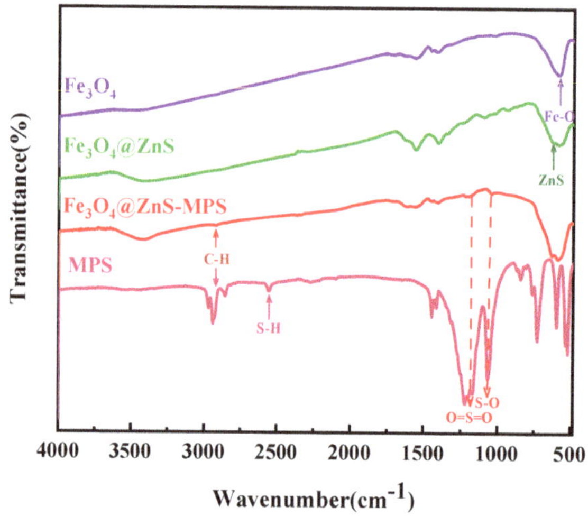

Figure 3. FT−IR spectra of commercial Fe_3O_4, Fe_3O_4@ZnS, Fe_3O_4@ZnS@MPS, and MPS.

3.1.4. XPS Analysis

The elemental composition of the Fe_3O_4@ZnS@MPS nanoparticle was explored by XPS analysis (Figure 4), The five peaks at 1019.2 eV, 709.8 eV, 529.6 eV, 285.6 eV, and 159.2 eV are composed of Zn 2p, Fe 2p, O 1s, C 1s, and S 2p, respectively, confirming the successful synthesis of Fe_3O_4@ZnS@MPS nanoparticles. Figure 4b shows that the binding energies of Fe^{3+} $2p_{3/2}$ and Fe^{3+} $2p_{1/2}$ are 710.8 and 725.0 eV, respectively, and the bimodal fitting of Fe 2p can obtain the binding energies of Fe^{2+} $2p_{3/2}$ and Fe^{2+} $2p_{1/2}$ at 708.7 and 721.6 eV, respectively. The results indicate that Fe_3O_4 exists in the nanoparticle. According to Figure 4c, the difference in binding energy between Zn $2p_{3/2}$ and Zn $2p_{1/2}$ is 22.5 eV, indicating that metallic Zn supported by the MFNPs mainly exists in the Zn^{2+} valence state. Peaks at 159.5 eV and 160.7 eV (Figure 4d) are attributed to metal sulfides S^{2-} ($2p_{3/2}$ and S $2p_{1/2}$) from ZnS, respectively [32,33].

3.1.5. Hysteresis Curve

The magnetic properties of the Fe_3O_4 and Fe_3O_4@ZnS@MPS MFNPs were studied using a VSM in an external magnetic field from −20,000 to +20,000 Oe, and results are shown in Figure 5. The saturation magnetization of Fe_3O_4 MNPs was 64.52 emu/g, and that of Fe_3O_4@ZnS@MPS nanocomposites was 47.09 emu/g. The magnetization in the nanocomposites is reduced due to the diamagnetic effect of the thick sulfhydryl-modified ZnS layer around the Fe_3O_4 MNPs. Ion redistribution of Zn^{2+} may also have contributed to the decrease in the saturation magnetization of the composite. However, it still possesses typical superparamagnetism, which can meet the experimental requirements, and still shows excellent magnetic properties after removal of Ag^+ (inset figure) [11,34].

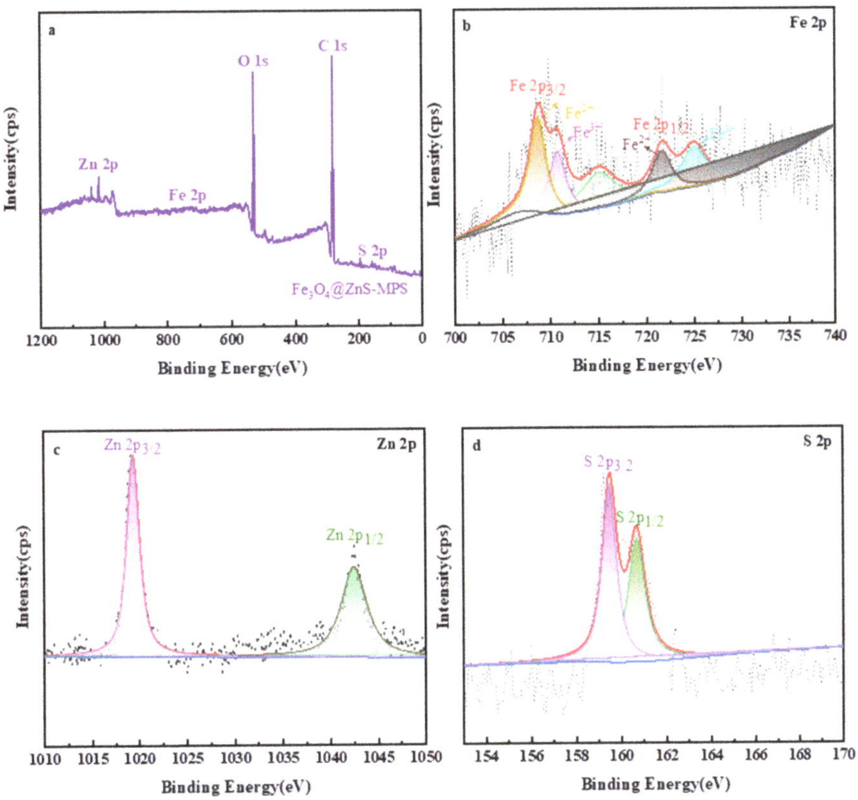

Figure 4. XPS analysis of Fe_3O_4@ZnS@MPS (**a**), high-resolution XPS spectra of Fe 2p (**b**), Zn 2p (**c**), S 2p (**d**) of Fe_3O_4@ZnS@MPS.

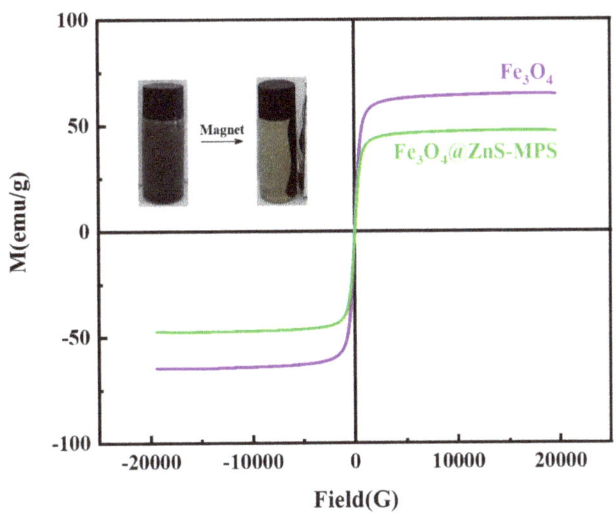

Figure 5. VSM measurements of Fe_3O_4 and Fe_3O_4@ZnS@MPS. Inset: The response of Fe_3O_4@ZnS@MPS to external magnetic field.

3.1.6. Thermogravimetric Analysis

In order to study the thermal properties of the samples, a thermal weight loss analyzer was used to analyze the samples. The thermal properties constitute an important index used to evaluate the hybrid materials, and differ with different synthesis methods. The TGA curves of MFNFs were measured at a heating rate of 10 °C/min under a nitrogen atmosphere, and the results are shown in Figure 6. As shown in Figure 6, the overall weight loss rate of the products is not very large, and the weight loss rate of Fe_3O_4 is only about 8%. After the modification of ZnS, its weight loss is reduced to 15%, and the extra weight loss is the water contained in ZnS. The weight loss rate reaches 18% after the modification of MPS, indicating that the polymer in the sample was removed. Therefore, the TGA analysis shows that the sulfhydryl group was modified on the surface of Fe_3O_4 magnetic microspheres.

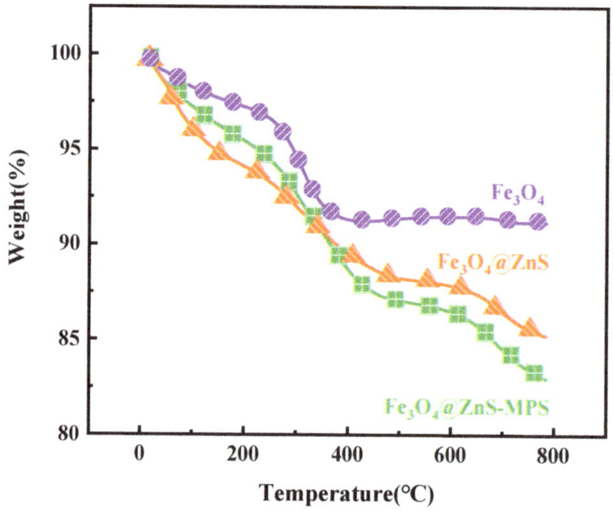

Figure 6. TGA curves of Fe_3O_4, Fe_3O_4@ZnS, and Fe_3O_4@ZnS@MPS.

Combined with the above data, it can be proven that the Fe_3O_4@ZnS@MPS magnetic fluorescent nanosensors (MFNPs) have been successfully synthesized.

3.2. Detection Performance Study of MFNPs

3.2.1. Performance Analysis of Magnetic Fluorescence Nanosensor MFNPs

Figure 7 shows the comparison of fluorescence characteristic spectra before and after the addition of Ag^+ to MFNPs. Figure 7 shows that the addition of Ag^+ can significantly quench the fluorescence intensity of MFNPs, and the inset shows the TEM images of MFNPs before and after the addition of Ag^+. It can be clearly seen that Ag^+ has been loaded and dispersed on the surface of the MFNPs. MFNPs have good dispersibility and are almost spherical. The complexation of Ag^+ with sodium 3-sulfhydryl-1-propane sulfonate on the surface of MFNPs leads to Ag^+ aggregation and obvious fluorescence quenching.

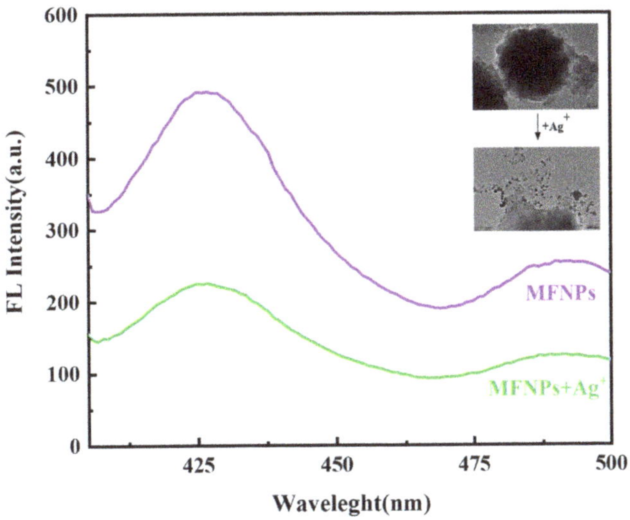

Figure 7. Comparison of fluorescence spectra of MFNPs before and after Ag^+ addition. Inset: TEM images of MFNPs before and after Ag^+ addition.

3.2.2. The Effect of pH on the Fluorescence of MFNPs

One of the crucial factors determining the sensor's capacity for detection is the pH level of the solution. As can be seen from Figure 8, the fluorescence intensity of the probe itself does not change greatly within the range of 4.8–9.0, and the degree of fluorescence intensity quenched by the addition of silver ions was not affected by the pH value. Therefore, we chose the common neutral liquid pH value of 7.0 as the experimental standard.

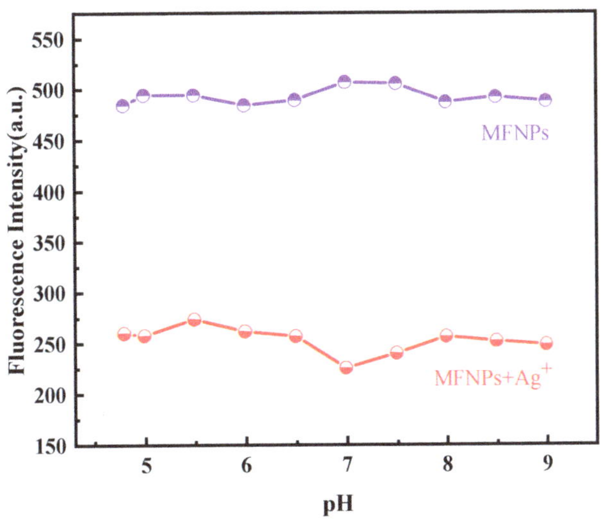

Figure 8. Fluorescence intensity of MFNPs at 425 nm in the absence and presence of Ag^+ at different pH.

3.2.3. The Response Range and Relation Curve of MFNPs to Ag^+

With the increase in the concentration of Ag^+, the fluorescence intensity gradually weakens in the range of 0–100 μM (Figure 9). There was a linear relationship between the concentration of Ag^+ and the fluorescence intensity $I/I_0 = -0.00426x + 1.00934$ ($R^2 = 0.9967$), where x is the concentrations of Ag^+ (μM), with the LOD of 7.04 μM. Therefore, the sensor of the presented invention can more accurately determine the content of Ag^+. Based on the data of the linear range and the detection limit, it can be seen that the probe showed good performance (Table 1).

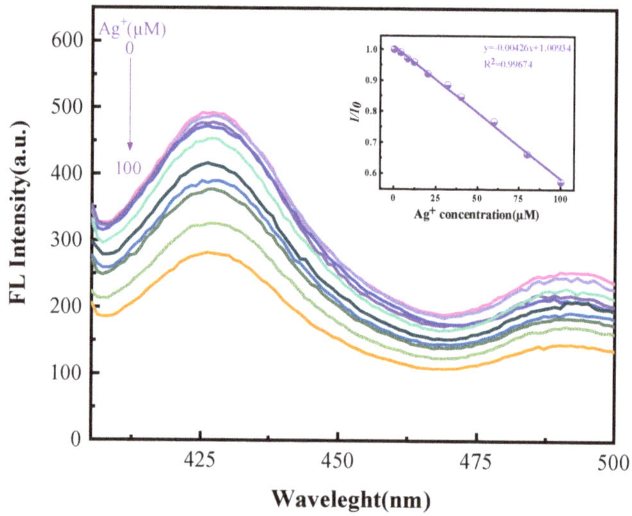

Figure 9. Emission spectra of MFNPs in the presence of increasing amounts of Ag^+ at room temperature. Inset: The curve of fluorescence intensity at 425 nm; the concentrations of Ag^+ are 0, 4, 8, 12, 20, 32, 40, 60, 80, 100 μM, respectively.

Table 1. Comparison of the present nanoprobe with other reported sensors for Ag^+ detection.

Type of Sensor	Detection Range	LOD	Reference
Eu-MIL-61(Ga(OH)(btec)·0.5H_2O)	0–1000 μM	0.23 μM	Dalton Transactions, 46 (2017)875 [35]
Triphenylamine–thiophene-pyridinium	5.0–9.0 μM	3.6 μM	Talanta, 255(2023)124222 [36]
Nitrogen and bromine co-doped carbon dots	0–6.0 μM	0.14 μM	Spectrochimica Acta Part A: Molecular and Biomolecular Spectroscopy, 296(2023)122642 [37]
NPCl-doped carbon quantum dots	15.89–27.66 μM	26.38 μM	Analytica Chimica Acta, 1144 (2021) 1–13 [38]
Cholesteric chiral artificial receptor L5	2–20 μM	0.13 μM	Microchemical Journal, 190(2023)108633 [39]
Fe_3O_4@ZnS@MPS	0–100 μM	7.04 μM	This work

3.2.4. Particle Selectivity and Interference Ion Determination

High selectivity is a necessary condition for the sensor. Therefore, under the same conditions, the Ag^+ selectivity of the prepared magnetic fluorescent sensor was investigated by detection of the reaction of the relevant analytes. The results show that the fluorescence quenching effect of Ag^+ is the best. Although the other ions will experience weak quenching, this is obviously negligible compared with that of silver ions (Figure 10a). To further investigate the ability of the magnetic fluorescent sensor to recognize silver ions in the presence of other metal ions, the anti-interference ability of the sensor was also investigated. When an equal amount of silver ions was added to an equal amount of other metal ions (400 μM Ag^+, Co^{2+}, Ni^{2+}, Al^{3+}, Cu^{2+}, Zn^{2+}, Cd^{2+}, Fe^{3+}, Fe^{2+}, K^+, Ca^{2+}, Na^+, Pb^{2+}, Hg^{2+}), the other ions did not affect the detection of silver ions by the magnetic fluorescent sensor. The bar graph clearly proves this point (Figure 10b); it is obvious that the common metal ions in the environment do not interfere with the qualitative and quantitative detection of silver ions by the particles.

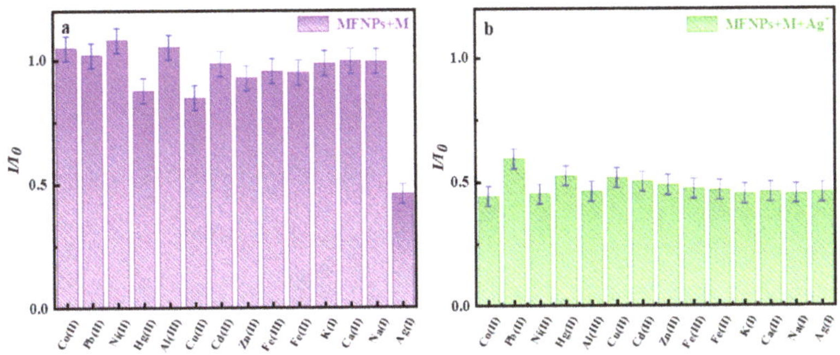

Figure 10. (**a**) Bar graph represents the ratio of fluorescence quenching of MFNPs in the presence of different metal ions (1, Co^{2+}; 2, Pb^{2+}; 3, Ni^{2+}; 4, Hg^{2+}; 5, Al^{3+}; 6, Cu^{2+}; 7, Zn^{2+}; 8, Cd^{2+}; 9, Fe^{3+}; 10, Fe^{2+}; 11, K^+; 12, Ca^{2+}; 13, Na^+; 14, Ag^+); (**b**) bar graph represents the ratio of fluorescence quenching of MFNPs upon the addition of Ag^+ (400 μM) to the solution containing other metal ions (400 μM, 1, Co^{2+}; 2, Pb^{2+}; 3, Ni^{2+}; 4, Hg^{2+}; 5, Al^{3+}; 6, Cu^{2+}; 7, Zn^{2+}; 8, Cd^{2+}; 9, Fe^{3+}; 10, Fe^{2+}; 11, K^+; 12, Ca^{2+}; 13, Na^+;14, Blank).

3.2.5. Removal of Ag^+

Adsorption capacity is considered to be one of the most important properties of nanomaterials, so adsorption tests were performed in conical flasks (100 mL) to determine the adsorption amount of Ag^+. The effect of the initial Ag^+ concentration on the adsorption efficiency is shown in Figure 11. The adsorption analysis showed that with the increase in the initial Ag^+ concentration, the adsorption capacity of MFNPs also gradually increased. This was due to the increase in the initial concentration, which improved the adsorption driving force and ion mass transfer rate, and the binding site on the adsorbents' surface was gradually occupied by Ag^+; thus, the adsorption capacity increased. The removal rate of Ag^+ in the solution gradually decreased after 100 μM, reaching 99.62%, which is because the amount of adsorbent in the solution is finite, and the adsorption site provided for Ag^+ is also finite. When the concentration of heavy metals is low, almost all Ag^+ can be combined with the adsorption site to achieve a higher removal rate, and then the adsorption gradually reaches equilibrium. Therefore, the adsorption effect of the adsorbent on the free metal ions in the solution is reduced, resulting in a lower removal rate. When the Ag^+ concentration changed from 300 μM to 600 μM, the removal rate decreased from 99.54% to 93.54%, and the adsorption capacity increased from 322.496 mg/g to 606.112 mg/g. In general, the optimal Ag^+ concentration at the intersection point of 400 μM could be obtained. Compared with other materials, the adsorption and transfer rates of this material are relatively high (Table 2).

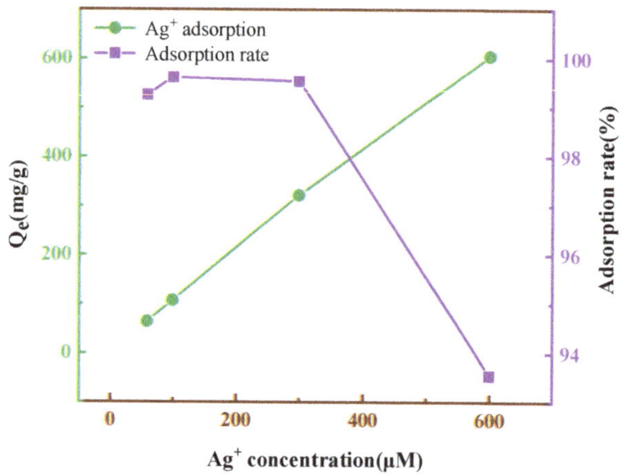

Figure 11. Effect of the initial concentration of Ag^+ on the adsorption capacity and removal efficiency of magnetic fluorescent nanoparticles.

Table 2. The adsorption capacity of reported adsorbents for Ag^+.

Type of Sensor	Adsorption Capacity (mg/g)	Reference
Bifunctional polysilsesquioxane microspheres	416.88	Journal of Hazardous Materials 442(2023) 130,121 [40]
ZMC-MAH-TEPA (grafted magnetic zeolite/chitosan)	70.12	International Journal of Biological Macromolecules 222 (2022) 2615–2627 [41]
DMTD-AP(2,5-dimercap-to-1,3,4-thiadiazole modified apple pomace)	196.9	Sustainable Chemistry and Pharmacy 26 (2022) 100,621 [42]
L-PRL(o -phenanthroline-based polymer)	325.8	Chinese Chemical Letters 34 (2023) 107,485 [43]
MFNPs	395.79	This work

3.3. Infrared Spectroscopy Analysis before and after Complexation of MFNPs with Ag^+

Infrared spectra of MFNPs before and after the addition of Ag^+ were analyzed using FT-IR, and the FT-IR spectra of MFNPs and MFNPs + Ag^+ complexes are shown in Figure 12. From the infrared spectra of MFNPs, the corresponding peaks of S = O telescopic vibration at 1042 cm^{-1} were observed.

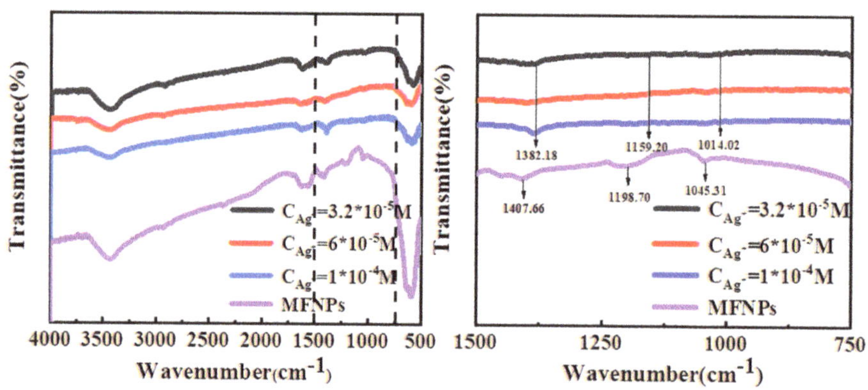

Figure 12. Infrared absorption spectra of MFNPs and inclusion complex MFNPs + Ag^+.

The peak shape of the absorption peaks at 1045.31 cm^{-1}, 1198.70 cm^{-1}, and 1407.66 cm^{-1} changed, indicating that S–O, O=S=O, and C–O might play a key role in the synergistic effect with Ag^+. The result can be clearly observed in Figure 8; when the Fe_3O_4@ZnS-MPS interacted with Ag+, the tensile peak of methoxy (S–O) was changed from 1045.31 cm^{-1} to 1014.02 cm^{-1}, while the characteristic absorption peak of O=S=O moved from 1198.70 cm^{-1} to 1159.20 cm^{-1} and the characteristic absorption peak of C–O moved from 1407.66 cm^{-1} to 1382.18 cm^{-1}. The reason for the change in the infrared absorption peak of the sulfonic acid group might be that the oxygen atoms in the sulfonic acid group cooperated with Ag^+ to form a synergistic complexation. The oxygen atoms of the sulfonic acid group had a negative charge, indicating that the S–O and O=S=O group could adsorb Ag^+ by electrostatic interaction as Ag^+ laced electrons. Thus, it was speculated that the oxygen atom of the S–O and O=S=O group donated electrons to the Ag^+, which led to the transfer of charge and resulted in changes in the infrared spectrum [43].

3.4. The Fluorescent Detection of Ag^+ in Actual Samples

The practicability of the MFNPs for the detection of Ag^+ was demonstrated by analyzing real water samples. The real samples were selected from tap water, water from Songhua River in Jilin, and electrolysis waste water. First, suspended particles were removed from these water samples using centrifugal separation. Then, the supernatant was analyzed according to the previous procedure of Ag^+. Certain volumes of the supernatant and Ag^+ solutions were added to 200 µL of the Fe_3O_4@ZnS@MPS suspension, and the final volume was tuned to 2 mL by PBS buffer solution. As shown in Table 3, Ag^+ was not detected in the real water samples. The recovery of Ag^+ detection was in the range of 85.20%–105.30%, with the RSD (relative standard deviation) in the range of 3.18%–5.32%. The result indicated that the MFNPs had great potential for the detection of Ag+ in real water samples.

Table 3. The performance of the MFNPs in real water samples.

Sample	Added (µM)	Measured (µM)	Recovery (%)	RSD (%)
Tap water	1	0.926	92.60	3.18
	5.0	4.859	97.18	4.25
	10.0	10.321	103.21	3.86
River water	1	1.053	105.30	4.34
	5.0	4.987	99.74	3.89
	10.0	9.842	98.42	4.45
Electrolysis waste water	1	0.852	85.20	5.32
	5.0	4.320	86.40	4.96
	10.0	8.620	86.20	4.85

4. Conclusions

In summary, we present the design, synthesis, and performance of a novel magnetic fluorescent nanoprobe (Fe_3O_4@ZnS@MPS(MFNPs)) modified with 3-sulfhydryl-1-propane sodium for simultaneous detection and removal of Ag^+ from water solutions. Our proposed Fe_3O_4@ZnS@MPS sensor exhibits significant quenching fluorescence intensity ability and high selectivity for Ag^+. It was shown that the Fe_3O_4@ZnS@MPS fluorescent sensor was a good adsorbent for the removal of Ag^+, due to the aggregation and magnetic separation of the sensor. In addition, it could be applied over a wide pH range. The above results confirmed that this method is an inexpensive, simple, convenient, and extremely sensitive method that enables highly specific recognition for the simultaneous detection and removal of Ag^+ from aqueous solutions. This study provided a new idea for the determination and removal of Ag^+ and other heavy metals.

Author Contributions: Conceptualization, Y.G. and X.C.; methodology, X.Z.; software, X.C.; validation, Z.L., S.Y. and J.C.; formal analysis, P.X.; investigation, X.Z.; resources, Z.L.; data curation, X.C.; writing—original draft preparation, X.C.; writing—review and editing, X.Z.; visualization, X.Z.;

supervision, X.Z.; project administration, X.Z.; funding acquisition, Z.L. All authors have read and agreed to the published version of the manuscript.

Funding: This research was funded by X.Z. grant number (20220203020SF) And The APC was funded by Jilin Provincial Department of Science and Technology. This research was funded by J.C. grant number (51902125) And The APC was funded by National Natural Science Foundation of China. This research was funded by S.Y. grant number (22106051) And The APC was funded by National Natural Science Foundation of China.

Data Availability Statement: Z.X. and G.Y. designed the work. C.X. performed the experiment and data analysis.

Acknowledgments: We gratefully acknowledge financial support from Jilin Provincial Department of Science and Technology (20220203020SF), the National Natural Science Foundation of China (51902125, 22106051).

Conflicts of Interest: The authors declare no conflict of interest.

References

1. Singha, S.; Kim, D.; Seo, H.; Cho, S.W.; Ahn, K.H. Fluorescence sensing systems for gold and silver species. *Chem. Soc. Rev.* **2015**, *44*, 4367–4399. [CrossRef]
2. Bahareh, R.; Hossein, R.M.; Mehdi, S.K. Colorimetric/fluorometric optical chemosensors based on oxazolidine for highly selective detection of Fe^{3+} and Ag^{+} in aqueous media: Development of ionochromic security papers. *J. Mol. Struct.* **2023**, *1271*, 134021.
3. Pal, C.; Majumder, S. Label-free electrochemical biosensor for ultra-low level detection of Ag(I) using ssDNA functionalized multi-walled carbon nanotube. *Colloids Surf. A Physicochem. Eng. Asp.* **2022**, *645*, 128927. [CrossRef]
4. Han, J.S.; Zhang, X.; Zhou, Y.B.; Ning, Y.; Wu, J.; Liang, S.; Sun, H.C.; Zhang, H.; Yang, B. Fabrication of CdTe nanoparticles-based superparticles for an improved detection of Cu^{2+} and Ag^{+}. *J. Mater. Chem.* **2012**, *22*, 2679–2686. [CrossRef]
5. Bian, J.; Xia, Y.X.; Sang, L.Y.; Zhu, C.X.; Li, Y.X.; Li, G.Y.; Liu, X.Y.; Wang, X.; Liu, Y. A recyclable colorimetric probe: In situ fabrication of highly stable HPEI–AuNPs for selective Ag^{+} detection. *New J. Chem.* **2020**, *44*, 5438–5447. [CrossRef]
6. Gumpu, M.B.; Sethuraman, S.; Krishnan, U.M. A review on detection of heavy metal ions in water an electrochemical approach. *Sens. Actuators B Chem.* **2015**, *213*, 515–533. [CrossRef]
7. Cai, C.; Cheng, H.Y.; Wang, Y.C.; Bao, H.F. Mercaptosuccinic acid modified CdTe quantum dots as a selective fluorescence sensor for Ag^{+} determination in aqueous solutions. *RSC Adv.* **2014**, *4*, 59157–59163. [CrossRef]
8. Pandey, R.; Kumar, A.; Xu, Q.; Pandey, D.S. Zinc(II), copper(II) and cadmium(II) complexes as fluorescent chemosensors for cations. *Dalton Trans.* **2020**, *49*, 542–568. [CrossRef] [PubMed]
9. Deng, P.Y.; Wang, W.; Liu, X.Q.; Wang, L.; Yan, Y.S. A hydrophobic polymer stabilized $CsPbBr_3$ sensor for environmental pollutant detection. *New J. Chem.* **2021**, *45*, 930–938. [CrossRef]
10. Wu, P.; Yan, X.P. Doped quantum dots for chemo/biosensing and bioimaging. *Chem. Soc. Rev.* **2013**, *42*, 5489–5521. [CrossRef]
11. Stefan, M.; Leostean, C.; Pana, O.; Suciu, M.; Popa, A.; Toloman, D.; Macavei, S.; Bele, C.; Tabaran, F.; Barbu-Tudoran, L. Synthesis and characterization of Fe_3O_4–ZnS: Mn nanocomposites for biomedical applications. *Mater. Chem. Phys.* **2021**, *264*, 124474. [CrossRef]
12. Arano-Martinez, J.A.; Martínez-González, C.L.; Salazar, M.L.; Torres-Torres, C. A Framework for Biosensors Assisted by Multiphoton Effects and Machine Learning. *Biosensors* **2022**, *12*, 710. [CrossRef]
13. Wang, C.W.; Shen, W.Z.; Rong, Z.; Liu, X.X.; Gu, B.; Xiao, R.; Wang, S.Q. Layer-by-layer assembly of magnetic-core dual quantum dot-shell nanocomposites for fluorescence lateral flow detection of bacteria. *Nanoscale* **2020**, *12*, 795–807. [CrossRef]
14. Shah, M.T.; Balouch, A.; Alveroglu, E. Sensitive Fluorescence Detection of Ni^{2+} ions Using Fluorescein Functionalized Fe_3O_4 Nanoparticles. *J. Mater. Chem. C* **2018**, *6*, 1105–1115. [CrossRef]
15. Chen, M.; Shao, L.L.; Li, J.J.; Pei, W.J.; Chen, M.K.; Xie, X.H. One-step hydrothermal synthesis of hydrophilic Fe_3O_4/carbon composites and their application in removing toxic chemicals. *RSC Adv.* **2016**, *6*, 35228–35238. [CrossRef]
16. Dulyasucharit, R.; Wongkasemjit, S.; Nanan, S.; Intharaksa, O.; Masingboon, C. Magnetic $Fe_3O_4/Bi_2O_2(OH)(NO_3)$ as a sunlight-driven photocatalyst for rhodamine B degradation. *J. Solid State Chem.* **2023**, *319*, 123784. [CrossRef]
17. Zhu, Y.Z.; Wu, D.P.; Chen, J.H.; Ma, N.; Dai, W. Boosting highly capture of trace tetracycline with a novel water-resistant and magnetic (ZIF-8)-on-(Cu-BTC@Fe_3O_4) composite. *J. Solid State Chem.* **2023**, *319*, 123797. [CrossRef]
18. Gao, F.; Gu, H.; Wang, H.; Wang, X.; Xiang, B. Magnetic amine-functionalized polyacrylic acid-nanomagnetite for hexavalent chromium removal from polluted water. *RSC Adv.* **2015**, *5*, 60208–60219. [CrossRef]
19. Duan, M.; Xia, F.F.; Li, T.L.; Shapter, J.G.; Yang, S.; Li, Y.Y.; Gao, G.; Cui, D.X. Matrix metalloproteinase-2-targeted superparamagnetic Fe_3O_4-PEG-G5-MMP2@Ce6 nanoprobes for dual-mode imaging and photodynamic therapy. *Nanoscale* **2019**, *11*, 18426. [CrossRef]
20. Li, G.Y.; Jiang, Y.R.; Huang, K.L.; Ding, P.D.; Chen, J. Preparation and properties of magnetic Fe_3O_4-chitosan nanoparticles. *J. Alloys Compd.* **2008**, *466*, 451–456. [CrossRef]

21. Liu, M.; Tao, Z.; Wang, H.; Zhao, F.; Sun, Q. Study on the adsorption of Hg(II) by one-pot synthesis of amino-functionalized graphene oxide decorated with Fe_3O_4 microspheres nanocomposite. *RSC Adv.* **2016**, *6*, 84573–84586. [CrossRef]
22. Chen, X.; Chen, J.; Ma, M.S.; Yu, S.H.; Liu, Z.G.; Zeng, X.D. An Ethyl-Thioglycolate-Functionalized Fe_3O_4@ZnS Magnetic Fluorescent Nanoprobe for the Detection of Ag+ and Its Applications in Real Water Solutions. *Nanomaterials* **2023**, *13*, 1992. [CrossRef] [PubMed]
23. Dafeh, S.R.; Iranmanesh, P.; Salarizadeh, P. Fabrication, optimization, and characterization of ultra-small superparamagnetic Fe_3O_4 and biocompatible Fe_3O_4@ZnS core/shell magnetic nanoparticles: Ready for biomedicine applications. *Mater. Sci. Eng. C* **2019**, *98*, 205–212. [CrossRef] [PubMed]
24. Khawla, M.; Zouhour, H.; Yves, C.; Souhaira, H.; Rym, M. ZnS quantum dots as fluorescence sensor for quantitative detection of tetracycline. *Opt. Mater.* **2022**, *125*, 112103. [CrossRef]
25. Nawaz, T.; Zulfiqar, S.; Sarwar, M.I.; Iqbal, M. Synthesis of diglycolic acid functionalized core-shell silica coated Fe_3O_4 nanomaterials for magnetic extraction of pb(II) and cr(VI) ions. *Sci. Rep.* **2020**, *10*, 10076. [CrossRef]
26. Stefan, M.; Leostean, C.; Pana, O.; Soran, M.; Suciu, R.; Gautron, E.; Chauvet, O.; Stefan, M. Synthesis and characterization of Fe_3O_4@ZnS and Fe_3O_4@Au@ZnS core–shell nanoparticles. *Appl. Surf. Sci.* **2014**, *288*, 180–192. [CrossRef]
27. Zavidovskiy, I.A.; Streletskiy, O.A.; Nuriahmetov, I.F.; Nishchak, O.U.; Savchenko, N.F.; Tatarintsev, A.A.; Pavlikov, A.V. Highly Selective Polyene-Polyyne Resistive Gas Sensors: Response Tuning by Low-Energy Ion Irradiation. *J. Compos. Sci.* **2023**, *7*, 156. [CrossRef]
28. Mondal, D.K.; Phukan, G.; Paul, N.; Borah, J.P. Improved self heating and optical properties of bifunctional Fe_3O_4/ZnS nanocomposites for magnetic hyperthermia application. *J. Magn. Magn. Mater.* **2021**, *528*, 167809. [CrossRef]
29. Koc, K.; Karakus, B.; Rajar, K.; Alveroglu, E. Synthesis and Characterization of ZnS@Fe_3O_4 Fluorescent-Magnetic Bifunctional Nanospheres. *Superlattices Microstruct.* **2017**, *110*, 198–204. [CrossRef]
30. Wen, A.; Li, G.L.; Wu, D.; Yu, Y.X.; Yang, Y.; Hu, N.; Wang, H.L.; Chen, J.; Wu, Y.N. Sulphonate functionalized covalent organic framework-based magnetic sorbent for effective solid phase extraction and determination of fluoroquinolones. *J. Chromatogr. A* **2020**, *1612*, 460651. [CrossRef]
31. Liu, Q.; Yang, D.L.; Shi, Y.L.; Pan, Q.Q.; Wang, T. Shape-selective separation of copper nanowires and copper-based nanoparticles by a ligand exchange strategy. *Appl. Surf. Sci.* **2023**, *611*, 155597. [CrossRef]
32. Masoud, H.Z.; Arjomand, M.Z. Photocatalytic activity of ZrO_2/TiO_2/Fe_3O_4 ternary nanocomposite for the degradation of naproxen: Characterization and optimization using response surface methodology. *Sci. Rep.* **2022**, *12*, 10388.
33. Zangmeister, R.A.; Morris, T.A.; Tarlov, M.J. Characterization of Polydopamine Thin Films Deposited at Short Times by Autoxidation of Dopamine. *Langmuir.* **2013**, *29*, 8619–8628. [CrossRef]
34. Chang, C.J.; Tsai, W.C. CuS-ZnS decorated Fe_3O_4 nanoparticles as magnetically separable composite photocatalysts with excellent hydrogen production activity. *Int. J. Hydrog. Energy* **2019**, *44*, 20872–20880. [CrossRef]
35. Sun, N.N.; Yan, B. A reliable amplified fluorescence-enhanced chemosensor (Eu-MIL-61) for the directional detection of Ag^+ in an aqueous solution. *Dalton Trans.* **2017**, *46*, 875–881. [CrossRef]
36. Li, J.R.; Huang, H.; Zhang, C.; Chen, X.L.; Hu, Y.J.; Huang, X.H. Dual-key-and-lock AIE probe for thiosulfate and Ag^+ detection in mitochondria. *Talanta* **2023**, *255*, 124222. [CrossRef] [PubMed]
37. Zhu, J.T.; Shen, M.X.; Shen, J.W.; Wang, C.Z.; Wei, Y.M. Nitrogen and bromine co-doped carbon dots with red fluorescence for sensing of Ag^+ and visual monitoring of glutathione in cells. *Spectrochim. Acta A Mol. Biomol. Spectrosc* **2023**, *296*, 122642. [CrossRef]
38. Wang, Z.H.; Zhang, L.; Hao, Y.M.; Dong, W.J.; Liu, Y.; Song, S.M.; Shuang, S.M.; Dong, C.; Gong, X.J. Ratiometric fluorescent sensors for sequential on-off-on determination of riboflavin, Agt and L-cysteine based on NPCl-doped carbon quantum dots. *Anal. Chim. Acta* **2021**, *1144*, 1–13. [CrossRef]
39. Luo, L.X.; Ye, Y.; Wang, H.; Liu, Z.; Li, J.Q. Rapid Detection of Ag^+ in Food Using Cholesteric Chiral Artificial Receptor L5. *Microchem. J.* **2023**, *190*, 108633.
40. Wang, B.X.; Wu, K.Y.; Liu, T.H.; Cheng, Z.K.; Liu, Y.; Liu, Y.F.; Niu, Y.Z. Feasible synthesis of bifunctional polysilsesquioxane microspheres for robust adsorption of Hg(II) and Ag(I): Behavior and mechanism. *J. Hazard. Mater.* **2023**, *442*, 130121. [CrossRef]
41. Liu, X.Q.; Zhang, Y.Y.; Liu, Y.; Zhang, T.A. Removal of Cr(VI) and Ag(I) by grafted magnetic zeolite/chitosan for water purification: Synthesis and adsorption mechanism. *Int. J. Biol. Macromol.* **2022**, *222*, 2615–2627. [CrossRef] [PubMed]
42. Zhang, S.J.; Lin, J.; Chen, S.L.; Xie, H.D.; Qi, Y.B. Selective recovery of Ag(I) from effluent using 2,5-dimercap-to-1,3,4-thiadiazole modified apple pomace. *Sustain. Chem. Pharm.* **2022**, *26*, 100621. [CrossRef]
43. Ding, X.; Yu, W.J.; Sheng, X.; Shi, H.; You, D.; Peng, M.M.; Shao, P.H.; Yang, L.M.; Liu, L.M.; Luo, X.B. Feasible fabrication of o -phenanthroline-based polymer adsorbent for selective capture of aqueous Ag(I). *Chin. Chem. Lett.* **2023**, *34*, 107485. [CrossRef]

coatings

Article

β-Cyclodextrin-Modified Mesoporous Silica Nanoparticles with Photo-Responsive Gatekeepers for Controlled Release of Hexaconazole

Hua Pan [†], Wenjing Li [†], Litao Wu, Weilan Huang and Fang Zhang *

Faculty of Environment and Life, Beijing University of Technology, Beijing 100124, China; panh@emails.bjut.edu.cn (H.P.); liwenjing9511@126.com (W.L.); wltuiao@163.com (L.W.); huangweilan1226@163.com (W.H.)

* Correspondence: zhangfang2000@bjut.edu.cn

† These authors contributed equally to this work.

Abstract: In the present research, photo-responsive controlled-release hexaconazole (Hex) nanoparticles (Nps) were successfully prepared with azobenzene (Azo)-modified bimodal mesoporous silica (BMMs), in which β-cyclodextrin (β-CD) was capped onto the BMMs-Azo surface via host–guest interactions (Hex@BMMs/Azo/β-CD). Scanning electron microscopy (SEM) confirmed that the nanoparticles had a spherical structure, and their average diameter determined by dynamic light scattering (DLS) was found to be 387.2 ± 3.8 nm. X-ray powder-diffraction analysis and N_2-adsorption measurements indicated that Hex was loaded into the pores of the mesoporous structure, but the structure of the mesoporous composite was not destroyed. The loading capacity of Hex@BMMs/Azo/β-CD nanoparticles for Hex was approximately 27.3%. Elemental components of the nanoparticles were characterized by X-ray photoelectron spectroscopy (XPS) and electron dispersive spectroscopy (EDS). Ultraviolet–visible-light (UV–Vis) absorption spectroscopy tests showed that the azophenyl group in BMMs-Azo undergoes effective and reversible *cis-trans* isomerization under UV–Vis irradiation. Hex@BMMs/Azo/β-CD Nps exhibited excellent light-sensitive controlled-release performance. The release of Hex was much higher under UV irradiation than that in the dark, which could be demonstrated by the bioactivity test. The nanoparticles also displayed excellent pH-responsive properties, and the sustained-release curves were described by the Ritger–Peppas release kinetic model. BMMs nanocarriers had good biological safety and provided a basis for the development of sustainable agriculture in the future.

Keywords: nanoparticle; hexaconazole; photo-responsibility; mesoporous silicas; controlled-release

Citation: Pan, H.; Li, W.; Wu, L.; Huang, W.; Zhang, F. β-Cyclodextrin-Modified Mesoporous Silica Nanoparticles with Photo-Responsive Gatekeepers for Controlled Release of Hexaconazole. *Coatings* **2021**, *11*, 1489. https://doi.org/10.3390/coatings11121489

Academic Editor: Yu Shen

Received: 14 October 2021
Accepted: 22 November 2021
Published: 2 December 2021

1. Introduction

Pesticides are considered the most effective way to control pests and diseases in farmland, and improve agricultural productivity [1]. Currently, traditionally formulated pesticides are dominant in China. However, more than 90% of the pesticides cannot accurately act on specific targets, and their effective utilization rate is low due to poor water dispersibility, dust drift, volatilization, evaporation, and degradation, which has led to serious environmental pollution [2–4]. With the rise of nanotechnology in the agricultural field, slow- and controlled-release pesticides have been rapidly developed [5–7]. Environmental stimulus-responsive systems, such as pH- [8–10], redox- [11,12], temperature- [13,14], enzyme- [15,16], and ultraviolet (UV)-light-responsive [17,18] materials have been developed for triggered pesticide release, which can improve the utilization rate of pesticides and reduce environmental pollution.

Among the various stimulative-responsive systems, light-responsive materials are particularly attractive and more achievable in practical applications [19,20]. UV light, as an external stimulative factor, is widely used in intelligent nano slow/controlled-release systems, due to its good controllability, remote control, and clean energy. Unlike the traditional "gatekeeper" system, a light-response system is reversible [21]. Pesticides can be accurately released under the illumination of a specific light source at the specified location.

Azobenzene (Azo) is a novel UV-light-responsive molecule, and shows *trans/cis*-isomers. It has been reported that *trans*- and *cis*-isomers of azobenzene (Azo) can be reversibly converted between each other. *Trans-to-cis* isomerization can occur under UV-light (300–400 nm) irradiation, while *cis-to-trans* isomerization can be induced by visible light (435 nm) or heating [22,23]. The host–guest interaction of *trans*-Azo and β-cyclodextrin (β-CD) was well recognized by hydrophobic and van der Waals forces, while *cis*-Azo cannot be, which is fully reversible. No side reaction or decomposition after long-term irradiation occurs, which is one of the ideal photochemical reactions in the materials application field [12,24].

Bimodal mesoporous silica (BMMs) with a double-channel structure was synthesized by the sol–gel method [25,26]. Compared with traditional silicon mesoporous materials, the large pore volume of BMMs can significantly improve the loading capacity of pesticides. BMMs are widely used as drug carriers due to their adjustable structure, good stability, excellent biocompatibility, and low toxicity [27,28]. Hexaconazole (Hex) is a triazole fungicide with broad-spectrum protection and is used as a treatment for plant diseases caused by fungi [29,30]. However, the usage of Hex is limited due to its poor water solubility and utilization efficiency. In addition, Hex has a long residual life in the environment and is very difficult to degrade in the agricultural soil and water environment. The residual Hex reduces soil and water quality, and causes irreversible harm and toxicity to soil microorganisms and animals and plants in water [31,32]. Therefore, it is desirable to design and prepare an intelligent, controlled-release system to improve the utilization rate and biological safety of Hex.

Herein, aminoazobenzene modified with a silane coupling agent was grafted onto the surface of BMMs, and Hex was encapsulated in the mesopore surface through physical adsorption. A host–guest supramolecular valve was formed after *trans*-Azo was recognized by β-cyclodextrin (β-CD), which was further wrapped on a BMMs carrier to prepare the light-responsive pesticide-release system (Hex@BMMs/Azo/β-CD). The procedure is shown in Figure 1. The physico-chemical characteristics, release behavior, and biological activity of nanoparticles (Nps) were investigated. This study provides a new method of improving the utilization efficiency of pesticides and reducing environmental pollution.

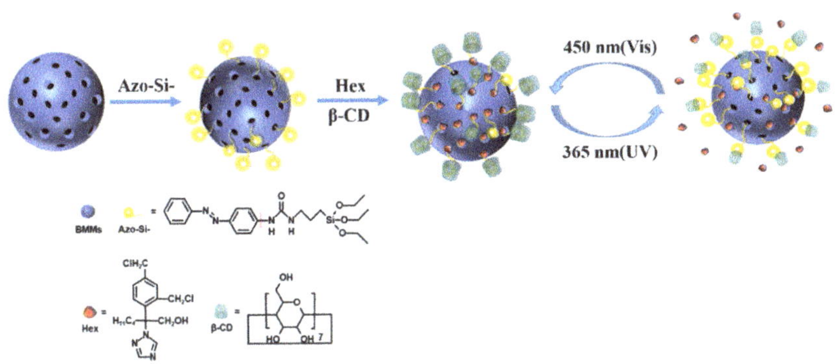

Figure 1. Schematic diagram for the preparation of Hex@BMMs/Azo/β-CD and its light-responsive release mechanism.

2. Materials and Methods

2.1. Materials

Hexaconazole (95%) was purchased from Beijing Jinyue Biotechnology Co., Ltd. (Beijing, China). P-aminoazobenzene, 3-isocyanatopropyltriethoxysilane, and β-cyclodextrin (β-CD) were supplied by Sigma Aldrich (Saint Louis, MO, USA). Cetyltrimethy-lammonium bromide (CTAB), tetraethyl orthosilicate (TEOS), and ammonia (25%, analytical reagent grade) were obtained from Sinopharm Chemical Reagent Co., Ltd. (Beijing, China). Anhydrous tetrahydrofuran (THF) was purchased from Tianjin Fuchen Chemical Reagent Factory (Tianjin, China). N-hexane was purchased from the Beijing Chemical Plant (Beijing, China). Deionized water was obtained with a Milli-Q water purification system (Millipore Co., Burlington, MA, USA) and used for all reactions and treatment processes in the laboratory.

2.2. Preparation of BMMs

CTAB (1.0448 g) was dissolved in 41.6 mL of distilled water in a flask and TEOS (3.2 mL) was slowly added dropwise to the solution. Subsequently, 0.96 mL of ammonia was quickly added into the flask with continued stirring until the solution turned into a white gel. The white gel was then suction-filtered, washed with distilled water, and dried at 100 °C for 8 h. The obtained product was heated to 550 °C at 5 °C/min for 5 h to remove CTAB to attain BMMs.

2.3. Modification of Azo-Si

Isocyanate propyltriethoxysilane (1.0255 g), p-aminoazobenzene (0.794 g), and anhydrous tetrahydrofuran (THF, 7.5 mL) were added to a three-necked flask. The reaction mixture was stirred at 70 °C for 10 h under a nitrogen atmosphere. To obtain Azo-Si, 40 mL of nhexane was used as a precipitator. After THF was removed by rotary evaporation, the solid precipitate was frozen overnight at −20 °C and collected by vacuum filtration. The product was washed with n-hexane and dried at 35 °C for 8 h to yield orange needle-like crystals.

2.4. Preparation of BMMs/Azo Nps

After drying at 110 °C for 2 h, 300 mg of BMMs were dispersed in 30 mL of anhydrous methanol and sonicated for 10 min. Next, 30 mg of Azo was added into the BMMs solution with nitrogen protection. The final reaction mixture was stirred at 65 °C for 12 h, and then filtered. The resulting yellow solid precipitate was washed with THF and methanol, and dried in a vacuum at 35 °C for 10 h to provide the product (BMMs/Azo).

2.5. Preparation of Hex@BMMs/Azo Nps

BMMs/Azo (100 mg) and hexaconazole (20 mg/mL) were added to a flask, and the resulting solution was stirred at 35 °C for 24 h, centrifuged at a speed of 5000 rpm, washed with methanol three times, and dried at 35 °C for 8 h. The obtained products were denoted Hex@BMMs/Azo Nps.

2.6. Preparation of Hex@BMMs/Azo/β-CD Nps

Hex@BMMs/Azo (100 mg) was immersed in phosphate buffer solution (PBS, pH = 7.4), and then 50 mg of β-CD was added. The mixture was placed in the magnetic stirrer for 24 h. Next, the solution was centrifuged for 8 min at 5000 rpm, rinsed with PBS three times, and dried at 35 °C for 8 h to obtain H@BMMs/Azo/β-CD Nps.

2.7. Azo Loading Capacity

Nps were collected by centrifugation, and the concentrations of Azo were detected by high-performance liquid chromatography (HPLC, 1200 Series, Agilent Corp., Wilmington, DE, USA) with a C18 bonded reverse-phase column. The mobile phase of the methanol–water mix (70:30, *v/v*) was programmed with a flow rate of 1.0 mL/min at 210 nm with a UV detector. The injection volume was 10 μL and the retention time was 12 min at 30 °C. The loading capacities (LC) were calculated as follows:

$$LC(\%) = \frac{C_{loading} \times V_{loading} - C_{residual}V_{residual} - C_1V_1 - C_2V_2 - C_3V_3}{m}$$

where LC is the Hex loading capacity (%), m is the total amount of Nps (mg), $C_{loading}$, $C_{residual}$, and C_1, C_2, C_3 are the Hex concentrations of the original loading, residual, and eluted solutions for the first, second, and third time respectively (mg/mL) and $V_{loading}$, $V_{residual}$, V_1, V_2, and V_3 are the volumes of the original loading, residual, and eluted solutions for the first, second, and third time, respectively (mL).

2.8. Characterization of Hex@BMMs/Azo/β-CD Nps

The morphologies and the electron dispersive spectroscopy (EDS) maps of nanoparticles were observed by scanning electron microscopy (SEM) (Zeiss, Oberkochen, Germany). Particle size, polydispersity index (PDI), and zeta potential were analyzed by a dynamic light scattering (DLS) detector (Malvern Instruments Ltd., Malvern, UK). The obtained nanoparticles were suspended in deionized water at pH 7.0 ± 0.1.

The structure and interaction analyses of samples were conducted using X-ray powder diffraction (XRPD, D8 ADVANCE X, Bruker/AXS, Inc., Karlsruhe, Germany) and Fourier-transform infrared spectroscopy (FTIR, Nicolet Nexus 470, Nicolet Instrument Corp., Concord, CA, USA). The nitrogen adsorption–desorption isotherms were determined using an Autosorb-iQ pore analyzer (Quantachrome, Boynton Beach, FL, USA). The surface areas, pore-size distributions, and pore volumes were calculated using the Brunauer–Emmett–Teller (BET) and Barrett–Joyner–Halenda methods. Thermogravimetric analyses (TGA) were carried out by a thermal analyzer (PerkinElmer, Waltham, MA, USA). Elements were analyzed by using X-ray photoelectron spectroscopy (ESCALAB 250Xi, Thermo Fisher Scientific, Waltham, MA, USA). Photoisomerization of azophenyl groups was measured using a UV–Vis spectrophotometer (UV-2600 Shimadzu Co., Ltd., Tokyo, Japan) at wavelength of 250–600 nm.

2.9. Adhesion Test

The contact angles of Hex@BMMs/Azo/β-CD Nps on leaves were measured using a contact-angle apparatus (JC2000D2M, Powereach Ltd., Shanghai, China). Technical Hex, Hex@BMMs/Azo, and deionized water were used as controls. Two-week-old tomato leaves were selected as model plants, and carefully washed with deionized water. After the leaves were air-dried, Hex@BMMs/Azo/β-CD dispersion, Hex@BMMs/Azo dispersion, technical Hex solution, and water were dropped onto the leaves on a glass slide, respectively. Images of the droplet on the leaf surface were taken, and the corresponding contact angles were detected using a contact-angle meter (Dataphysics-TP50, DataPhysics Instruments, Filderstadt, Germany).

2.10. Release Behavior of Hex@BMMs/Azo/β-CD Nps

Hex@BMMs/Azo/β-CD Nps (15 mg) were suspended in 3 mL of methanol–water solvents (50:50, *v/v*). The suspension was transferred to a dialysis bag (molecular cutoff capacity 3500 kDa) and then put into flasks with 60 mL of methanol/H_2O (1:1, *v/v*) solution. The release medium was then shaken at a speed of 100 rpm at 37 °C and irradiated continuously with 150, 300, and 500 W UV lamps (365 nm). Five milliliters of the sample were collected at fixed time intervals, and the same amount of fresh medium was replenished in the system. Hex contents were measured by HPLC, and the release ratio of Hex from the nano-delivery sample was calculated.

To investigate the pH sensitivity of Hex, the release of Hex from Nps at different pH levels (4.0, 7.0, and 9.0) under 300 W, 365 nm UV light was measured using the same method as described above. The cumulative release rate of Hex in Nps was then calculated using the following equation:

$$Q(\%) = \frac{V_0 \times C_T + V \times \sum_{N-1}^{T-1} C}{W} \times 100\%$$

where W is the total mass concentration of Hex in the Pickering emulsion, V_0 is the volume of the sustained-release solution, and V is the volume of the slow-release solution taken at a specific interval time. C_T and C are the concentrations of T and N samples, respectively. The release kinetics of Hex from Hex@BMMs/Azo/β-CD Nps was separately investigated using four mathematical models, namely the pseudo-zero-order equation, pseudo-first-order equation, Higuchi model, and Ritger–Peppas model.

2.11. Bioactivity Test

Hex@BMMs/Azo/β-CD Nps (40 mg) were dispersed in 8 mL of sterile distilled water with 0.1% Tween 80, and then 4 mL of the aqueous suspension was exposed to UV light (365 nm) under stirring at 25 °C. Technical Hex solution, non-irradiated Nps, and distilled water were used as controls. *Rhizoctonia solani* cakes (5 mm in diameter) were placed into the center of potato dextrose agar (PDA) petri dishes with Nps (50, 100, and 200 mg/L). The dishes were incubated for 14 days at 28 °C and each colony diameter was measured every 24 h. The relative inhibition rates against the fungi were calculated according to the colony diameters.

2.12. Biosafety Evaluation

Human embryo skin fibroblast cells (CCC-ESF-1 cells) were seeded in 96-well plates in triplicate and cultured in DMEM medium supplemented with 10% FBS and 1% penicillin/streptomycin at 37 °C for 24 h. Then, CCC-ESF-1 cells were treated with different concentrations (0, 31.25, 62.5, 125, 250, and 500 μg/mL) of BMMs/Azo/β-CD for 24 h, and the cell viability was evaluated using a cell counting kit (CCK8, Dojindo, Japan). *E. coli* was cultured in LB culture medium with different concentrations (0, 31.25, 62.5, 125, 250, and 500 μg/mL) of BMMs/Azo/β-CD Nps at 37 °C for 24 h, and the *E. coli* concentrations were determined by UV–Vis absorptiometry at 600 nm. All tests were carried out at least in triplicate.

2.13. Data Analysis

The data were analyzed using SPSS 20.0 statistical analysis software (SPSS, Chicago, IL, USA). All experiments were performed three times. Statistical significance was determined as $p < 0.05$.

3. Results and Discussion

3.1. Morphological Observations

Figure 2 shows SEM images of BMMs, BMMs/Azo, Hex@BMMs/Azo, and Hex@BMMs/Azo/β-CD. The Nps had spherical shapes and were not destroyed due to the encapsulations of Hex, Azo, and β-CD. The particle sizes, zeta-potential values, and PDIs were further measured by DLS analysis (Table 1). After Hex was loaded and the mesopore surface was grafted with Azo/β-CD, the average particle size of Nps increased from 269.8 ± 6.8 nm (BMMs) to 387.2 ± 3.8 nm (Hex@BMMs/Azo/β-CD). The highest PDI value, i.e., 0.153 ± 0.02, indicated that the Nps had better mono-dispersity and stability. Owing to the presence of -OH on the surface of the mesoporous silica, the zeta potential of BMMs was −13.93 ± 1.57 mV. After the modification of azobenzene, the zeta-potential value of BMMs/Azo increased to −8.42 ± 1.71 mV because of the neutralization of the amino groups present on the surface of the modified azobenzene. After loading Hex, the zeta-potential value of Hex@BMMs/Azo/β-CD Nps decreased to −16.97 ± 0.95 mV, due to the negative charge of Hex, indicating that the pesticide Hex was successfully loaded into the Nps.

3.2. Structure and Interaction Analysis

The structure of the Nps was also tested by XRPD and the results are shown in Figure 3A. BMMs had an obvious (100) crystal-plane diffraction peak at 2θ = 1.86°, which is the characteristic peak of BMMs, indicating that it had a highly ordered, double-hole structure. After grafting the modified azobenzene, BMMs/Azo exhibited the same (100) crystal-plane diffraction peak, indicating that the BMMs/Azo Nps still maintained an ordered mesoporous structure. After Hex was loaded into the BMMs, the peak intensity decreased from 1.99° to 1.97°, and the d value moved from 44.09 to 44.85 nm. This is because the Hex in the channel reduced the scattering between the mesoporous channels and the pore wall. The strength of the XRPD peaks (2θ = 2°) of the β-CD-coated Hex@BMMs/Azo Nps decreased significantly, and the shape broadened, showing that the mesoporous structure was affected, and β-CD was successfully encapsulated into the system.

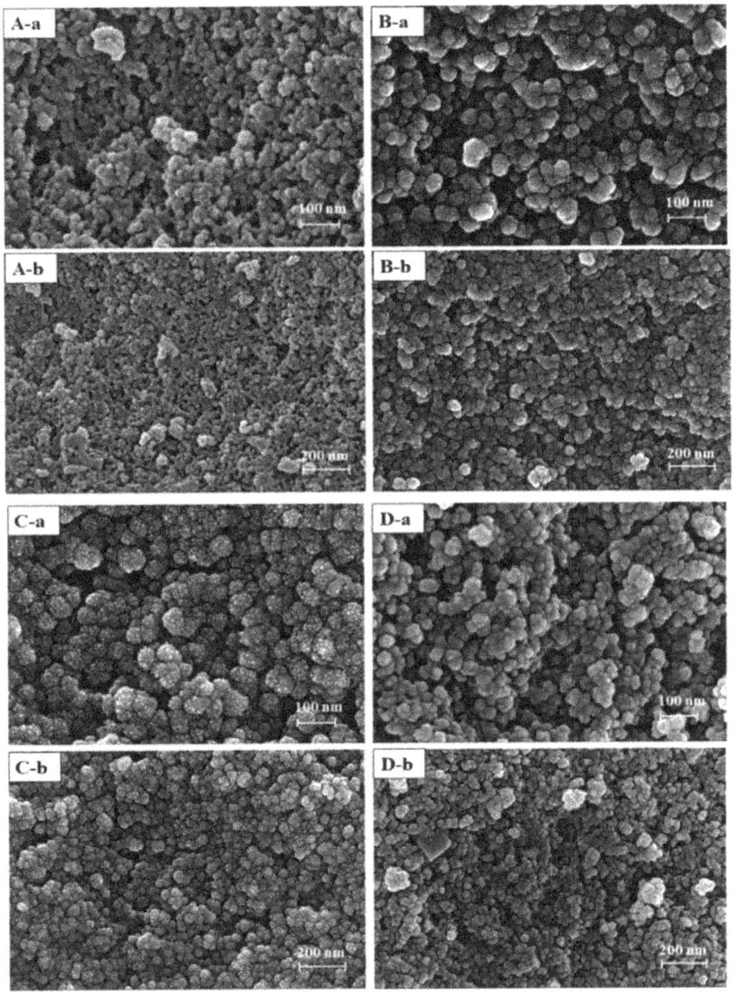

Figure 2. SEM images of different nanoparticles. (**A-a**,**A-b**) BMMs, (**B-a**,**B-b**) BMMs/Azo, (**C-a**,**C-b**) Hex@BMMs/Azo, (**D-a**,**D-b**) Hex@BMMs/Azo/β-CD.

Table 1. The mean diameters and distributions of nanoparticles based on BMMs.

Nps	Mean Size (nm)	PDI	Zeta Potential (mV)
BMMs	269.8 ± 6.8	0.062 ± 0.03	−13.93 ± 1.57
BMMs/Azo	367.4 ± 3.3	0.144 ± 0.01	−8.42 ± 1.71
Hex@BMMs/Azo	553.4 ± 4.1	0.123 ± 0.04	−16.77 ± 1.60
Hex@BMMs/Azo/β-CD	387.2 ± 3.8	0.153 ± 0.02	−16.97 ± 0.95

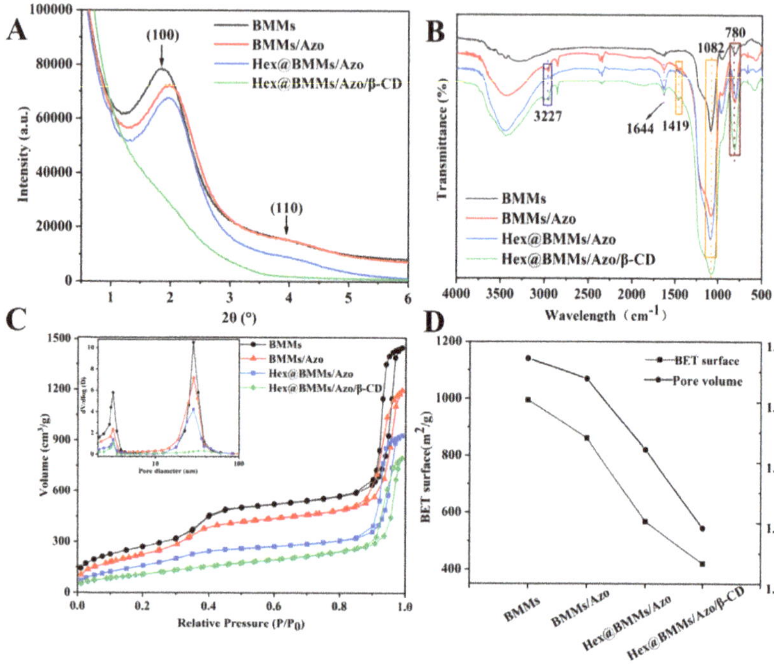

Figure 3. XRPD patterns (**A**), FTIR (**B**), N2 adsorption/desorption isotherms (**C**), and BET surface areas and pore volumes (**D**) of BMMs, BMMs/Azo, Hex@BMMs/Azo, and Hex@BMMs/Azo/β-CD.

The FTIR spectra of BMMs, BMMs/Azo, Hex@BMMs/Azo, and Hex@BMMs/Azo/β-CD were determined to evaluate the Nps' structural changes with various functional groups (Figure 3B). BMMs exhibited three characteristic peaks at 1082, 780, and 810 cm^{-1} that were anti-symmetrical and symmetrical stretching-vibration peaks of Si–O–Si groups. The absorption bands at 1644 and 1419 cm^{-1} were the stretching-vibration peaks of the –CONH– group and the vibration peaks of the C=C group in the aromatic ring respectively, indicating that the modified Azo was successfully grafted onto the surface of BMMs. The characteristic absorption peak of Hex at 3227 cm^{-1} and the absorption peak with increasing intensity at 1082 cm^{-1} suggested that the pesticide Hex was adsorbed in the pores of mesoporous silica through van der Waals forces and hydrogen bonds. The loading content of Hex in Nps was also tested by HPLC, and the loading ratios of Hex (27.3%) in Hex@BMMs/Azo/β-CD were obtained. To investigate the porosity, pore surface areas, and pore volumes of the nano-delivery system, nitrogen adsorption/desorption measurements were performed. As shown in Figure 4C, the N_2 adsorption/desorption isotherms of BMMs, BMMs/Azo, Hex@BMMs/Azo, and Hex@BMMs/Azo/β-CD belong to the Langmuir IV isotherm with two hysteresis loops. The first hysteresis loop, at $0.2 < P/P_0 < 0.4$, rose rapidly owing to monolayer adsorption of nitrogen. The second hysteresis loop appeared at P/P_0 = 0.8–0.95, indicating that the capillary tube of the particle accumulation pore had been condensed. The corresponding pore-size distribution revealed that the Nps had a dual-model structure and two pore sizes (Figure 3C, inset). After the loading of Hex molecules and the modification of Azo/β-CD in BMMs, the shape of the adsorption isotherm remained basically unchanged compared with BMMs, indicating that the mesoporous structure of the sample still existed. However, the BET specific surface area and pore volume of Hex@BMMs/Azo/β-CD significantly decreased, implying that Hex molecules occupied a significant number of pores and surface sites of Nps (Figure 3D). These observations further suggested that the modified Azo interacted with Si–OH groups, and that Hex was successfully loaded into Hex@BMMs/Azo/β-CD Nps.

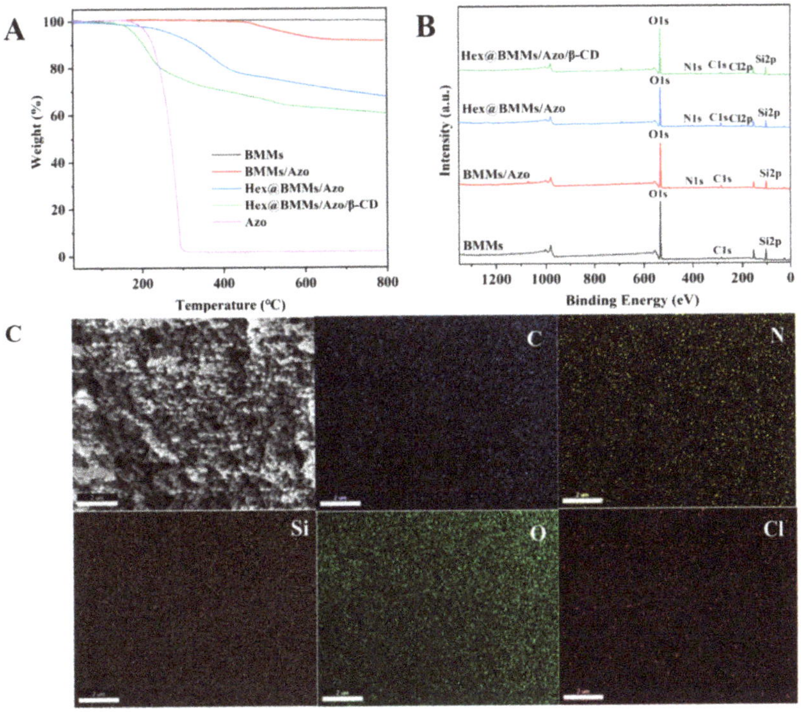

Figure 4. TG (**A**), XPS (**B**), and element distributions (**C**) of Hex@BMMs/AZO/β-CD and the control samples.

The loading rate of the Hex@BMMs/Azo/β-CD Nps was determined by TG analysis. The TG curves of nanoparticles are shown in Figure 4A. The mass loss of all samples mainly included two stages. The first stage was the evaporation of water in the samples before 150 °C, and the second weight-loss peak occurred at the stage from 150 to 800 °C due to the decomposition of organic components incorporated in the samples. The weight losses of BMMs/Azo and Hex@BMMs/Azo were 8.1% and 32.8%, respectively. Hence, the loading rate of Hex@BMMs/Azo was about 24.7%, which was approximately consistent with the result of HPLC. After β-CD was grafted to the surface of Hex@BMMs/Azo Nps, the weight loss was 39.5%, proving that Hex was successfully loaded into Azo-functionalized BMMs and β-CD played a good encapsulation role.

To analyze the chemical elements on the surface of nanoparticles, the samples were characterized by XPS. As shown in Figure 4B, the surface of BMMs nanoparticles mainly contains two elements, Si and O. The binding energies of approximately 533.08 and 104.06 eV belong to O1s and Si2p, respectively. The weak emerging signal at 285.08 eV was attributed to C1s, which was a residual component after the calcination of the CTAB template. In the BMMs/AZO spectrum, the C1s peak was more intense than that of BMMs and a new signal of N1s was observed at the binding energy of 401.08 eV, confirming that Azo was successfully modified on the surface of BMMs. Furthermore, a new peak of Cl2p appeared at 201.08 eV in the spectrum of Hex@BMMs/Azo, compared with BMMs/Azo, indicating that BMMs/Azo nanoparticles were successfully loaded with Hex. The element components of Hex@BMMs/Azo/β-CD were completely consistent with those of Hex@BMMs/Azo. Due to the coating of β-CD, the thickness of the surface of Hex@BMMs/Azo increased, resulting in the weakening of the absorption peak intensity of Hex@BMMs/Azo/β-CD nanoparticles. In addition, the result of EDS analysis (Figure 4C) showed that the elements of Si, O, C, N, and Cl were all distributed in Hex@BMMs/Azo/β-CD, which further proved the successful preparation of the nano-pesticide system.

3.3. Foliage Adhesion of Hex

Adhesion experiments were conducted to prove that the Hex@BMMs/Azo/β-CD nano-delivery system had better adhesion behavior on the leaf surfaces. As shown in Figure 5, the contact-angle values of Hex@BMMs/Azo Nps and Hex@BMMs/Azo/β-CD Nps were 83.89° ± 0.36° and 71.64° + 0.41° respectively, which were obviously lower than those of Hex technical solution (103.62° ± 0.37°) and deionized water (105.57° ± 0.48°). The data showed that Azo/β-CD-coated Hex@BMMs microcapsules exhibited excellent adhesion properties. Compared with deionized water and technical Hex, the contact angles of the Nps were reduced because the hydroxyl groups on the surface of BMMs increased the infiltration of the nano-delivery system on the leaf surface. The smaller contact angle of Hex@BMMs/Azo/β-CD Nps was due to the properties of β-CD, including internal hydrophobicity and external hydrophilicity. After mesoporous silica was modified by β-CD, the hydrophilicity of the silica surface increased and the contact angle decreased, which improved the water solubility of the Nps and the adhesion of Nps on leaves.

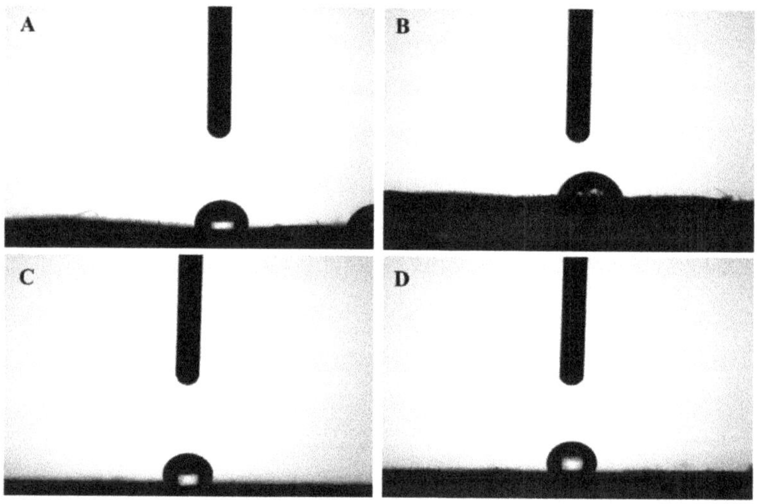

Figure 5. The contact angles of Hex@BMMs/Azo (**A**), Hex@BMMs/Azo/β-CD (**B**), technical Hex (**C**), and deionized water (**D**).

3.4. Photo-Responsive Property

To verify that β-CD could self-assemble with BMMs/Azo to form a gatekeeper for the controlled-release of pesticide, we detected the UV absorption spectra of BMMs, BMMs/Azo, and BMMs/Azo/β-CD under UV irradiation. As shown in Figure 6A, com-pared with BMMs, an obvious UV absorption peak appeared at 357 nm after the modification of Azo, indicating that Azo was successfully grafted on the surface of BMMs. When β-CD was capped onto the nanoparticles' surface, the UV peak intensity was higher than that of BMMs/Azo. This was mainly because the electron cloud density of the Azo-conjugated system was interfered by the higher electron cloud in the inner cavity of β-CD molecules, resulting in the decrease of electron transition energy and the increase of absorption intensity of the conjugated system, which led to the enhancement of BMMs/Azo/β-CD absorption spectrum intensity.

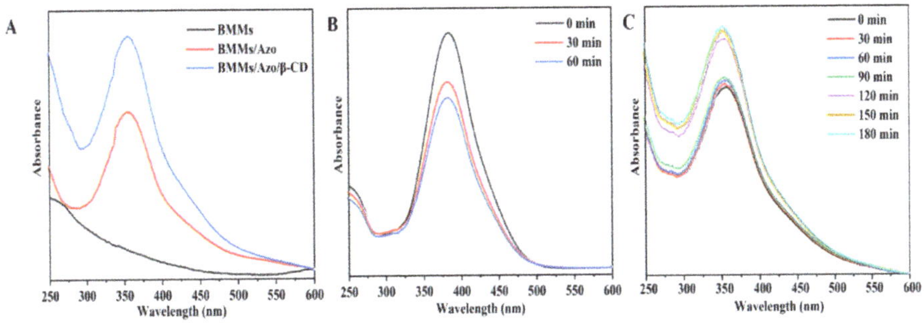

Figure 6. UV–Vis spectra of BMMs/Azo/β-CD and control samples (**A**). UV–Vis spectra of Azo after different radiation: UV (365 nm) light radiation for 0, 30, and 60 min (**B**), Vis (450 nm) light irradiation for 0, 30, 60, 90, 120, 150, and 180 min after UV irradiation (120 min) (**C**).

We further investigated the effect of light on the *trans/cis*-isomers of Azo molecules. The UV absorption spectra of BMMs, BMMs/Azo, and BMMs/Azo/β-CD nanoparticles were detected by UV–Vis spectroscopy. After nanoparticles were radiated by UV light, the peak at 357 nm decreased with increasing time from 0 to 60 min (Figure 6B), indicating that the UV light could induce *trans-cis* transformation of Azo molecules. Under the following Vis light radiation, the peak intensities showed contrasting trends (Figure 6C), suggesting that the Vis light radiation could induce *cis-trans* transformation of Azo molecules. Especially from 90 to 120 min, a significant transition appeared in the nano-system, implying that *cis*-Azo structures were mostly converted to *trans*-Azo structures under the irradiation of visible light, and the detached β-CD recombined with BMMs/Azo again. As a result, Azo in BMMs/Azo/β-CD Nps exhibited *trans-cis* reversible photoisomerization, which could be mutually transformed from a *trans* structure to a *cis* structure under the alternating irradiation of UV–Vis light, which was consistent with previous reports [22,33]. In addition, BMMs-Azo Nps were repeatedly irradiated many times, and no "fatigue" to light was observed, indicating that the nanomaterial had good reversible light-responsive capability.

3.5. Release Behavior

3.5.1. Effect of Light Intensity

To investigate the controlled-release performance of Hex from Hex@BMMs-Azo-CD under UV light, Hex@BMMs-Azo-CD Nps were irradiated under different UV-light intensities (150, 300, and 500 W at 365 nm). As shown in Figure 7A, the release of Hex from the microcapsules under UV irradiation was faster than that in darkness. This was because the Hex@BMMs-Azo-CD Nps were endowed with irreversible "gatekeeper" systems, and Azo underwent *trans*- to *cis*-isomerization under UV irradiation. The "cap" of the pesticide-carrying non-complex that blocked the mesopores was opened, and Hex molecules were released from the pores, indicating that the nano-pesticide delivery system responded to light stimulation. Compared with the control with a release rate of only 6.37% in darkness, Hex was released rapidly from the Nps within 100 min of illumination. After 360 min, the release ratios of Hex reached 41.5%, 58.3%, and 79.0% at light intensities of 150, 300, and 500 W, respectively. In addition, the Azo-modified BMMs carrier was more sensitive to UV light with increasing light intensity, and the Hex-release rate was obviously increased, which further showed that UV light was the dominant driving force for Hex release. In short, these results demonstrated that Hex could be released from Hex@BMMs-Azo-CD Nps via stimulation by UV light.

3.5.2. Effect of pH

The release behaviors of Hex@BMMs-Azo-CD Nps were investigated at different pH values (4.0, 7.0, and 9.0). The sustained-release curves of pesticide at various pH values are shown in Figure 7B. The cumulative release rate gradually increased as the pH value decreased. At pH 4, the Hex-release rate reached 66.58% at 360 min. After 360 min, the release rate of Hex@BMMs-Azo-CD Nps at pH 4 was 66.9%, and those at pH 7 and pH 9 were 58.3% and 40.4%, respectively. This was mainly due to the grafting of β-CD on the surface of the Nps. CD is relatively stable under alkaline and neutral conditions, but is easily hydrolyzed under acidic conditions. After the sealed "cap" was opened by hydrolysis, Hex molecules were easily released from the mesopores. In addition, excessive hydrogen ions in the acid solution will be added to the N=N bond of azobenzene, resulting in a pH response.

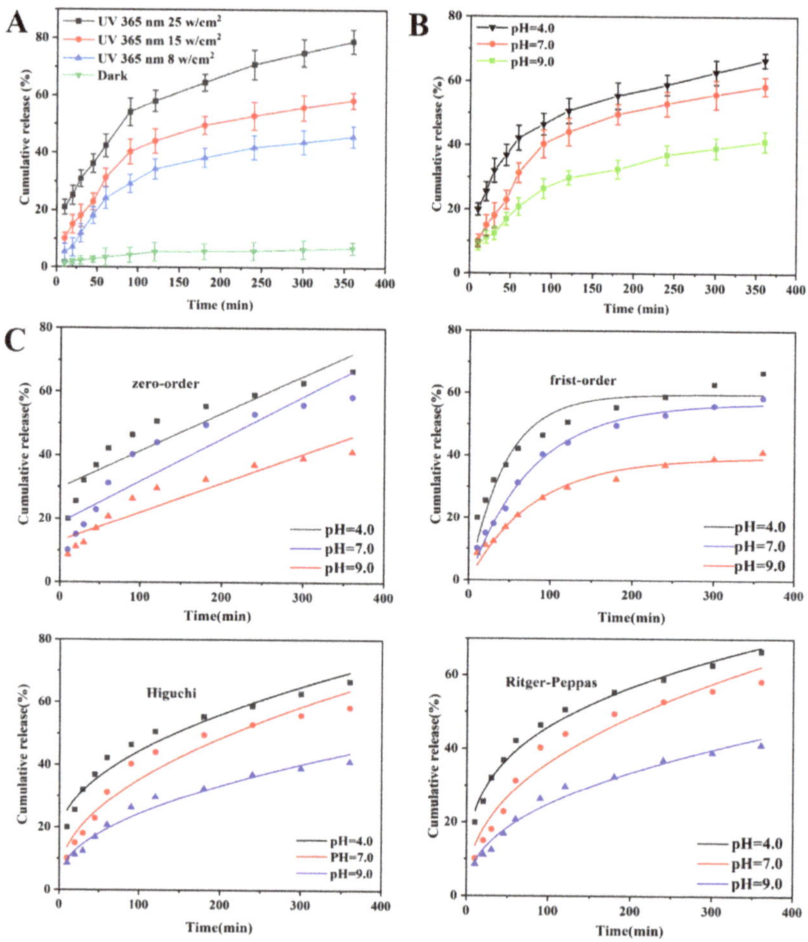

Figure 7. Cumulative release curves with different intensity UV light (**A**) and different pH (**B**), and kinetic fitting models of Hex@BMMs/Azo/β-CD nanoparticles (**C**). Error bars represent standard deviation from the mean (n = 3).

3.5.3. Release Kinetics Analysis

To further elucidate the effect of pH on the mechanism of Hex sustained release from Hex@BMMs-Azo-CD NPs, we studied the release kinetics using the zero-order kinetics model, first-order kinetics model, Higuchi kinetics model, and Ritger–Peppas kinetics model (Figure 7C). Table 2 presents the values of parameters and the regression coefficients (R^2).

Table 2. Release parameters of HEX at different pH values.

pH	Fitting Methods	Kinetic Equation	Determination Coefficient, R^2
4.0	Zero-order fitting	$Q = 0.1176t + 29.65$	0.8576
	First-order fitting	$Q = 59.58(1 - e^{-0.0224}t)$	0.9098
	Higuchi fitting	$Q = 2.79t^{1/2} + 16.425$	0.9577
	Ritger–Peppas fitting	$Q = 11.45x^{n_1}$	0.9827
7.0	Zero-order fitting	$Q = 0.1326t + 18.64$	0.8392
	First-order fitting	$Q = 56.74(1 - e^{-0.0129}t)$	0.9899
	Higuchi fitting	$Q = 3.17t^{1/2} + 3.596$	0.9457
	Ritger–Peppas fitting	$Q = 5.09x^{n_2}$	0.9509
9.0	Zero-order fitting	$Q = 0.0914t + 12.93$	0.8873
	First-order fitting	$Q = 39.36(1 - e^{-0.0125}t)$	0.9743
	Higuchi fitting	$Q = 2.14t^{1/2} + 2.879$	0.9687
	Ritger–Peppas fitting	$Q = 3.52x^{n_3}$	0.9789

Note: Q is the fractional release of pesticide, t is the elapsed time, R^2 is the high value of the linear regression coefficient, and n is the release exponent, where n_1, n_2, and n_3 are 0.308, 0.426, and 0.425, respectively.

Regardless of acidic or alkaline conditions, the R^2 value of the Ritger–Peppas kinetic equation was higher than those of the other three mathematical models, indicating that the Ritger–Peppas kinetic model was more suitable for the release behavior of Hex. The values of n in acidic and alkaline environments were 0.308 and 0.425 respectively, and both values, which were lower than 0.45, proved that the Hex release in Hex@BMMs/Azo/β-CD mainly followed Fick diffusion, and the concentration was the main influencing factor in the slow-release process. However, under neutral conditions, the R^2 value of the first-order kinetic equation was the highest, and the release of Hex conforms to the first-order kinetic model, implying that the release of Hex was closely related to concentration.

3.6. Bioactivity Test

Figure 8 shows the bioactivity of Hex@BMMs/Azo/β-CD plotted against concentrations of *Rhizoctonia solani* ranging from 50 to 200 mg/L. Compared with the negative control, the inhibition rates were 40.3% (non-irradiated Nps), 60.1% (irradiated Nps), and 56.8% (technical Hex) respectively, at the Hex-as-an-active-ingredient concentration of 50 mg/L at 14 days, due to the sustained release of Hex in the nano-delivery systems. In addition, the control efficiencies of UV-irradiated samples were significantly higher than those of non-UV-irradiated samples and technical controls, indicating that UV irradiation significantly improved the release of Hex in the nano-delivery systems. This result is consistent with previous reports [23,34]. In short, the UV-stimuli-responsive Nps exhibited a better and more sustained antibacterial activity against *Rhizoctonia solani* than technical Hex.

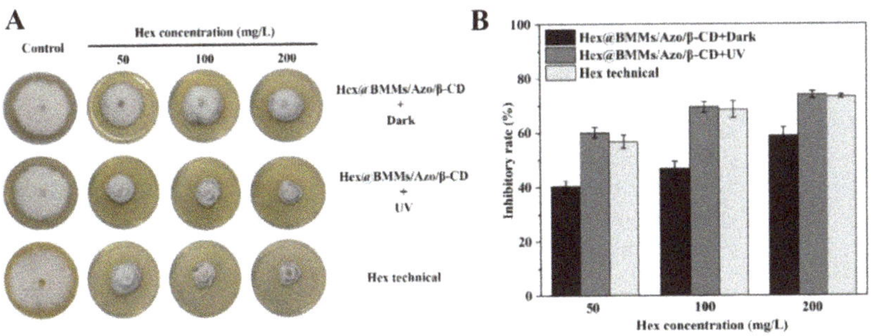

Figure 8. Digital images (**A**) and inhibitory rates (**B**) of Hex@BMMs/Azo/β-CD nanoparticles against *Rhizoctonia solani* at 14 days.

3.7. Biosafety Evaluation

To further evaluate the biological safety of nanocarriers, the toxicological effects of different concentrations of BMMs/Azo/β-CD on CCC-ESF-1 cells and *E. coli* were studied. Figure 9 shows that different concentrations of BMMs/Azo/β-CD Nps had little influence on the growth and metabolism of CCC-ESF-1 cells and *E. coli*, indicating that an Azo-modified nano-pesticide loading system had excellent biological safety.

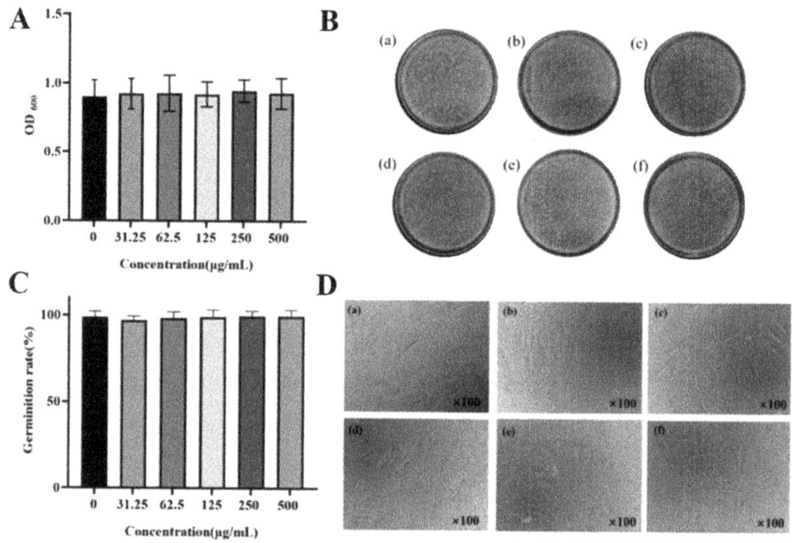

Figure 9. Biosafety evaluation of different concentrations of BMMs/Azo: (**A**) *Escherichia coli* suspension OD600, (**B**) photograph of *Escherichia coli* colony, (**C**) CCC-ESF-1 cell survival rate, and (**D**) photograph of CCC-ESF-1 cell. (**a**) Control, (**b**) 31.25 μg/mL, (**c**) 62.55 μg/mL, (**d**) 125 μg/mL, (**e**) 250 μg/mL, and (**f**) 500 μg/mL.

4. Conclusions

In this work, we prepared a novel UV-responsive nano-pesticide delivery system by the sol–gel method. Azo was modified and grafted with BMMs, and then the fungicide Hex was loaded into mesoporous silica by adsorption. The Hex@BMMs/Azo/β-CD Nps had a uniformly spherical morphology and good dispersibility in water. The nanocomplex showed excellent photo-responsive controlled-release performance owing to the strong host–guest complex between the *trans*-Azo and β-CD, which was used to control Hex release from BMMs under UV–Vis irradiation. In addition, the release of Hex in Nps was promoted in an acidic environment by varying the pH of the structure. The Ritger–Peppas kinetic model was a better fit for the release behavior of Hex. The carrier displayed good adhesion on leaf surfaces, and was biologically benign. The sustained fungicidal efficacy against *Rhizoctonia solani* indicated that Hex@BMMs/Azo/β-CD nanoparticles could effectively improve the utilization of Hex and decrease pesticide residue. Therefore, this work provides a promising approach to reduce the risk to the environment, and promote the development of green agriculture in the future.

Author Contributions: Data curation, L.W.; Formal analysis, W.H.; Funding acquisition, F.Z.; Investigation, H.P. and W.L.; Supervision, F.Z.; Writing—review & editing, F.Z. All authors have read and agreed to the published version of the manuscript.

Funding: The research was funded by the National Key R&D Program of China (2016YFD0200502-2).

Institutional Review Board Statement: Not applicable.

Informed Consent Statement: Not applicable.

Data Availability Statement: Not applicable.

Conflicts of Interest: The authors declare no conflict of interest.

References

1. Chamberlain, J.L.; Darr, D.; Meinhold, K. Rediscovering the contributions of forests and trees to transition global food systems. *Forests* **2020**, *11*, 1098. [CrossRef]
2. Singh, A.; Dhiman, N.; Kar, A.K.; Singh, D.; Purohit, M.P.; Ghosh, D.; Patnaik, S. Advances in controlled release pesticide formulations: Prospects to safer integrated pest management and sustainable agriculture. *J. Hazard. Mater.* **2020**, *385*, 121525–121601. [CrossRef] [PubMed]
3. Kumar, S.; Nehra, M.; Dilbaghi, N.; Marrazza, G.; Hassan, A.A.; Kim, K.-H. Nano-based smart pesticide formulations: Emerging opportunities for agriculture. *J. Control. Release* **2019**, *294*, 131–153. [CrossRef] [PubMed]
4. Kookana, R.S.; Boxall, A.B.; Reeves, P.T.; Ashauer, R.; Beulke, S.; Chaudhry, Q.; Cornelis, G.; Fernandes, T.F.; Gan, J.; Kah, M.; et al. Nanopesticides: Guiding principles for regulatory evaluation of environmental risks. *J. Agric. Food Chem.* **2014**, *62*, 4227–4240. [CrossRef]
5. Xia, X.; Shi, B.; Wang, L.; Liu, Y.; Zou, Y.; Zhou, Y.; Chen, Y.; Zheng, M.; Zhu, Y.; Duan, J.; et al. From mouse to mouse-ear cress: Nanomaterials as vehicles in plant biotechnology. *Exploration* **2021**, *1*, 9–20. [CrossRef]
6. Feng, J.; Chen, W.; Shen, Y.; Chen, Q.; Yang, J.; Zhang, M.; Yang, W.; Yuan, S. Fabrication of abamectin-loaded mesoporous silica nanoparticles by emulsion-solvent evaporation to improve photolysis stability and extend insecticidal activity. *Nanotechnology* **2020**, *31*, 345705. [CrossRef] [PubMed]
7. Mahajan, R.; Selim, A.; Neethu, K.M.; Sharma, S.; Shanmugam, V.; Jayamurugan, G. A systematic study to unravel the po-tential of using polysaccharides based organic-nanoparticles versus hybrid-nanoparticles for pesticide delivery. *Nanotechnolgy* **2021**, *32*, 475704. [CrossRef] [PubMed]
8. Gao, Y.; Zhang, Y.; He, S.; Xiao, Y.; Qin, X.; Zhang, Y.; Li, D.; Ma, H.; You, H.; Li, J. Fabrication of a hollow mesoporous silica hybrid to improve the targeting of a pesticide. *Chem. Eng. J.* **2019**, *364*, 361–369. [CrossRef]
9. Shan, Y.; Cao, L.; Muhammad, B.; Xu, B.; Zhao, P.; Cao, C.; Huang, Q. Iron-based porous metal-organic frameworks with crop nutritional function as carriers for controlled fungicide release. *J. Colloid Interface Sci.* **2020**, *566*, 383–393. [CrossRef] [PubMed]
10. Liang, Y.; Song, J.; Dong, H.; Huo, Z.; Gao, Y.; Zhou, Z.; Tian, Y.; Li, Y.; Cao, Y. Fabrication of pH-responsive nanoparticles for high efficiency pyraclostrobin delivery and reducing environmental impact. *Sci. Total Environ.* **2021**, *787*, 147422. [CrossRef]
11. Gao, Y.; Liang, Y.; Dong, H.; Niu, J.; Tang, J.; Yang, J.; Tang, G.; Zhou, Z.; Tang, R.; Shi, X.; et al. A bioresponsive system based on mesoporous organosilica nanoparticles for smart delivery of fungicide in response to pathogen presence. *ACS Sustain. Chem. Eng.* **2020**, *8*, 5716–5723. [CrossRef]
12. Liang, Y.; Gao, Y.; Wang, W.; Dong, H.; Tang, R.; Yang, J.; Niu, J.; Zhou, Z.; Jiang, N.; Cao, Y. Fabrication of smart stimu-li-responsive mesoporous organosilica nano-vehicles for targeted pesticide delivery. *J. Hazard. Mater.* **2020**, *389*, 122075–122085. [CrossRef]
13. Gao, Y.; Xiao, Y.; Mao, K.; Qin, X.; Zhang, Y.; Li, D.; Zhang, Y.; Li, J.; Wan, H.; He, S. Thermoresponsive polymer-encapsulated hollow mesoporous silica nanoparticles and their application in insecticide delivery. *Chem. Eng. J.* **2020**, *383*, 123169–123176. [CrossRef]
14. Zhang, L.; Ren, S.; Chen, C.; Wang, D.; Liu, B.; Cai, D.; Wu, Z. Near infrared light-driven release of pesticide with magnetic collectability using gel-based nanocomposite. *Chem. Eng. J.* **2021**, *411*, 127881. [CrossRef]
15. Liang, Y.; Fan, C.; Dong, H.; Zhang, W.; Tang, G.; Yang, J.; Jiang, N.; Cao, Y. Preparation of MSNs-chitosan@prochloraz nanoparticles for reducing toxicity and improving release properties of prochloraz. *ACS Sustain. Chem. Eng.* **2018**, *6*, 10211–10220. [CrossRef]
16. Abdelrahman, T.M.; Qin, X.; Li, D.; Senosy, I.A.; Mmby, M.; Wan, H.; Li, J.; He, S. Pectinase-responsive carriers based on mesoporous silica nanoparticles for improving the translocation and fungicidal activity of prochloraz in rice plants. *Chem. Eng. J.* **2021**, *404*, 126440. [CrossRef]
17. Chen, C.; Zhang, G.; Dai, Z.; Xiang, Y.; Liu, B.; Bian, P.; Zheng, K.; Wu, Z.; Cai, D. Fabrication of light-responsively con-trolled-release herbicide using a nanocomposite. *Chem. Eng. J.* **2018**, *349*, 101–110. [CrossRef]
18. Ye, Z.; Guo, J.; Wu, D.; Tan, M.; Xiong, X.; Yin, Y.; He, G. Photo-responsive shell cross-linked micelles based on carbox-ymethyl chitosan and their application in controlled release of pesticide. *Carbohydr. Polym.* **2015**, *132*, 520–528. [CrossRef]
19. Liang, W.; Xie, Z.; Cheng, J.; Xiao, D.; Xiong, Q.; Wang, Q.; Zhao, J.; Gui, W. A light-triggered ph-responsive met-al-organic framework for smart delivery of fungicide to control sclerotinia diseases of oilseed rape. *ACS Nano* **2021**, *15*, 6987–6997. [CrossRef]
20. Zheng, D.; Bai, B.; Zhao, H.; Xu, X.; Hu, N.; Wang, H. Stimuli-responsive Ca-alginate-based photothermal system with enhanced foliar adhesion for controlled pesticide release. *Colloids Surf. B Biointerfaces* **2021**, *207*, 112004. [CrossRef]
21. Bruneau, M.; Bennici, S.; Brendle, J.; Dutournie, P.; Limousy, L.; Pluchon, S. Systems for stimuli-controlled release: Materials and applications. *J. Control. Release* **2019**, *294*, 355–371. [CrossRef] [PubMed]

22. Mei, X.; Yang, S.; Chen, D.; Li, N.; Li, H.; Xu, Q.; Ge, J.; Lu, J. Light-triggered reversible assemblies of azobenzene-containing amphiphilic copolymer with beta-cyclodextrin-modified hollow mesoporous silica nanoparticles for controlled drug re-lease. *Chem. Commun.* **2012**, *48*, 10010–10012. [CrossRef] [PubMed]
23. Tarn, D.; Ferris, D.P.; Barnes, J.C.; Ambrogio, M.W.; Stoddart, J.F.; Zink, J.I. A reversible light-operated nanovalve on mesoporous silica nanoparticles. *Nanoscale* **2014**, *6*, 3335–3343. [CrossRef] [PubMed]
24. Wang, D.; Wu, S. Red-light-responsive supramolecular valves for photocontrolled drug release from mesoporous nanoparticles. *Langmuir* **2016**, *32*, 632–636. [CrossRef] [PubMed]
25. Gao, L.; Sun, J.; Li, Y.; Zhang, L. Bimodal mesoporous silicas functionalized with different level and species of the amino groups for adsorption and controlled release of aspirin. *J Nanosci. Nanotechnol.* **2011**, *11*, 6690–6697. [CrossRef]
26. Jin, X.; Wang, Q.; Sun, J.; Panezai, H.; Bai, S.; Wu, X. P(NIPAM-co-AA)@BMMs with mesoporous silica core and controlled copolymer shell and its fractal characteristics for dual pH- and temperature-responsive performance of ibuprofen release. *Int. J. Polym. Mater.* **2017**, *67*, 131–142. [CrossRef]
27. Li, W.; Wang, Q.; Zhang, F.; Shang, H.; Bai, S.; Sun, J. pH-sensitive thiamethoxam nanoparticles based on bimodal meso-porous silica for improving insecticidal efficiency. *R. Soc. Open Sci.* **2021**, *8*, 201967. [CrossRef]
28. Li, W.; Xu, X.; Pan, H.; Wu, L.; Bai, S.; Sun, J.; Zhang, F. Comparative study on two different methods for fabrication of sus-tained release boscalid based on mesoporous silica. *Mater. Res. Express* **2021**, *8*, 045018. [CrossRef]
29. Chauhan, N.; Dilbaghi, N.; Gopal, M.; Kumar, R.; Kim, K.H.; Kumar, S. Development of chitosan nanocapsules for the controlled release of hexaconazole. *Int. J. Biol. Macromol.* **2017**, *97*, 616–624. [CrossRef]
30. Ju, C.; Xu, J.; Wu, X.; Dong, F.; Liu, X.; Tian, C.; Zheng, Y. Effects of hexaconazole application on soil microbes community and nitrogen transformations in paddy soils. *Sci. Total Environ.* **2017**, *609*, 655–663. [CrossRef]
31. Yu, L.; Chen, M.; Liu, Y.; Gui, W.; Zhu, G. Thyroid endocrine disruption in zebrafish larvae following exposure to hexa-conazole and tebuconazole. *Aquat. Toxicol.* **2013**, *138–139*, 35–42. [CrossRef] [PubMed]
32. Liang, H.; Li, L.; Li, W.; Wu, Y.; Liu, F. The decline and residues of hexaconazole in tomato and soil. *Environ. Monit. Assess.* **2012**, *184*, 1573–1579. [CrossRef] [PubMed]
33. Liu, B.; Zhang, J.; Chen, C.; Wang, D.; Tian, G.; Zhang, G.; Cai, D.; Wu, Z. Infrared-light-responsive controlled-release pesticide using hollow carbon microspheres@polyethylene glycol/alpha-cyclodextrin. *Gel J. Agric. Food Chem.* **2021**, *69*, 6981–6988. [CrossRef] [PubMed]
34. Li, H.; Tan, L.-L.; Jia, P.; Li, Q.-L.; Sun, Y.-L.; Zhang, J.; Ning, Y.-Q.; Yu, J.; Yang, Y.-W. Near-infrared light-responsive supramolecular nanovalve based on mesoporous silica-coated gold nanorods. *Chem. Sci.* **2014**, *5*, 2804. [CrossRef]

coatings

Article

Heterogeneous Fenton-like Catalyzation of Nanoscale Schwertmannite for Sulfamethoxazole Degradation

Xiaoqing Meng, Lin Wang, Ying Yang, Yuqi Song and Cansheng Yuan *

College of Rural Revitalization, Jiangsu Open University, Nanjing 210036, China; mengxq@jsou.edu.cn (X.M.); wanglin@jsou.edu.cn (L.W.); yangy@jsou.edu.cn (Y.Y.); songyq@jsou.edu.cn (Y.S.)
* Correspondence: csyuan2023@163.com; Tel.: +86-(025)86265520

Abstract: Sulfamethoxazole (SMX) contamination in large quantities of wastewater can cause potential environmental problems. Due to difficulty in degrading SMX by natural processes, it is necessary to develop a novel technology to solve this problem. Advanced oxidation processes (AOPs) have been identified as methods with a high potential to treat recalcitrant organic pollutants. The nanoscale schwertmannite (nano-SWT) was prepared with an indoor-temperature synthesis method facilitated by polyvinylpyrrolidone (PVP). In this study, we performed a reaction of the nano-SWT materials with Fenton-like catalysts for SMX degradation in hydrogen peroxide (H_2O_2) media. The findings showed that the nano-SWT prepared by addition of 0.1 $g{\cdot}L^{-1}$ PVP (nano-SWT-n, n = 0.1) could degrade 92.5% of the SMX within 90 min at indoor temperature, which was due to the nano-SWT providing abundant reaction sites at the solid/solution interfaces. Additionally, SMX could be highly mineralized with 75% TOC removal and H_2O_2 was efficiently utilized during the nano-SWT/H_2O_2 process. In addition, after six cycles of Fenton-like degradation, the nano-SWT remained stable and reusable as a Fenton-like catalyst for SMX degradation. The nano-SWT performed well as a catalyst for SMX degradation. Additionally, this work provides a feasible environmental purification approach for the efficient degradation of SMX through the use of nanoscale schwertmannite as a catalyst in heterogeneous Fenton-like systems.

Keywords: Fenton-like; sulfamethoxazole; nanoscale schwertmannite; degradation

Citation: Meng, X.; Wang, L.; Yang, Y.; Song, Y.; Yuan, C. Heterogeneous Fenton-like Catalyzation of Nanoscale Schwertmannite for Sulfamethoxazole Degradation. *Coatings* **2023**, *13*, 1097. https://doi.org/10.3390/coatings13061097

Academic Editor: M. Shaheer Akhtar

Received: 24 May 2023
Revised: 9 June 2023
Accepted: 12 June 2023
Published: 14 June 2023

1. Introduction

A crucial sulfonamide antibacterial agent known as sulfamethoxazole (SMX) is effective in treating people and animals with illnesses and infections [1,2]. As SMX is widely used in the areas of industry and home activities, large quantities of wastewater that contain SMX will be produced each year. However, due to its limited biodegradability, even minute levels of SMX will lead to severe environmental problems and its degradation in the environment is difficult [1,3]. For example, SMX can impede aquatic organism activity or cause changes in microorganisms that lead to antibiotic resistance [4]. The German Federal Environment Agency suggests a threshold of 100 $ng{\cdot}L^{-1}$ for the introduction of antibiotics into groundwater [5,6]. Hence, an effective strategy to remove SMX from wastewater or groundwater needs to be developed.

Advanced oxidation processes (AOPs) have been deemed a very popular method for treating recalcitrant organic compounds through generating abundant, highly active hydroxyl radicals (•OH), as it has a rapid rate of oxidation, a broad range of applications, and a high level of mineralization [7–9]. In the presence of Fe(II) and H_2O_2, the Fenton reaction is regarded as one of the most efficient AOPs [9,10], and is used for its simple operation, high efficiency, and cost-effective production of •OH. However, as a wastewater treatment process, the Fenton reaction is severely limited by the high H_2O_2 consumption, the amount of ferric hydroxide sludge, and the difficulty in recovering the catalyst after it has been used [11–13]. Heterogeneous catalysts for Fenton-like reactions have been thoroughly investigated in order to solve these problems [14,15].

Schwertmannite ($Fe_8O_8(OH)_{8-2x}(SO_4)_x$, $1.75 \geq x \geq 1$) is a secondary Fe(III)-oxyhydroxysulfate mineral that is frequently found in acidic mining effluent, acid sulphate soils, or the leachate of acidic sludge [2,16,17]. Generally, weakly crystalline mineral particles with spherical shapes that range in size from nanometers to microns are seen in schwertmannite [16,17]. In recent years, among the heterogeneous Fenton-like reactions, schwertmannite can serve as a catalyst and shows excellent catalytic performance [17–20]. Furthermore, both chemogenic and biogenic schwertmannites have been used as catalysts to degrade organic pollutants in heterogeneous Fenton-like reactions under difficult conditions [21,22]. It is important to note that chemically synthesized schwertmannite (ferrous ions are readily oxidized by hydrogen peroxide) typically takes the shape of micron-sized spherical aggregates with diminished catalytic activity and specific surface area. Higher H_2O_2 concentrations or catalyst dosages are necessary to ensure the effective removal of organic contaminants since the aggregation of schwertmannite particles may significantly impede oxidation processes [23]. Hence, on the basis of synthetic design methodologies, it is extremely important to investigate the connection between the morphology/micro-environment and the catalytic capabilities of schwertmannite in Fenton-like systems.

Therefore, the aim of this study was to investigate the feasibility of nanoscale schwertmannite (nano-SWT) as Fenton-like catalysts for SMX degradation in the presence of H_2O_2 (nano-SWT system). The main objectives were to: (1) prepare nano-SWT through the addition of polyvinylpyrrolidone (PVP), which is a type of organic surfactant that is inexpensive, and characterize it; (2) investigate the SMX degradation efficiency using nano-SWT as a Fenton-like catalyst in different conditions (such as nano-SWT dosage, H_2O_2 concentration, or initial pH); (3) explore the SMX mineralization degree and the utilization efficiency of H_2O_2 during the nano-SWT/H_2O_2 process; (4) study the stability and reusability of the nano-SWT as a Fenton-like catalyst for SMX degradation after six cycles of usage. This work can provide a feasible environmental purification approach for the efficient degradation of SMX through the use of nanoscale schwertmannite as a catalyst in heterogeneous Fenton-like systems.

2. Materials and Methods

2.1. Chemicals

Sulfamethoxazole (SMX, >99%) was obtained from Aladdin Co., Ltd. (Shanghai, China). Polyvinylpyrrolidone (PVP, K30) was purchased from Solarbio Co., Ltd. (Beijing, China). Methanol, acetonitrile, acetic acid, and formic acid were supplied by Tedia Co., Ltd. (Shanghai, China). Hydrogen peroxide (H_2O_2, >99.9%), sulfuric acid (H_2SO_4, >99%), sodium hydroxide (NaOH, >99%), ferrous sulfate heptahydrate (Fe_2SO_4•$7H_2O$, >99%), and other analytical-grade chemicals were purchased from Shanghai Sinopharm Chemical Reagent Co. Ltd. based in China, without further purification. Milli-Q water was employed in this work to prepare the solutions and suspensions that were used.

2.2. Preparation and Characterization of Nano-SWT

Schwertmannite was prepared using the technique outlined in the literature [1]. Simply, in 100 mL deionized water, 5.56 g of $FeSO_4$•$7H_2O$ and PVP (0, 0.1, 0.25 g) were dissolved. Then, the solution pH was raised to 3.0 using 1.0 M H_2SO_4. After that, 3.0 mL of H_2O_2 (v/v 30%) was gradually added to the aforementioned mixture at a rate of 0.3 mL/min using an injection pump while stirring for 24 h at room temperature. The resulting mixture was centrifuged at 10,000 rpm for five minutes, and water and ethanol were used to rinse the precipitates three times. After that, the precipitates were dried under vacuum at 60 °C to achieve a constant weight. Then, the sample were recorded as nano-SWT-*n* (*n* = 0, 0.1, and 0.25), where *n* is the amount of PVP used.

Using an X-ray diffractometer (XRD, 40 kV, 40 mA, ThermoFisher XTRA, Waltham, MA, USA), the crystal structure of the nano-SWT was analyzed. Using a scanning electron microscope (SEM, Hitachi S-4700, Tokyo, Japan) with an accelerating voltage of 20 kV, the morphology of the nano-SWT was determined. The functional groups of the nano-SWT were examined using Fourier transform infrared spectroscopy (FTIR, Nicolet iS50, ThermoFisher, Waltham, MA, USA). The specific surface area of the nano-SWT was measured using the Brunauer–Emmett–Teller (BET, TriStarII 3020, Micromeritics, Norcross, GA, USA) via N_2 adsorption–desorption method. The sulfur (S) and iron (Fe) contents of the nano-SWT were examined through inductively coupled plasma atomic emission spectrometry (ICP–OES, Agilent 720, Santa Clara, CA, USA).

2.3. Catalytic Degradation of SMX

The SMX degradation experiments were conducted in a 250 mL flask without energy input, and this procedure was regulated using the methods of previous research [2]. Briefly, a specific amount of the nano-SWT was added into 100 mL of a SMX solution. Then, NaOH (0.1 mM) or H_2SO_4 (0.1 mM) were used to bring the original pH levels of the suspensions to the appropriate levels. After that, all suspensions in flasks were mixed under magnetic stirring at 180 rpm, and the reaction temperature was controlled at 25 °C for half an hour to achieve adsorption equilibrium. By adding a specific amount of H_2O_2 (v/v 30%) to the aforementioned solution, the reaction was then started. To filter 0.5 mL suspension samples from each flask at predetermined intervals, a 0.22 μm membrane filter was used. To stop SMX from reacting with the reactive species, 0.5 mL of methanol was added to the filtrate. Afterward, the concentration of SMX was determined using the treated filtrates. No quenching agent was added before the TOC analysis; the filtrates were quickly cooled down using ice packs for the TOC measurement. After the SMX degradation, suspensions of the nano-SWT were gathered by filtration via a 0.22 μm filter paper and subjected to reusability experiments using Milli-Q water. Duplicates of each degradation experiment were run.

The following equation was used to compute the SMX degradation efficiency:

$$R = \frac{(C_0 - C_t)}{C_0} \times 100\%$$

In this equation, C_t represents the concentration of SMX at sampling time t, C_0 represents the initial concentration of SMX, and R represents the degradation efficiency.

2.4. Analytical Methods

High Performance Liquid Chromatography (HPLC, Agilent, Santa Clara, CA, USA), outfitted with a UV detector and a C18 chromatographic column (250 mm × 4.6 mm, 5 μm), was used to determine the concentration of SMX. The SMX concentration was analyzed in an eluent comprised of acetonitrile and 0.1% formic acid with volume ratio of 55:45 (v/v) at UV 270 nm. The eluent flowed at a rate of 1 mL per minute [2]. Shimadzu TOC-LCPH/CPN analyzers were used to calculate the total organic carbon (TOC) [17]. The titanium sulfate method was used to measure the concentration of H_2O_2 by combining H_2SO_4 with $K_2TiO(C_2O_4)_2$ to create an orange complex with a maximum absorbance at 400 nm [23].

3. Results

3.1. Characterizations of Prepared Schwertmannite

According to Figure 1, the XRD patterns of the three prepared iron precipitates revealed that seven typical broad characteristic peaks were observed in the diffraction spectrum of the weak crystal structure (2θ: 18.2, 26.3, 35.2, 39.5, 46.5, 55.3, 61.3°), which are all in agreement with the standard diffraction data of schwertmannite (PDF#47-1775) [23]. This result indicated the three iron precipitates that were created consisted of pure schwertmannite particles. The surface morphology of the three schwertmannite materials (nano-SWT-n, n = 0, 0.1, 0.25) was investigated using scanning electron microscopy (SEM) techniques (Figure 2). The SEM showed that the nano-SWT-0 surface was relatively smooth, while the nano-SWT-0.1 and nano-SWT-0.25 showed rather rough hierarchical structures. Moreover, at diameters of about 1.0 μm, the nano-SWT-0 particles were considerably agglomerated. Nevertheless, with sizes ranging from 50 to 200 nm, the nano-SWT-0.1 and nano-SWT-0.25 particles were evenly scattered nanoparticles, which may be due to the lowered surface energy of the nano-SWT and the surfactant-like behavior of PVP [23]. Evidently, the PVP coating decreased the size of the nanoparticles, causing them to aggregate and form a loose spherical structure that produced a lot of active sites in the nano-SWT.

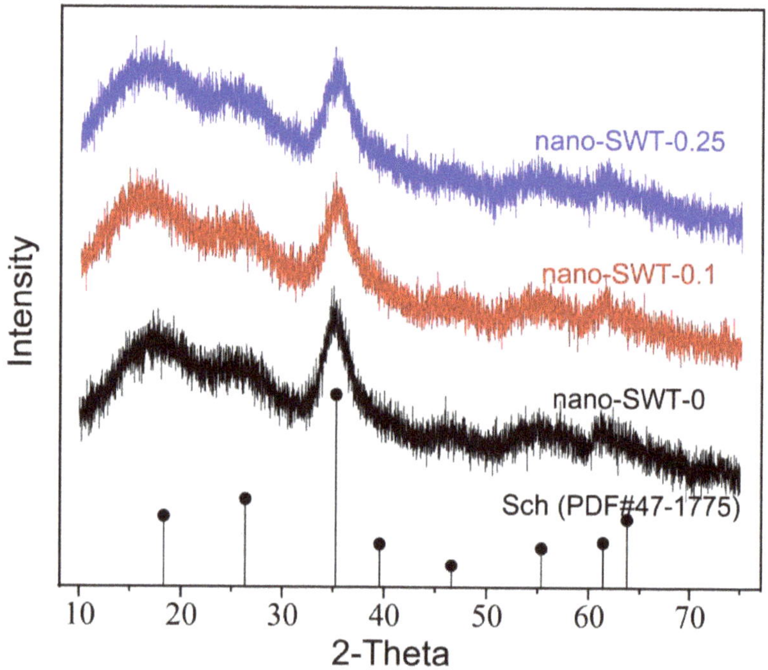

Figure 1. XRD analyses of nano-SWT-*n* (*n* = 0, 0.1, 0.25).

The chemical formula of the nano-SWT-*n* (*n* = 0~0.25) were determined as $Fe_8O_8(OH)_{5.92}(SO_4)_{1.14}$, $Fe_8O_8(OH)_{5.84}(SO_4)_{1.08}$, and $Fe_8O_8(OH)_{5.78}(SO_4)_{1.11}$, respectively (Table 1), which are consistent with those reported by previous studies [17,24–26]. In these schwertmannite samples, the Fe/S molar ratio ranged from 7.03 to 7.44 with nano-SWT-0.1 having the highest Fe/S ratio of 7.44 and having more hydroxyl (OH) and fewer sulphate (SO_4^{2-}) groups in its structure (Table 1). These findings indicated that nano-SWT-0.1 had a greater surface area available since the aggregation of nanoparticles was reduced [23,27].

Table 1. Sulfur (S) and iron (Fe) contents, as well as BET surface areas in nano-SWT.

Sample	Fe (wt.%)	S (wt.%)	Molar Ratio (Fe/S)	*n*	Formulae	Bet (m²/g)
Nano-SWT-0	46.88	6.67	7.03	1.14	$Fe_8O_8(OH)_{5.92}(SO_4)_{1.14}$	4.58
Nano-SWT-0.1	39.73	5.34	7.44	1.08	$Fe_8O_8(OH)_{5.84}(SO_4)_{1.08}$	130.12
Nano-SWT-0.25	30.08	4.17	7.21	1.11	$Fe_8O_8(OH)_{5.78}(SO_4)_{1.11}$	85.40

Indeed, the study showed that the specific surface areas of the nano-SWT-*n* (*n* = 0~0.25) were 4.58, 130.12, and 85.40 $m^2 \cdot g^{-1}$, respectively. It is clear that the introduction of PVP is conducive to increasing the BET surface area. Meanwhile, the optimal nano-SWT-0.1 would be produced if the amount of the PVP surfactant could be controlled. More catalytic active sites might be provided by the relatively high BET surface area of the nano-SWT, which would be advantageous for increasing catalytic efficiency [28–30]. Moreover, the N_2 adsorption–desorption isotherms of the nano-SWT-*n* (n = 0~0.25) were determined, and these three samples represented type IV isotherms with hysteresis loops (Figure 3), indicating that these nanoparticles formed a porous structure [31,32].

Figure 2. SEM images of nano-SWT-0 ((**a**): × 50,000, (**b**): ×150,000), nano-SWT-0.1 ((**c**): ×50,000, (**d**): ×150,000), and nano-SWT-0.25 ((**e**): ×50,000, (**f**): ×150,000).

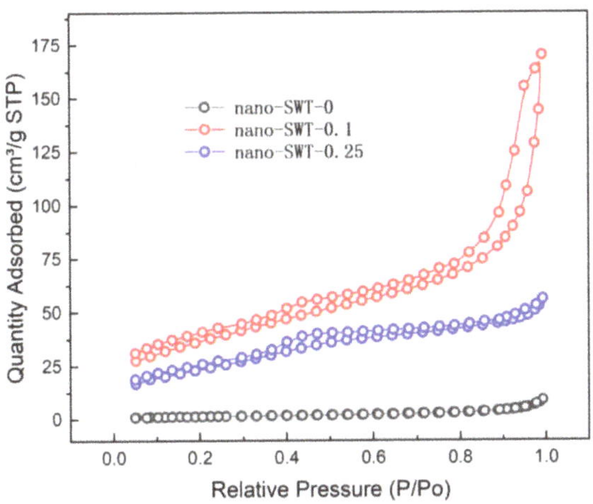

Figure 3. N_2 adsorption–desorption isotherms of nano-SWT–*n* (n = 0, 0.1, 0.25).

3.2. Catalytic Performances of Catalyst for SMX Degradation

The SMX degradation conditions in the nano-SWT/H_2O_2 system were optimized under a variety of significant factors, such as nano-SWT dosage, H_2O_2 concentration, and initial pH of the reaction solution. The impact of nano-SWT dosage on the catalytic degradation of SMX was examined. According to Figure 4a, when the dosage of the nano-SWT was increased from 0.1 g·L^{-1} to 0.2 g·L^{-1} within 90 min of the reaction, the degradation efficiency of SMX rose from 72.1% to 92.5%. The findings showed that the increased number of higher dosages of $\equiv Fe(OH)^{2+}$ on nano-SWT led to the production of more $\equiv Fe^{2+}$ and •OH [2,17,33], which was the cause of the better SMX degrading efficiency. However, when the nano-SWT loading was raised to 0.5 g·L^{-1}, the considerable "screen effect" on SMX degradation in the nano-SWT/H_2O_2 system prevented a significant increase in SMX degradation efficiency [34,35]. Thus, 0.2 g·L^{-1} of nano-SWT was appropriate for SMX degradation in the nano-SWT/H_2O_2 system.

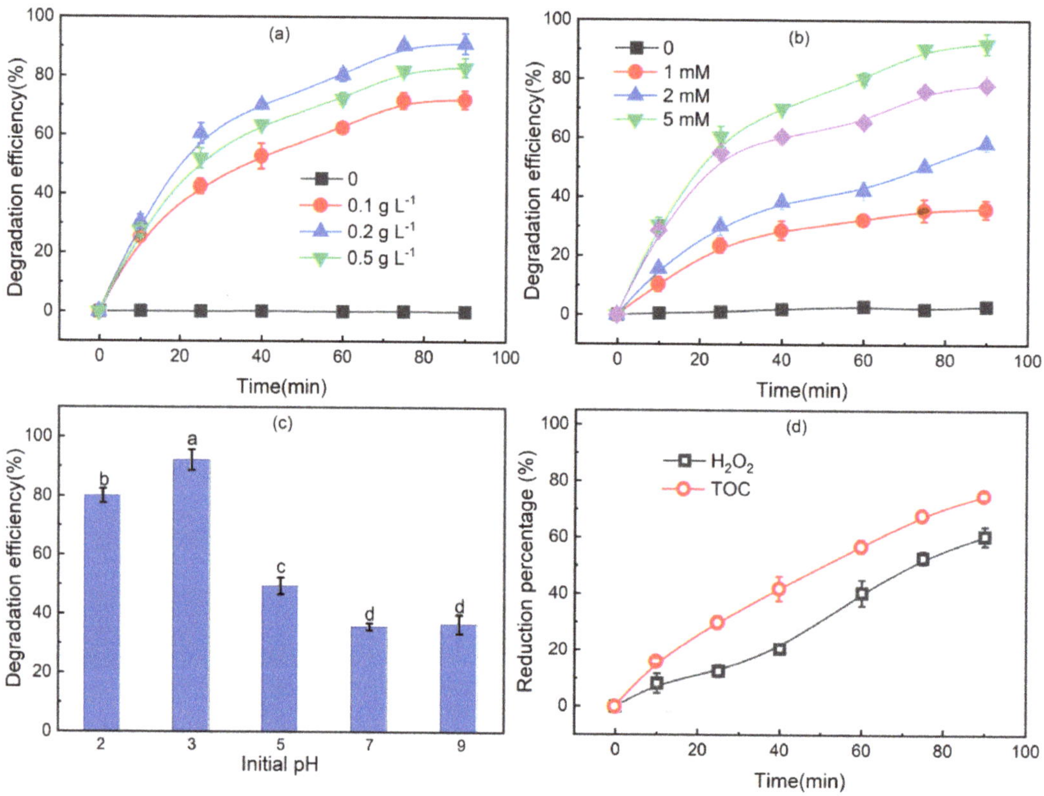

Figure 4. The degradation efficiency of SMX at various catalyst (nano-SWT-0.1) dosages (**a**), H_2O_2 concentrations (**b**), initial pH values (**c**), and columns marked with the same letter in the group of SMX degradation efficiency do not differ significantly from each other at $p < 0.05$, and TOC removal and H_2O_2 decomposition (**d**).

Investigations were conducted into how H_2O_2 concentration affected the rate of SMX degradation in the nano-SWT/H_2O_2 system. Figure 4b demonstrates that the degradation efficiency of SMX noticeably increased from 36.1% to 92.2% in the nano-SWT/H_2O_2 system as the H_2O_2 concentration was increased from 1 mM to 5 mM. This phenomenon signified that more reactive species might be generated at higher H_2O_2 concentrations in the nano-

SWT/H_2O_2 system. Nevertheless, the subsequent increase in H_2O_2 concentration from 5 mM to 10 mM caused the SMX degradation efficiency in the nano-SWT/H_2O_2 system to drop from 92.5% to 83.1%. The result can be explained by the fact that reactive species were mixed again or quenched by excess H_2O_2 [2,18,36]; similar phenomena appeared in previous studies [2,17]. The ideal H_2O_2 concentration for the SMX degradation in the nano-SWT/H_2O_2 system was 40 μM because too much H_2O_2 would undoubtedly increase the treatment cost.

Because pH affects the surface charge of catalysts and SMX molecules, as well as the conversion of reactive species [37,38], the nano-SWT was also investigated for catalytic activity in the pH range of 2 to 9. Figure 4c shows that the catalytic activities for the degradation of SMX in acidic conditions (pH = 2, 3, 5) was more efficient than that in neutral (pH = 7) or alkaline (pH = 9) conditions. The efficiency of SMX degradation reached a high at pH 3. The following points may help to explain the findings. Previous studies have shown that Sch has a pHzpc of 5.3, and SMX has a pKa,1 and pKa,2 of 1.6 and 5.7, respectively [33,39]. Hence, the surface charge of the nano-SWT increased when the pH was lowered from 5 to 2, and the SMX molecule mostly existed in the neutral phase (pKa,1 < pH < pKa,2). Based on the above theory, the nano-SWT electrostatically adsorbs SMX and H_2O_2, thus facilitating the heterogeneous activation of H_2O_2 to the reactive species for SMX degradation. The electrostatic repulsion between the nano-SWT and the SMX molecule occurred as the pH rose from 5 to 7 or 9, resulting in the nano-SWT becoming negatively charged (pH > pHzpc) and SMX entering an anionic phase (pH > pKa,2). This prevented the heterogeneous activation of H_2O_2 to the reactive species for SMX degradation. It is worth noting that at an initial pH of 7 or 9, the SMX degradation efficiencies were 35.8% and 36.8% in the nano-SWT/H_2O_2 process, respectively, and there was no difference ($p > 0.05$) between the two conditions. This result can be explained that, in the nano-SWT/H_2O_2 process, when the pH value drops from 7 or 9 to close to 4 [2], it creates a more acidic environment for the nano-SWT to adsorb and activate H_2O_2 into the reactive species for the degradation of SMX. Meanwhile, the consumption of OH^- from H_2O by the nano-SWT at pH > pHzpc may be the source of the pH drop in the solution.

Moreover, the degradation efficiency of SMX at pH 2 (80.1%) was slightly lower than that of pH 3 (92.2%). A possible reason is that the degradation of SMX may be converted to reactive species by various solution pH levels [40,41]. At pH 2, •OH may be scavenged more readily by H^+ in the nano-SWT/H_2O_2 system [42,43], whereas at pH 5 to 9, the conversion of •OH was facilitated, but the redox potential of •OH was reduced [44]. Based on the above results, the nano-SWT demonstrated excellent catalytic activity under acidic conditions (pH 2–5) with its SMX degradation efficiency ranging from 49.5% to 92.5% after only 90 min. Moreover, the optimal environments for SMX degradation in this nano-SWT/H_2O_2 system were H_2O_2 5.0 mM, schwertmannite 0.2 $g{\cdot}L^{-1}$, and pH 3. The SMX mineralization degree in the nano-SWT/H_2O_2 system was assessed by TOC removal. Figure 4d shows that the nano-SWT/H_2O_2 system removed 75% of the TOC within 90 min, and this result further proved that the nano-SWT/H_2O_2 system has the ability to degrade SMX and mineralize it efficiently. The result implied that the •OH from the nano-SWT/H_2O_2 system promoted the degradation of SMX and its intermediates, such as 4-aminobenzenesulfonic acid, 4-nitro sulfamethoxazole, and nitrosobenzene [2]. The concentration of H_2O_2 steadily dropped in line with the decline in TOC. H_2O_2 utilization efficiency can be reflected by the ratio of the stoichiometric consumption of H_2O_2 to the actual consumption for the degradation of SMX during the nano-SWT/H_2O_2 process. After almost all the SMX disappeared from the nano-SWT/H_2O_2 system, the utilization efficiency of H_2O_2 reached up to 60.5%, which was much higher than of Sch in the degradation of some organic pollutants, such as nitrobenzene and phenanthrene, through Fenton-like reactions [17,22].

3.3. Stability and Reusability of the Nano-SWT in the SMX Degradation Process

The long-term stability and reusability of the nano-SWT in the SMX degradation process should also be taken into account from an application standpoint. Therefore, the catalytic performance of the nano-SWT was examined via multi-cycle experiments. The initial concentrations of SMX, H_2O_2, and nano-SWT were 50 μM, 5.0 mM, and 0.2 g·L^{-1}, respectively, in each cycle, the reaction time was 90 min, and the initial pH of the solution was 3. Figure 5 shows that when the nano-SWT was reused for six cycles, the SMX degradation efficiency decreased from 92.5% to 85.2%. The slight decrease in performance of the nano-SWT/H_2O_2 system may be attributed to the decrease in surface iron content on the catalyst after each cycle [2]. This result demonstrated that the nano-SWT can be reused as a Fenton-like catalyst for SMX degradation because it still maintained up to 85.2% of its catalytic capacity after six cycles of reuse. The stability and reusability properties of the nano-SWT in the nano-SWT/H_2O_2 system are similar to those for the previous reports using schwertmannite as a catalyst [45,46].

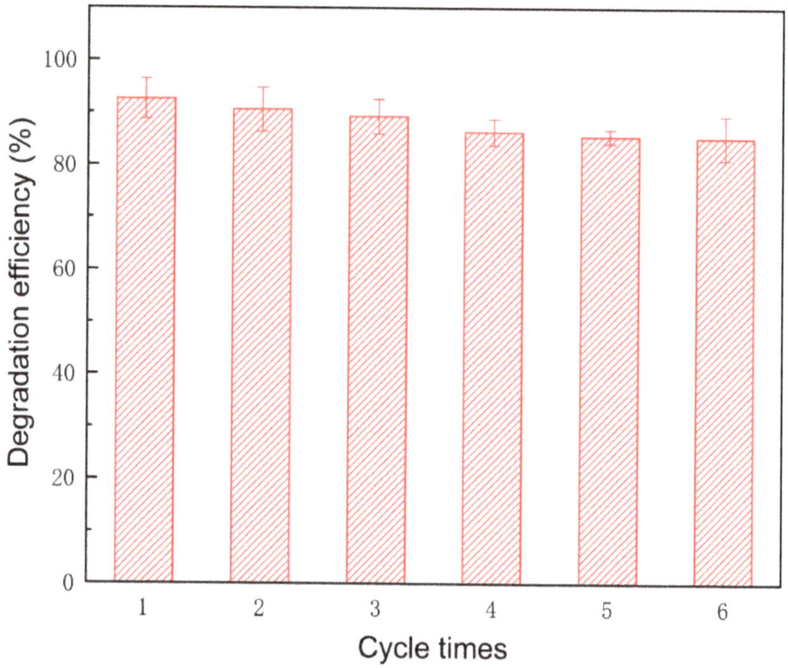

Figure 5. SMX degradation efficiency during repeated nano-SWT applications. Experimental conditions: H_2O_2 concentration = 5 mM, nano-SWT dosage = 0.2 g·L^{-1}, SMX concentration = 50 μM, initial pH =3.0, and reaction time = 90 min.

The FTIR spectra of newly synthesized nano-SWT and nano-SWT after being used for six cycles are shown in Figure 6. It was found that the absorption peaks were caused by the stretching of free or bound hydroxyl groups at a wavelength of 3217 cm^{-1}. The structural OH groups or the O-H vibrations of H_2O molecules were responsible for the absorption peaks at 1631 cm^{-1}. The absorption peaks at 609 cm^{-1} were attributed to the vibration of SO_4^{2-}. The wavelengths of 1117 and 982 cm^{-1} corresponded to the γ1 and γ3 vibrations of SO_4^{2-}, respectively. The vibration of the Fe-O bond was a contributing factor to the absorption peaks at 483 cm^{-1}. The FTIR spectra of all functional groups matched those of Sch as stated in previous investigations [17,24,47,48]. The FTIR spectrum of the reused nano-SWT was consistent with that of the newly synthesized nano-SWT. The findings showed that the structure and functional groups of the nano-SWT were minimally affected by the nano-SWT/H_2O_2 reaction. This result suggests that the mineral structure of the nano-SWT, a great Fenton-like catalyst for SMX degradation, remained unchanged over the course of multiple uses. Above all, after six cycles of Fenton-like degradation, the nano-SWT maintain stability and was reusable as a Fenton-like catalyst for SMX degradation.

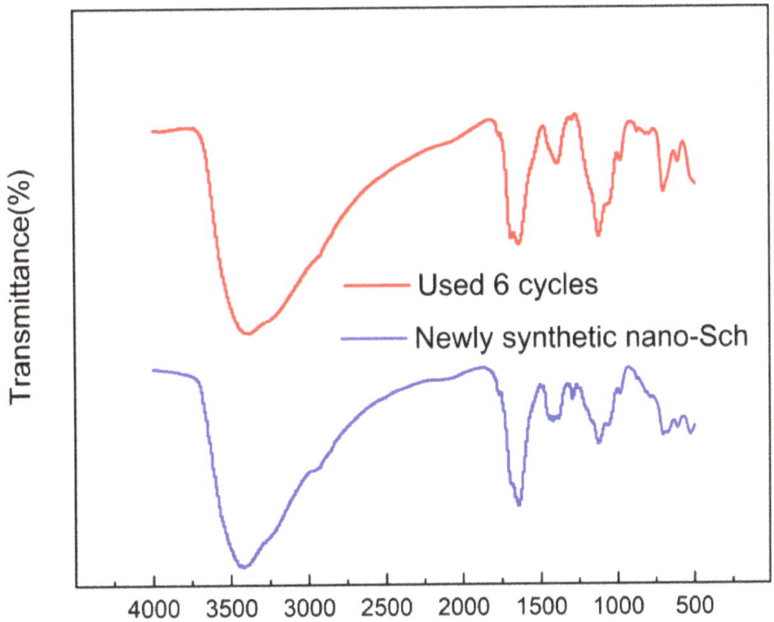

Figure 6. FTIR spectra of newly synthesized nano-SWT and nano-SWT after six cycles of use. Experimental conditions: H_2O_2 concentration = 5 mM, nano-SWT dosage = 0.2 g·L^{-1}, SMX concentration = 50 μM, initial pH =3.0, and reaction time = 90 min.

4. Conclusions

A simple PVP-assisted synthetic method was used to synthesize nano-SWT. By adding a small amount of PVP, the structure and surface characteristics of synthesized nano-SWT can be engineered. The nano-SWT can be used as a catalyst to efficiently degrade SMX in a heterogeneous Fenton-like reaction (nano-SWT/H_2O_2 reaction). The degradation efficiency of SMX was more than 92.5% within 90 min of reaction time under suitable reaction conditions, which is due to the nano-SWT providing abundant reaction sites at the solid/solution interfaces. Moreover, the SMX could be highly mineralized with removal of 75% of the TOC within 90 min, and the utilization efficiency of H_2O_2 could reach up to 60.5% after almost all the SMX disappeared from the nano-SWT/H_2O_2 system. In addition, the nano-SWT had good stability and reusability as a Fenton-like catalyst for SMX degradation, as evidenced by the fact that repeated use did not appreciably alter its catalytic capacity and function groups. As a result, in heterogeneous Fenton-like reactions, the nano-SWT is a great catalyst for the degradation of SMX. Hence, the nano-SWT/H_2O_2 system was verified to be a promising method for the efficient degradation of SMX.

Author Contributions: Conceptualization, X.M.; methodology, X.M.; software, X.M.; validation, X.M.; formal analysis, X.M.; investigation, X.M.; resources, C.Y. and L.W.; data curation, X.M.; writing—original draft preparation, X.M.; writing—review and editing, Y.Y. and Y.S.; visualization, X.M.; supervision, C.Y. and L.W.; project administration, X.M.; funding acquisition, X.M. All authors have read and agreed to the published version of the manuscript.

Funding: The authors would like to thank the Jiangsu Open University Youth Research Program (2022XK010) for financial support.

Institutional Review Board Statement: Not applicable.

Informed Consent Statement: Not applicable.

Data Availability Statement: The authors declare that the data are unavailable due to privacy or ethical restrictions in this paper.

Conflicts of Interest: The authors declare no conflict of interest.

References

1. Lee, K.; Kim, T.H.; Kim, T.H.; Lee, J.; Yu, S. Enhancement of TOC removal efficiency of sulfamethoxazole using catalysts in the radiation treatment: Effects of band structure and electrical properties of radiocatalysts. *Sep. Purif. Technol.* **2023**, *312*, 123390. [CrossRef]
2. Yan, S.; Zhan, L.; Meng, X.; Wang, D.; Wang, X.; Zheng, G.; Lu, J.; Zhou, L. Role of schwertmannite or jarosite in photocatalytic degradation of sulfamethoxazole in ultraviolet/peroxydisulfate system. *Sep. Purif. Technol.* **2021**, *274*, 118991. [CrossRef]
3. Gong, H.; Chu, W. Determination and toxicity evaluation of the generated products in sulfamethoxazole degradation by UV/$CoFe_2O_4$/TiO_2. *J. Hazard. Mater.* **2016**, *314*, 197–203. [CrossRef] [PubMed]
4. Oh, D.; Lee, C.S.; Kang, Y.G.; Chang, Y.S. Hydroxylamine-assisted peroxymonosulfate activation using cobalt ferrite for sulfamethoxazole degradation. *Chem. Eng. J.* **2020**, *386*, 123751. [CrossRef]
5. Meffe, R.; de Bustamante, I. Emerging organic contaminants in surface water and groundwater: A first overview of the situation in Italy. *Sci. Total Environ.* **2014**, *481*, 280–295. [CrossRef] [PubMed]
6. Burke, V.; Richter, D.; Greskowiak, J.; Mehrtens, A.; Schulz, L.; Massmann, G. Occurrence of antibiotics in surface and groundwater of a drinking water catchment area in Germany. *Water Environ. Res.* **2016**, *88*, 652–659. [CrossRef]
7. Pisharody, L.; Gopinath, A.; Malhotra, M.; Nidheesh, P.V.; Kumar, M.S. Occurrence of organic micropollutants in municipal landfill leachate and its effective treatment by advanced oxidation processes. *Chemosphere* **2022**, *287*, 132216. [CrossRef]
8. Lama, G.; Meijide, J.; Sanromán, A.; Pazos, M. Heterogeneous Advanced Oxidation Processes: Current Approaches for Wastewater Treatment. *Catalysts* **2022**, *12*, 344. [CrossRef]
9. Munoz, M.; De Pedro, Z.M.; Casas, J.A.; Rodriguez, J.J. Preparation of magnetite-based catalysts and their application in heterogeneous Fenton oxidation—A review. *Appl. Catal. B Environ.* **2015**, *176*, 249–265. [CrossRef]
10. Xu, X.R.; Zhao, Z.Y.; Li, X.Y.; Gu, J.D. Chemical oxidative degradation of methyl tert-butyl ether in aqueous solution by Fenton's reagent. *Chemosphere* **2004**, *55*, 73–79. [CrossRef]
11. Neyens, E.; Baeyens, J. A review of classic Fenton's peroxidation as an advanced oxidation technique. *J. Hazard. Mater.* **2003**, *98*, 33–50. [CrossRef]
12. Zhu, Y.; Zhu, R.; Xi, Y.; Zhu, J.; Zhu, G.; He, H. Strategies for enhancing the heterogeneous Fenton catalytic reactivity: A review. *Appl. Catal. B Environ.* **2019**, *255*, 117739. [CrossRef]
13. Hartmann, M.; Kullmann, S.; Keller, H. Wastewater treatment with heterogeneous Fenton-type catalysts based on porous materials. *J. Mater. Chem.* **2010**, *20*, 9002–9017. [CrossRef]
14. Babuponnusami, A.; Muthukumar, K. Removal of phenol by heterogenous photo electro Fenton-like process using nano-zero valent iron. *Sep. Purif. Technol.* **2012**, *98*, 130–135. [CrossRef]
15. Ganiyu, S.O.; Zhou, M.; Martínez-Huitle, C.A. Heterogeneous electro-Fenton and photoelectro-Fenton processes: A critical review of fundamental principles and application for water/wastewater treatment. *Appl. Catal. B Environ.* **2018**, *235*, 103–129. [CrossRef]
16. Jin, D.; Wang, X.; Liu, L.; Liang, J.; Zhou, L. A novel approach for treating acid mine drainage through forming schwertmannite driven by a mixed culture of *Acidiphilium multivorum* and *Acidithiobacillus ferrooxidans* prior to lime neutralization. *J. Hazard. Mater.* **2020**, *400*, 123108. [CrossRef]
17. Meng, X.; Yan, S.; Wu, W.; Zheng, G.; Zhou, L. Heterogeneous Fenton-like degradation of phenanthrene catalyzed by schwertmannite biosynthesized using *Acidithiobacillus ferrooxidans*. *Rsc. Adv.* **2017**, *7*, 21638–21648. [CrossRef]
18. Wang, W.M.; Song, J.; Han, X. Schwertmannite as a new Fenton-like catalyst in the oxidation of phenol by H_2O_2. *J. Hazard. Mater.* **2013**, *262*, 412–419. [CrossRef]
19. Cui, Q.; Li, Y.; Chai, S.; Zhang, W.; Zuo, Q.; He, C. Enhanced catalytic activation of H_2O_2 by CNTs/SCH through rapid Fe (III)/Fe (II) redox couple circulation: Insights into the role of functionalized multiwalled CNTs. *Sep. Purif. Technol.* **2022**, *282*, 120000. [CrossRef]

20. Yang, Z.; Zhu, P.; Yan, C.; Wang, D.; Fang, D.; Zhou, L. Biosynthesized Schwertmannite@ Biochar composite as a heterogeneous Fenton-like catalyst for the degradation of sulfanilamide antibiotics. *Chemosphere* **2021**, *266*, 129175. [CrossRef]
21. Guo, J.; Dong, C.; Zhang, J.; Lan, Y. Biogenic synthetic schwertmannite photocatalytic degradation of acid orange 7 (AO7) assisted by citric acid. *Sep. Purif. Technol.* **2015**, *143*, 27–31. [CrossRef]
22. Duan, H.; Liu, Y.; Yin, X.; Bai, J.; Qi, J. Degradation of nitrobenzene by Fenton-like reaction in a H_2O_2/schwertmannite system. *Chem. Eng. J.* **2016**, *283*, 873–879. [CrossRef]
23. Qiao, X.X.; Yu, K.; Xu, J.Y.; Cai, Y.L.; Li, Y.F.; Cao, H.L.; Lü, J. Engineered nanoscale schwertmannites as Fenton–like catalysts for highly efficient degradation of nitrophenols. *Appl. Surf. Sci.* **2021**, *548*, 149248. [CrossRef]
24. Bigham, J.M.; Schwertmann, U.; Traina, S.J.; Winland, R.L.; Wolf, M. Schwertmannite and the chemical modeling of iron in acid sulfate waters. *Geochim. Et Cosmochim. Acta* **1996**, *60*, 2111–2121. [CrossRef]
25. Liao, Y.; Zhou, L.; Liang, J.; Xiong, H. Biosynthesis of schwertmannite by *Acidithiobacillus ferrooxidans* cell suspensions under different pH condition. *Mater. Sci. Eng. C* **2009**, *29*, 211–215. [CrossRef]
26. Fan, C.; Guo, C.; Zeng, Y.; Tu, Z.; Ji, Y.; Reinfelder, J.R.; Chen, M.; Huang, W.; Lu, G.; Yi, X.; et al. The behavior of chromium and arsenic associated with redox transformation of schwertmannite in AMD environment. *Chemosphere* **2019**, *222*, 945–953. [CrossRef]
27. Jönsson, J.; Persson, P.; Sjöberg, S.; Lövgren, L. Schwertmannite precipitated from acid mine drainage: Phase transformation, sulphate release and surface properties. *Appl. Geochem.* **2005**, *20*, 179–191. [CrossRef]
28. Wu, M.; Gong, Y.; Nie, T.; Zhang, J.; Wang, R.; Wang, H.; He, B. Template-free synthesis of nanocage-like gC_3N_4 with high surface area and nitrogen defects for enhanced photocatalytic activity. *J. Mater. Chem. A* **2019**, *7*, 5324–5332. [CrossRef]
29. Mian, M.M.; Liu, G. Activation of peroxymonosulfate by chemically modified sludge biochar for the removal of organic pollutants: Understanding the role of active sites and mechanism. *Chem. Eng. J.* **2020**, *392*, 123681. [CrossRef]
30. Leonard, N.; Ju, W.; Sinev, I.; Steinberg, J.; Luo, F.; Varela, A.S.; Cuenya, B.R.; Strasser, P. The chemical identity, state and structure of catalytically active centers during the electrochemical CO_2 reduction on porous Fe–nitrogen–carbon (Fe–N–C) materials. *Chem. Sci.* **2018**, *9*, 5064–5073. [CrossRef]
31. Huo, C.; Ouyang, J.; Yang, H. CuO nanoparticles encapsulated inside Al-MCM-41 mesoporous materials via direct synthetic route. *Sci. Rep.* **2014**, *4*, 1–9. [CrossRef]
32. Vradman, L.; Landau, M.V.; Kantorovich, D.; Koltypin, Y.; Gedanken, A. Evaluation of metal oxide phase assembling mode inside the nanotubular pores of mesostructured silica. *Microporous Mesoporous Mater.* **2005**, *79*, 307–318. [CrossRef]
33. Yan, S.; Zheng, G.; Meng, X.; Zhou, L. Assessment of catalytic activities of selected iron hydroxysulphates biosynthesized using *Acidithiobacillus ferrooxidans* for the degradation of phenol in heterogeneous Fenton-like reactions. *Sep. Purif. Technol.* **2017**, *185*, 83–93. [CrossRef]
34. Chen, H.; Gao, B.; Li, H. Removal of sulfamethoxazole and ciprofloxacin from aqueous solutions by graphene oxide. *J. Hazard. Mater.* **2015**, *282*, 201–207. [CrossRef]
35. Tab, A.; Bellal, B.; Belabed, C.; Dahmane, M.; Trari, M. Visible light assisted photocatalytic degradation and mineralization of Rhodamine B in aqueous solution by Ag_3PO_4. *Optik* **2020**, *214*, 164858. [CrossRef]
36. Daneshvar, N.; Salari, D.; Khataee, A.R. Photocatalytic degradation of azo dye acid red 14 in water on ZnO as an alternative catalyst to TiO_2. *J. Photochem. Photobiol. A Chem.* **2004**, *162*, 317–322. [CrossRef]
37. Mao, S.; Sun, X.; Qi, H.; Sun, Z. Cu_2O nanoparticles anchored on 3D bifunctional CNTs/copper foam cathode for electrocatalytic degradation of sulfamethoxazole over a broad pH range. *Sci. Total Environ.* **2021**, *793*, 148492. [CrossRef]
38. Oyekunle, D.T.; Gendy, E.A.; Ifthikar, J.; Chen, Z. Heterogeneous activation of persulfate by metal and non-metal catalyst for the degradation of sulfamethoxazole: A review. *Chem. Eng. J.* **2022**, *437*, 135277. [CrossRef]
39. Nie, M.; Yan, C.; Xiong, X.; Wen, X.; Yang, X.; Dong, W. Degradation of chloramphenicol using a combination system of simulated solar light, Fe^{2+} and persulfate. *Chem. Eng. J.* **2018**, *348*, 455–463. [CrossRef]
40. Shang, K.; Wang, X.; Li, J.; Wang, H.; Lu, N.; Jiang, N.; Wu, Y. Synergetic degradation of Acid Orange 7 (AO7) dye by DBD plasma and persulfate. *Chem. Eng. J.* **2017**, *311*, 378–384. [CrossRef]
41. Zhou, Q.; Song, C.; Wang, P.; Zhao, Z.; Li, Y.; Zhan, S. Generating dual-active species by triple-atom sites through peroxymonosulfate activation for treating micropollutants in complex water. *Proc. Natl. Acad. Sci. USA* **2023**, *120*, e2300085120. [CrossRef] [PubMed]
42. Yang, L.; Xu, L.; Bai, X.; Jin, P. Enhanced visible-light activation of persulfate by Ti^{3+} self-doped TiO_2/graphene nanocomposite for the rapid and efficient degradation of micropollutants in water. *J. Hazard. Mater.* **2019**, *365*, 107–117. [CrossRef] [PubMed]
43. Mo, F.; Song, C.; Zhou, Q.; Xue, W.; Ouyang, S.; Wang, Q.; Wang, J. The optimized Fenton-like activity of Fe single-atom sites by Fe atomic clusters–mediated electronic configuration modulation. *Proc. Natl. Acad. Sci. USA* **2023**, *120*, e2300281120. [CrossRef] [PubMed]
44. Burbano, A.A.; Dionysiou, D.D.; Suidan, M.T.; Richardson, T.L. Oxidation kinetics and effect of pH on the degradation of MTBE with Fenton reagent. *Water Res.* **2005**, *39*, 107–118. [CrossRef]
45. Meng, X.; Wang, X.; Zhang, C.; Yan, S.; Zheng, G.; Zhou, L. Co-adsorption of As (III) and phenanthrene by schwertmannite and Fenton-like regeneration of spent schwertmannite to realize phenanthrene degradation and As (III) oxidation. *Environ. Res.* **2021**, *195*, 110855. [CrossRef]

46. Meng, X.; Zhang, C.; Zhuang, J.; Zheng, G.; Zhou, L. Assessment of schwertmannite, jarosite and goethite as adsorbents for efficient adsorption of phenanthrene in water and the regeneration of spent adsorbents by heterogeneous fenton-like reaction. *Chemosphere* **2020**, *244*, 125523. [CrossRef]
47. Regenspurg, S.; Brand, A.; Peiffer, S. Formation and stability of schwertmannite in acidic mining lakes. *Geochim. Cosmochim. Acta* **2004**, *68*, 1185–1197. [CrossRef]
48. Xu, Z.; Liang, J.; Zhou, L. Photo-Fenton-like degradation of azo dye methyl orange using synthetic ammonium and hydronium jarosite. *J. Alloys Compd.* **2013**, *546*, 112–118. [CrossRef]

MDPI
St. Alban-Anlage 66
4052 Basel
Switzerland
www.mdpi.com

Coatings Editorial Office
E-mail: coatings@mdpi.com
www.mdpi.com/journal/coatings